Digital Signal Processing
World Class Designs

Newnes World Class Designs Series

Analog Circuits: World Class Designs
Robert A. Pease
ISBN: 978-0-7506-8627-3

Embedded Systems: World Class Designs
Jack Ganssle
ISBN: 978-0-7506-8625-9

Power Sources and Supplies: World Class Designs
Marty Brown
ISBN: 978-0-7506-8626-6

FPGAs: World Class Designs
Clive "Max" Maxfield
ISBN: 978-1-85617-621-7

Digital Signal Processing: World Class Designs
Kenton Williston
ISBN: 978-1-85617-623-1

Portable Electronics: World Class Designs
John Donovan
ISBN: 978-1-85617-624-8

RF Front-End: World Class Designs
Janine Sullivan Love
ISBN: 978-1-85617-622-4

For more information on these and other Newnes titles visit: www.newnespress.com

Digital Signal Processing World Class Designs

Kenton Williston

with

Rick Gentile
Keith Jack
David Katz
Nasser Kehtarnavaz
Walt Kester
Dake Liu
Robert Meddins
Robert Oshana
Ian Poole
Khalid Sayood
Li Tan

AMSTERDAM • BOSTON • HEIDELBERG • LONDON
NEW YORK • OXFORD • PARIS • SAN DIEGO
SAN FRANCISCO • SINGAPORE • SYDNEY • TOKYO

Newnes is an imprint of Elsevier

Newnes is an imprint of Elsevier
30 Corporate Drive, Suite 400, Burlington, MA 01803, USA
Linacre House, Jordan Hill, Oxford OX2 8DP, UK

Library of Congress Cataloging-in-Publication Data
Application submitted.

British Library Cataloguing-in-Publication Data
A catalogue record for this book is available from the British Library.

ISBN: 978-1-85617-623-1

For information on all Newnes publications
visit our Web site at www.elsevierdirect.com

Transferred to Digital Printing, 2011

Printed and bound in the United Kingdom

Contents

Preface

What is this DSP thing I keep hearing about? What's so great about it? What do DSP engineers do? And how do I get in on the action? Great questions! Take a journey with me and all will be revealed…

Defining DSP

First, let's figure out what DSP is. The acronym DSP has two alternate meanings: Digital Signal Processing and Digital Signal Processor. Let's look at each, starting with digital signal processing. The meaning of "digital" is obvious—it means we are working in the world of 1's and 0's, and not in the analog world. The idea of a "signal" is a bit trickier. Our friend Wikipedia defines the term as "any time-varying or spatial-varying quantity." Speech is an example of a time-varying quantity; the pitch and volume of a voice changes from one moment to the next. A photograph is an example of a space-varying quantity; the color and brightness of an image are different in different areas of the photo. Now we are left to define "processing." This is a broad concept, but it generally involves analysis and manipulation using mathematical algorithms. For example, we could analyze a voice recording to determine its pitch, and we could manipulate a photograph by adjusting its colors.

DSP applications fall into four main categories:

- Communications

- Audio, video, and imaging (sometimes referred to as media processing)

- Motion control

- Military and aerospace

Of these areas, communications and video receive the most attention. Both areas are evolving rapidly, and both impose high computational loads. In addition, both areas include systems with severely limited power budgets.

With this background, it's easy to define the term digital signal processor. A digital signal processor is a simply a processor with specialized features for signal processing. For example, many signal processing algorithms involve multiplication followed by addition, an operation commonly referred to as a multiply accumulate or MAC. Since the MAC operation is so common in signal processing, all digital signal processors include special MAC hardware. As another example. many DSP applications have limited power budgets. Thus, many DSPs offer advanced power-saving features, such as the ability to change speeds and voltages n the fly—a feature known as dynamic frequency and voltage scaling.

It is important to note that digital signal *processing* does not require digital signal *processors*. Many signal processing systems use general-purpose processors (such as those available from ARM and MIPS) or custom hardware (built using ASICs and FPGAs). Many systems use a mix of hardware. For example, many systems contain both a DSP and a general-purpose processor (GPP).

The Role of the DSP Engineer

So much for the basics. What do DSP engineers actually do? My friend Shiv Balakrishnan recently wrote an article on this topic, and I agree with his conclusion that DSP engineers typically fall into three categories:

1. System designers create algorithms and (in some cases) the overall system.

2. Hardware designers implement (1) in hardware.

3. Programmers implement (1) in software, either by using hardware created by (2) or by using off-the-shelf hardware.

This book is intended mainly for programmers. This is the most common role for DSP engineers, and the role that is easiest to address without getting into graduate-level concepts. Nonetheless, it is worth looking at each of these jobs in detail, and exploring the skill sets for each.

System Designer

The system designer focuses on the big picture. This engineer may design the overall functionality of the system, including all of the attendant algorithms, or they may focus on specific subsystems and algorithms. The latter case is common for products such as cell phones where the system complexity is too great for a single engineer.

The system designer is referred to as a domain expert because they need an expert understanding of the system requirements and how to meet them. This includes expertise on the analog world, because most DSP systems have analog inputs and outputs. For example, wireless system designers must know how signals degrade as they propagate through the air.

The system designer also needs top-notch algorithmic expertise. To meet these demands, the system designer often needs a master's degree or even a PhD.

The system designer builds a functional model of the system using graphical tools such as *Simulink* and *LabVIEW* as well as text-based tools like MATLAB and C. The system designer often does not get into the details of the hardware and software design, but they must understand the basics of these disciplines. Even the most brilliant design is worthless if no one can build it! System designers must also take great care to ensure that hardware designers and programmers fully understand the functional model. Among other things, this means that the system designer must provide a means of testing the hardware and software against the system model. To meet these goals, system designers are increasingly turning to Electronic System Level (ESL) tools. ESL tools perform a number of functions, including automatic generation of reference hardware and software, as well as generation of test vectors.

Hardware Designer

Once the system is designed, it's the hardware designer's turn. The role of the hardware designer varies widely. As with the system designer, the hardware designer may work on the entire system or may focus on specific subsystems. The hardware designer may create custom hardware, or they may build a system using off-the-shelf parts. For custom hardware, designers once turned to *ASICs*, but ASIC design has become prohibitively expensive for all but the highest-volume products. As a result, hardware designers often use *FPGAs* instead. *Structured ASICs* are also an option, particularly for medium-volume applications.

In any of these cases, the hardware designer realizes the hardware as a set of blocks. Traditionally, the designer would implement each block in hand-coded *RTL* (either *VHDL* or *Verilog*), verify it, and optimize it. While hand-coded RTL is still in use, hardware engineers increasingly rely on ESL tools to generate hardware. This is particularly true for key DSP algorithms like FFTs and FIR filters. ESL tools have become quite proficient at generating hardware for these algorithms. In addition to the custom-built blocks, most applications also include one or more programmable processors. The processors may be implemented inside an ASIC or FPGA, or the hardware designer may use off-the shelf processors.

For systems that use off-the-shelf hardware, the hardware designer's job is much simpler. However, the hardware designer still has to make many careful choices in order to meet the system requirements. The processors must have enough performance to handle the workload, the buses must have enough bandwidth to handle the data, and so on. In many cases, it is impossible to find off-the-shelf hardware that fully meets the system requirements, so the hardware designer must create a small amount of custom hardware.

For example, many systems use an FPGA to implement I/O that is not available in the off-the-shelf processor.

The upshot of all of this is that hardware designers need a variety of skills. Although ESL tools are lightening the workload, the hardware designer must have good RTL coding skills. Obviously, hardware designers must understand the algorithms they implement. They must also understand the requirements of the software so they can make wise design decisions. Although it is possible to meet all of these requirements with a bachelor's-level education, a master's degree is quite helpful.

Programmer

Next up: the programmer. The programmer writes code to implement the remaining functionality on the systems processor or processors. Like programmers in any other field, the DSP programmer generally writes in C/C++. However, DSP code is unusual for two reasons. First, DSP code often begins as a MATLAB or Simulink model. Thus, tools that convert *MATLAB to C* are drawing a great deal of interest. Second, DSP code requires heavy optimization for performance, size, and power. In many cases, this requires the programmer to optimize key sections of their code using assembly language. To achieve these extreme levels of optimization, the DSP engineer must be intimately familiar with the details of their hardware.

As mentioned earlier, many DSP systems include multiple processors. For example, many systems include both a GPP and a DSP. In recent years, multicore processors have also become commonplace. (*Multicore* processors combine two or more processor cores on a single chip.) Thus, multiprocessor programming is a critical skill for many DSP programmers.

Until recently, DSP programmers wrote most of their own software. Today, DSP programmers often use off-the-shelf software for large parts of the system. This is particularly true for speech- and media-processing *codecs*. These codecs have become highly standardized and commoditized.

Obviously, the programmer must understand the algorithms they implement. Programmers who use off-the-shelf DSP software only need to know the basics of the underlying algorithms. In either case, a bachelor's-level education is generally sufficient, but a master's-level education is helpful.

DSP Project Flow

So far we have described the three DSP roles as separate disciplines, but the three roles tend to overlap in practice. For example, a single engineer may fill all three roles in a smaller project. It is also important to note that we have implied a linear project handoff from one discipline to the next. Projects rarely work this way in practice. Instead, most projects follow an iterative process with extensive feedback between the various roles. In addition, the system design, hardware design, and programming often proceed in parallel.

The Future of DSP

DSP was once a narrow, highly specialized field. Five years ago, DSP was nearly synonymous with telecom. If someone told you they were a DSP engineer, you had a good idea of what they did. You could be certain that they had good math skills and could explain exactly how a FIR filter works. Today, DSP is everywhere, often disguised under a moniker like "media processing." Many of the engineers working on DSP systems have only a general understanding of the underlying algorithms. With all these changes, it is hard to clearly define exactly what a DSP engineer is, or what they do—and this confusion is likely to get worse as DSP diffuses into an ever-growing list of applications. However, one thing is certain: DSP engineering will remain an important and in-demand skill for many years to come.

About the Editor

Kenton Williston is the owner of Cabral Consulting. He has been writing about Digital Signal Processing (DSP) technology and business trends since 2001. Kenton is currently the editor of the *DSP DesignLine*, and he was previously the editor and the senior contributor for *Inside DSP*. He is an author of numerous technology reports, including the *Buyer's Guide to DSP Processors*. Kenton is a popular presenter at the Embedded Systems Conference and other venues, and he has advised the leading DSP semiconductor firms on their marketing and product-planning strategies.

In previous lives, Kenton has worked for a major telecommunications company, an engineering services firm, and (believe it or not) an oil company. Kenton received his degree in Electrical Engineering from the University of Missouri-Rolla.

About the Contributors

James Bryant (Chapter 1) is a contributor to *Mixed Signal* and *DSP Design Techniques*.

Rick Gentile (Chapter 9) joined Analog Devices Inc. (ADI) in 2000 as a Senior DSP Applications Engineer. He currently leads the Applications Engineering Group, where he is responsible for applications engineering work on the Blackfin, SHARC and TigerSHARC processors . Prior to joining ADI, Rick was a Member of the Technical Staff at MIT Lincoln Laboratory, where he designed several signal processors used in a wide range of radar sensors. He received a B.S. in 1987 from the University of Massachusetts at Amherst and an M.S. in 1994 from Northeastern University, both in Electrical and Computer Engineering.

Keith Jack (Chapter 5) is Director of Product Marketing at Sigma Designs, a leading supplier of high-performance System-on-Chip (SoC) solutions for the IPTV, Blu-ray, and HDTV markets. Previously, he was Director of Product Marketing at Innovision, focused on solutions for digital televisions. Mr. Jack has also served as Strategic Marketing Manager at Harris Semiconductor and Brooktree Corporation. He has architected and introduced to market over 35 multimedia SoCs for the consumer markets, and is the author of "Video Demystified".

David Katz (Chapter 9) has over 15 years of experience in circuit and system design. Currently, he is Blackfin Applications Manager at Analog Devices, Inc., where he focuses on specifying new processors. He has published over 100 embedded processing articles domestically and internationally, and he has presented several conference papers in the field. Additionally, he is co-author of Embedded Media Processing (Newnes 2005). Previously, he worked at Motorola, Inc., as a senior design engineer in cable modem and automation groups. David holds both a B.S. and M. Eng. in Electrical Engineering from Cornell University.

Nasser Kehtarnavaz (Chapter 3) is the author of *Digital Signal Processing System Design* and *Real-Time Digital Signal Processing*.

Walt Kester (Chapter 1) is a corporate staff applications engineer at Analog Devices. For over 35 years at Analog Devices, he has designed, developed, and given applications support for high-speed ADCs, DACs, SHAs, op amps, and analog multiplexers. Besides writing many papers and articles, he prepared and edited eleven major applications books which form the basis for the Analog Devices world-wide technical seminar series including the topics of op amps, data conversion, power management, sensor signal conditioning, mixed-signal, and practical analog design techniques. He also is the editor of *The Data Conversion Handbook*, a 900+ page comprehensive book on data conversion published in 2005 by Elsevier. Walt has a BSEE from NC State University and MSEE from Duke University.

Dake Liu (Chapter 7) is a Professor and the Director of the Computer Engineering Division in the Department of Electrical Engineering at Linköping University, Linköping, Sweden. His research focuses on architectures of application specific instruction set processors (ASIP) and on-chip multi-core integrations based on VLSI. His research goal is to explore different processor architectures and inter-processor architectures and parallel programming for DSP firmware. On the processor level, his research focus is on integrated DSP-MCU solution based on homogeneous instruction set and heterogeneous architecture. On ILP (Instruction Level Parallelism) level research, his current focus is on architecture and toolchain for template based programming, especially for conflict free parallel memory accesses. On CMP (on Chip Multi Processor) level research, his current focus is on flexible low latency and low silicon cost system for real-time computing. Applications behind his research are embedded DSP computing, communications, and media signal processing. He published about 100 papers on international journals and reviewed conferences.

Robert Meddins (Chapter 2) is the author of *Introduction to Digital Signal Processing*.

Robert Oshana (Chapter 10) has over 25 years of experience in embedded and real-time systems and DSP systems development. He has numerous articles and books in this area. He is a member of the Embedded Systems Advisory board and speaks frequently at Embedded Systems Conferences worldwide. Rob is a senior member of IEEE, a licensed professional engineer (PE), and an adjunct professor at Southern Methodist University.

Ian Poole (Chapter 6) is an established electronics engineering consultant with considerable experience in the communications and cellular markets. He is the author of a number of books on radio and electronics and he has contributed to many magazines in the UK and worldwide. He is also winner of the inaugural Bill Orr Award for technical writing from the ARRL.

Khalid Sayood (Chapter 4) received his BS and MS in Electrical Engineering from the University of Rochester in 1977 and 1979, respectively, and his Ph.D. in Electrical Engineering from Texas A&M University in 1982. In 1982, he joined the University of Nebraska, where he is the Henson Professor of Engineering. His research interests include data compression, joint source channel coding, and bioinformatics.

Dan Sheingold (Chapter 1) is a contributor to *Mixed Signal and DSP Design Techniques*.

Li Tan (Chapter 8) is a Senior Professor and Curriculum Coordinator in Electronics Engineering Technology at DeVry University.

ADCs, DACs, and Sampling Theory

Walt Kester
Dan Sheingold
James Bryant

A chapter on analog? What's this doing in a DSP book? And at the very front of the book, no less!

Relax. You don't need to be an analog expert to do DSP. However, a little analog knowledge is a big help! Nearly all DSP systems have analog inputs. Before we can manipulate this analog data, we need to pass it through an analog-to-digital converter (ADC). This conversion process is always flawed—the digital data cannot capture the analog signal with perfect precision. ADCs also change the data by adding noise and distortion. What's worse, ADCs present the possibility of aliasing—a phenomenon where two signals that are obviously different in the analog world become indistinguishable in the digital world.

Things are just as bad on the output side. Most DSP systems have analog outputs, so we have to pipe the output through digital-to-analog converters (DACs). DACs cannot reproduce our digital signal with perfect precision—they always introduce noise and distortion.

If all of this sounds frightening, don't worry. As long as you understand the limitations of ADCs and DACs, these limitations are usually easy to manage. This chapter by Walt Kester will help you do just that. The author explains the principles behind ADCs and DACs, including the all-important Nyquist sampling theorem. He explains the basic operation of ADCs and DACs, and shows how even "ideal" ADCs and DACs introduce errors. He then goes on to explain the flaws of real-world ADCs and DACs.

The information in this chapter will be more than enough for most DSP engineers. If you want to dive deeper, I heartily recommend these series:

ADCs for DSP
http://www.dspdesignline.com/howto/202200877

DACs for DSP
http://www.dspdesignline.com/howto/205601725

—Kenton Williston

1.1 Coding and Quantizing

Analog-to-digital converters (ADCs) translate analog quantities, which are characteristic of most phenomena in the "real world," to digital language, used in information processing, computing, data transmission, and control systems. Digital-to-analog converters (DACs) are used in transforming transmitted or stored data, or the results of digital processing, back to "real-world" variables for control, information display, or further analog processing. The relationships between inputs and outputs of DACs and ADCs are shown in Figure 1.1.

Analog input variables, whatever their origin, are most frequently converted by transducers into voltages or currents. These electrical quantities may appear (1) as fast or slow "DC" continuous direct measurements of a phenomenon in the time domain, (2) as modulated AC waveforms (using a wide variety of modulation techniques), (3) or in some combination, with a spatial configuration of related variables to represent shaft angles. Examples of the first are outputs of thermocouples, potentiometers on DC references, and analog computing circuitry; of the second, "chopped" optical measurements, AC strain gage or bridge outputs, and digital signals buried in noise; and of the third, synchros and resolvers.

The analog variables to be dealt with in this chapter are those involving voltages or currents representing the actual analog phenomena. They may be either wideband or narrowband. They may be either scaled from the direct measurement, or subjected to some form of analog preprocessing, such as linearization, combination, demodulation, filtering, sample-hold, etc.

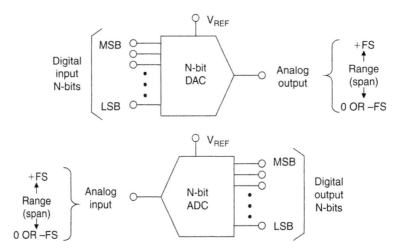

Figure 1.1: Digital-to-analog converter (DAC) and analog-to-digital converter (ADC) input and output definitions

As part of the process, the voltages and currents are "normalized" to ranges compatible with assigned ADC input ranges. Analog output voltages or currents from DACs are direct and in normalized form, but they may be subsequently post-processed (e.g., scaled, filtered, amplified, etc.).

Information in digital form is normally represented by arbitrarily fixed voltage levels referred to "ground," either occurring at the outputs of logic gates, or applied to their inputs. The digital numbers used are all basically binary; that is, each "bit," or unit of information has one of two possible states. These states are "off," "false," or "0," and "on," "true," or "1." It is also possible to represent the two logic states by two different levels of current; however, this is much less popular than using voltages. There is also no particular reason why the voltages need be referenced to ground—as in the case of emitter-coupled-logic (ECL), positive-emitter-coupled-logic (PECL) or low-voltage-differential-signaling logic (LVDS) for example.

Words are groups of levels representing digital numbers; the levels may appear simultaneously in *parallel,* on a bus or groups of gate inputs or outputs, *serially* (or in a time sequence) on a single line, or as a sequence of parallel bytes (i.e., "byte-serial") or nibbles (small bytes). For example, a 16-bit word may occupy the 16 bits of a 16-bit bus, or it may be divided into two sequential bytes for an 8-bit bus, or four 4-bit nibbles for a 4-bit bus.

Although there are several systems of logic, the most widely used choice of levels are those used in TTL (transistor-transistor logic) and, in which positive *true,* or 1, corresponds to a minimum output level of 2.4 V (inputs respond unequivocally to "1" for levels greater than 2.0 V); and *false,* or 0, corresponds to a maximum output level of 0.4 V (inputs respond unequivocally to "0" for anything less than 0.8 V). It should be noted that even though CMOS is more popular today than TTL, CMOS logic levels are generally made to be compatible with the older TTL logic standard.

A unique parallel or serial grouping of digital levels, or a *number,* or *code,* is assigned to each analog level which is quantized (i.e., represents a unique portion of the analog range). A typical digital code would be this array:

$$a_7 a_6 a_5 a_4 a_3 a_2 a_1 = 10111001$$

It is composed of 8 bits. The "1" at the extreme left is called the *most significant bit* (MSB, or bit 1), and the one at the right is called the *least significant bit* (LSB, or bit *N:* 8 in this case). The meaning of the code, as either a number, a character, or a representation of an analog variable, is unknown until the *code* and the *conversion relationship* have been defined. It is important not to confuse the designation of a particular bit (i.e., bit 1, bit 2, and so on) with the subscripts associated with the "a" array. The subscripts correspond to the power of 2 associated with the weight of a particular bit in the sequence.

The best-known code (other than base-10) is *natural or straight binary* (base-2). Binary codes are most familiar in representing integers; i.e., in a natural binary integer code having N bits, the LSB has a weight of 2^0 (i.e., 1), the next bit has a weight of 2^1 (i.e., 2), and so on up to the MSB, which has a weight of 2^{N-1} (i.e., $2^N/2$). The value of a binary number is obtained by adding up the weights of all non-zero bits. When the weighted bits are added up, they form a unique number having any value from 0 to $2^N - 1$. Each additional trailing zero bit, if present, essentially doubles the size of the number.

In converter technology, full-scale (abbreviated *FS*) is independent of the number of bits of resolution, N. A more useful coding is *fractional* binary, which is always normalized to full-scale. Integer binary can be interpreted as fractional binary if all integer values are divided by 2^N. For example, the MSB has a weight of ½ (i.e., $2^{(N-1)}/2^N = 2^{-1}$), the next bit has a weight of ¼ (i.e., 2^{-2}), and so forth down to the LSB, which has a weight of $1/2^N$ (i.e., 2^{-N}). When the weighted bits are added up, they form a number with any of 2^N values, from 0 to $(1-2^{-N})$ of full-scale. Additional bits simply provide more fine structure without affecting full-scale range. The relationship between base-10 numbers and binary numbers (base-2) are shown in Figure 1.2 along with examples of each.

1.1.1 Unipolar Codes

In data conversion systems, the coding method must be related to the analog input range (or span) of an ADC or the analog output range (or span) of a DAC. The simplest case is when the input to the ADC or the output of the DAC is always a unipolar positive voltage (current outputs are very popular for DAC outputs, much less for ADC inputs). The most popular

Whole numbers:

$$\text{Number}_{10} = a_{N-1}2^{N-1} + a_{N-1}2^{N-2} + \dots + a_1 2^1 + a_0 2^0$$

MSB LSB

Example: $1011_2 = (1 \times 2^3) + (0 \times 2^2) + (1 \times 2^1) + (1 \times 2^0)$
$= 8 + 0 + 2 + 1 = 11_{10}$

Fractional numbers:

$$\text{Number}_{10} = a_{N-1}2^{-1} + a_{N-2}2^{-2} + \dots + a_1 2^{-(N-1)} + a_0 2^{-N}$$

MSB LSB

Example: $0.1011_2 = (1 \times 0.5) + (0 \times 0.25) + (1 \times 0.125) + (1 \times 0.0625)$
$= 0.5 + 0 + 0.125 + 0.0625 = 0.6875_{10}$

Figure 1.2: Representing a base-10 number with a binary number (base-2)

code for this type of signal is *straight binary* and is shown in Figure 1.3 for a 4-bit converter. Notice that there are 16 distinct possible levels, ranging from the all-zeros code 0000, to the all-ones code 1111. It is important to note that the analog value represented by the all-ones code is not full-scale (abbreviated FS), but FS − 1 LSB. This is a common convention in data conversion notation and applies to both ADCs and DACs. Figure 1.3 gives the base-10 equivalent number, the value of the base-2 binary code relative to full-scale (FS), and also the corresponding voltage level for each code (assuming a 10 V full-scale converter. The Gray code equivalent is also shown, and will be discussed shortly.

Figure 1.4 shows the transfer function for an ideal 3-bit DAC with straight binary input coding. Notice that the analog output is zero for the all-zeros input code. As the digital input code increases, the analog output increases 1 LSB (1/8 scale in this example) per code. The most positive output voltage is 7/8 FS, corresponding to a value equal to FS − 1 LSB. The midscale output of 1/2 FS is generated when the digital input code is 100.

The transfer function of an ideal 3-bit ADC is shown in Figure 1.5. There is a range of analog input voltage over which the ADC will produce a given output code; this range is the *quantization uncertainty* and is equal to 1 LSB. Note that the width of the transition regions between adjacent codes is zero for an ideal ADC. In practice, however, there is always transition noise associated with these levels, and therefore the width is non-zero. It is customary to define the analog input corresponding to a given code by the *code center* which lies halfway between two adjacent transition regions (illustrated by the black dots in the

Base 10 number	Scale		+10V FS	Binary	Gray
+15	+FS −1LSB =	+15/16 FS	9.375	1 1 1 1	1 0 0 0
+14		+7/8 FS	8.750	1 1 1 0	1 0 0 1
+13		+13/16 FS	8.125	1 1 0 1	1 0 1 1
+12		+3/4 FS	7.500	1 1 0 0	1 0 1 0
+11		+11/16 FS	6.875	1 0 1 1	1 1 1 0
+10		+5/8 FS	6.250	1 0 1 0	1 1 1 1
+9		+9/16 FS	5.625	1 0 0 1	1 1 0 1
+8		+1/2 FS	5.000	1 0 0 0	1 1 0 0
+7		+7/16 FS	4.375	0 1 1 1	0 1 0 0
+6		+3/8 FS	3.750	0 1 1 0	0 1 0 1
+5		+5/16 FS	3.125	0 1 0 1	0 1 1 1
+4		+1/4 FS	2.500	0 1 0 0	0 1 1 0
+3		+3/16 FS	1.875	0 0 1 1	0 0 1 0
+2		+1/8 FS	1.250	0 0 1 0	0 0 1 1
+1	1LSB =	+1/16 FS	0.625	0 0 0 1	0 0 0 1
0		0	0.000	0 0 0 0	0 0 0 0

Figure 1.3: Unipolar binary codes, 4-bit converter

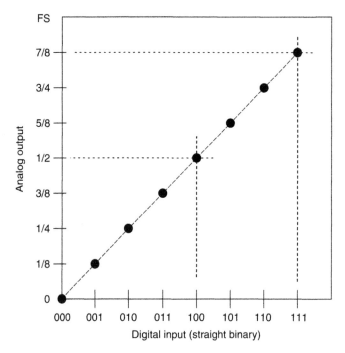

Figure 1.4: Transfer function for ideal unipolar 3-bit DAC

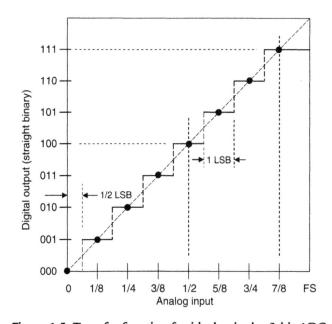

Figure 1.5: Transfer function for ideal unipolar 3-bit ADC

diagram). This requires that the first transition region occur at ½ LSB. The full-scale analog input voltage is defined by 7/8 FS, (FS − 1 LSB).

1.1.2 Gray Code

Another code worth mentioning at this point is the *Gray* code (or *reflective-binary*), which was invented by Elisha Gray in 1878 (Reference 1) and later re-invented by Frank Gray in 1949 (see Reference 2). The Gray code equivalent of the 4-bit straight binary code is also shown in Figure 1.3. Although it is rarely used in computer arithmetic, it has some useful properties which make it attractive to A/D conversion. Notice that in Gray code, as the number value changes, the transitions from one code to the next involve only one bit at a time. Contrast this to the binary code where all the bits change when making the transition between 0111 and 1000. Some ADCs make use of it internally and then convert the Gray code to a binary code for external use.

One of the earliest practical ADCs to use the Gray code was a 7-bit, 100 kSPS electron beam encoder developed by Bell Labs and described in a 1948 reference (Reference 3).

The basic electron beam coder concepts for a 4-bit device are shown in Figure 1.6. The early tubes operated in the serial mode (a). The analog signal is first passed through

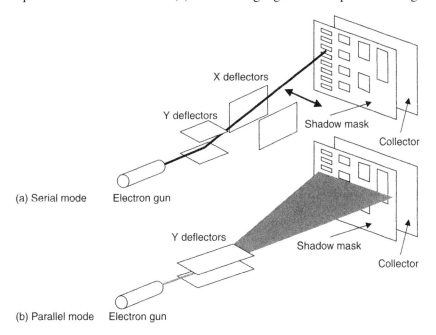

Figure 1.6: The electron beam coder: (a) serial mode and (b) parallel or "Flash" mode

a sample-and-hold, and during the "hold" interval, the beam is swept horizontally across the tube. The Y-deflection for a single sweep therefore corresponds to the value of the analog signal from the sample-and-hold. The shadow mask is coded to produce the proper binary code, depending on the vertical deflection. The code is registered by the collector, and the bits are generated in serial format. Later tubes used a fan-shaped beam (shown in Figure 1.6b), creating a "Flash" converter delivering a parallel output word.

Early electron tube coders used a binary-coded shadow mask, and large errors can occur if the beam straddles two adjacent codes and illuminates both of them. The way these errors occur is illustrated in Figure 1.7a, where the horizontal line represents the beam sweep at the midscale transition point (transition between code 0111 and code 1000). For example, an error in the most significant bit (MSB) produces an error of ½ scale. These errors were minimized by placing fine horizontal sensing wires across the boundaries of each of the quantization levels. If the beam initially fell on one of the wires, a small voltage was added to the vertical deflection voltage which moved the beam away from the transition region.

The errors associated with binary shadow masks were eliminated by using a Gray code shadow mask as shown in Figure 1.7b. As mentioned above, the Gray code has the property that adjacent levels differ by only one digit in the corresponding Gray-coded word. Therefore, if there is an error in a bit decision for a particular level, the corresponding error after conversion to binary code is only one least significant bit (LSB). In the case of midscale, note that only the MSB changes. It is interesting to note that this same phenomenon can

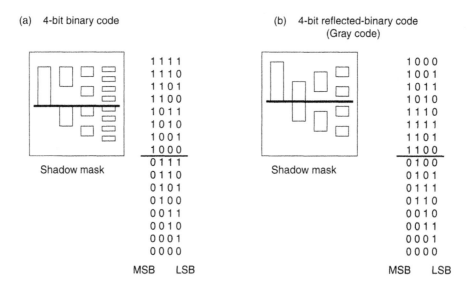

Figure 1.7: Electron beam coder shadow masks for (a) binary code and (b) Gray code

occur in modern comparator-based Flash converters due to comparator metastability. With small overdrive, there is a finite probability that the output of a comparator will generate the wrong decision in its latched output, producing the same effect if straight binary decoding techniques are used. In many cases, Gray code, or "pseudo-Gray" codes are used to decode the comparator bank. The Gray code output is then latched, converted to binary, and latched again at the final output.

As a historical note, in spite of the many mechanical and electrical problems relating to beam alignment, electron tube coding technology reached its peak in the mid-l960s with an experimental 9-bit coder capable of 12 MSPS sampling rates (Reference 4). Shortly thereafter, however, advances in all solid-state ADC techniques made the electron tube technology obsolete.

Other examples where Gray code is often used in the conversion process to minimize errors are shaft encoders (angle-to-digital) and optical encoders.

ADCs that use the Gray code internally almost always convert the Gray code output to binary for external use. The conversion from Gray-to-binary and binary-to-Gray is easily accomplished with the exclusive-or logic function as shown in Figure 1.8.

1.1.3 Bipolar Codes

In many systems, it is desirable to represent both positive and negative analog quantities with binary codes. Either *offset binary*, *two's complement*, *one's complement*, or *sign magnitude* codes will accomplish this, but offset binary and two's complement are by far the most

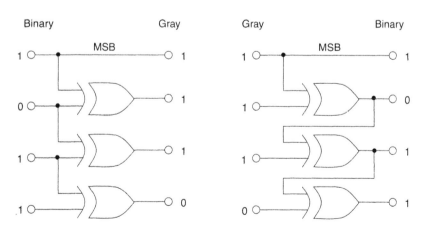

Figure 1.8: Binary-to-Gray and Gray-to-binary conversion using the exclusive-or logic function

Base 10 number	Scale		±5V FS	Offset binary	Twos comp.	Ones comp.	Sign mag.
+7	+FS −1LSB =	+7/8 FS	+4.375	1 1 1 1	0 1 1 1	0 1 1 1	0 1 1 1
+6		+3/4 FS	+3.750	1 1 1 0	0 1 1 0	0 1 1 0	0 1 1 0
+5		+5/8 FS	+3.125	1 1 0 1	0 1 0 1	0 1 0 1	0 1 0 1
+4		+1/2 FS	+2.500	1 1 0 0	0 1 0 0	0 1 0 0	0 1 0 0
+3		+3/8 FS	+1.875	1 0 1 1	0 0 1 1	0 0 1 1	0 0 1 1
+2		+1/4 FS	+1.250	1 0 1 0	0 0 1 0	0 0 1 0	0 0 1 0
+1		+1/8 FS	+0.625	1 0 0 1	0 0 0 1	0 0 0 1	0 0 0 1
0		0	0.000	1 0 0 0	0 0 0 0	*0 0 0 0	*1 0 0 0
−1		−1/8 FS	−0.625	0 1 1 1	1 1 1 1	1 1 1 0	1 0 0 1
−2		−1/4 FS	−1.250	0 1 1 0	1 1 1 0	1 1 0 1	1 0 1 0
−3		−3/8 FS	−1.875	0 1 0 1	1 1 0 1	1 1 0 0	1 0 1 1
−4		−1/2 FS	−2.500	0 1 0 0	1 1 0 0	1 0 1 1	1 1 0 0
−5		−5/8 FS	−3.125	0 0 1 1	1 0 1 1	1 0 1 0	1 1 0 1
−6		−3/4 FS	−3.750	0 0 1 0	1 0 1 0	1 0 0 1	1 1 1 0
−7	−FS + 1LSB =	−7/8 FS	−4.375	0 0 0 1	1 0 0 1	1 0 0 0	1 1 1 1
−8		−FS	−5.000	0 0 0 0	1 0 0 0		

	Ones comp.	Sign mag.
0+	0 0 0 0	0 0 0 0
0−	1 1 1 1	1 0 0 0

Not normally used
in computations (see text)

*

Figure 1.9: Bipolar codes, 4-bit converter

popular. The relationships between these codes for a 4-bit system is shown in Figure 1.9. Note that the values are scaled for a ±5 V full-scale input/output voltage range.

For *offset binary*, the zero signal value is assigned the code 1000. The sequence of codes is identical to that of straight binary.

The only difference between a straight and offset binary system is the half-scale offset associated with analog signal. The most negative value (−FS +1 LSB) is assigned the code 0001, and the most positive value (+FS −1 LSB) is assigned the code 1111. Note that in order to maintain perfect symmetry about midscale, the all-zeros code (0000) representing negative full-scale (−FS) is not normally used in computation. It can be used to represent a negative off-range condition or simply assigned the value of the 0001 (−FS +1 LSB).

The relationship between the offset binary code and the analog output range of a bipolar 3-bit DAC is shown in Figure 1.10. The analog output of the DAC is zero for the zero-value input code 100. The most negative output voltage is generally defined by the 001 code (−FS +1 LSB), and the most positive by 111 (+FS −1 LSB). The output voltage for the 000 input code is available for use if desired, but makes the output nonsymmetrical about zero and complicates the mathematics.

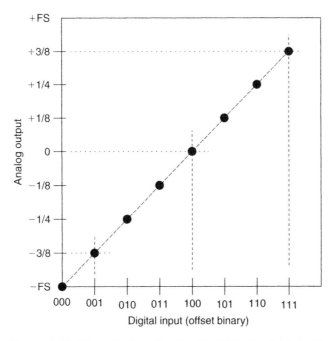

Figure 1.10: Transfer function for ideal bipolar 3-bit DAC

The offset binary output code for a bipolar 3-bit ADC as a function of its analog input is shown in Figure 1.11. Note that zero analog input defines the center of the midscale code 100. As in the case of bipolar DACs, the most negative input voltage is generally defined by the 001 code ($-$FS $+$1 LSB), and the most positive by 111 ($+$FS$-$1 LSB). As discussed above, the 000 output code is available for use if desired, but makes the output nonsymmetrical about zero and complicates the mathematics.

Two's complement is identical to offset binary with the most-significant-bit (MSB) complemented (inverted). This is obviously very easy to accomplish in a data converter, using a simple inverter or taking the complementary output of a "D" flip-flop. The popularity of two's complement coding lies in the ease with which mathematical operations can be performed in computers and DSPs. Two's complement, for conversion purposes, consists of a binary code for positive magnitudes (0 sign bit), and the two's complement of each positive number to represent its negative. The two's complement is formed arithmetically by complementing the number and adding 1 LSB. For example, $-$3/8 FS is obtained by taking the two's complement of $+$3/8 FS. This is done by first complementing $+$3/8 FS, 0011 obtaining 1100. Adding 1 LSB, we obtain 1101.

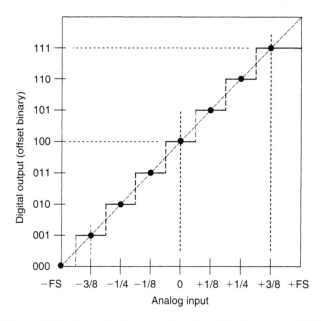

Figure 1.11: Transfer function for ideal bipolar 3-bit ADC

Two's complement makes subtraction easy. For example, to subtract 3/8 FS from 4/8 FS, add 4/8 to −3/8, or 0100 to 1101. The result is 0001, or 1/8, disregarding the extra carry.

One's complement can also be used to represent negative numbers, although it is much less popular than two's complement and rarely used today. The one's complement is obtained by simply complementing all of a positive number's digits. For instance, the one's complement of 3/8 FS (0011) is 1100. A one's complemented code can be formed by complementing each positive value to obtain its corresponding negative value. This includes zero, which is then represented by either of two codes, 0000 (referred to as 0+) or 1111 (referred to as 0−). This ambiguity must be dealt with mathematically, and presents obvious problems relating to ADCs and DACs for which there is a single code that represents zero.

Sign-magnitude would appear to be the most straightforward way of expressing signed analog quantities digitally. Simply determine the code appropriate for the magnitude and add a polarity bit. Sign-magnitude BCD is popular in bipolar digital voltmeters, but has the problem of two allowable codes for zero. It is therefore unpopular for most applications involving ADCs or DACs.

Figure 1.12 summarizes the relationships between the various bipolar codes: offset binary, two's complement, one's complement, and sign-magnitude, and shows how to convert between them.

To convert from → To ↓	Sign magnitude	Two's complement	Offset binary	One's complement
Sign magnitude	No change	If MSB = 1, complement other bits, add 00…01	Complement MSB If new MSB = 1, complement other bits, add 00…01	If MSB = 1, complement other bits
Two's complement	If MSB = 1, complement other bits, add 00…01	No Change	Complement MSB	If MSB = 1, add 00…01
Offset binary	Complement MSB If new MSB = 0 complement other bits, add 00…01	Complement MSB	No change	Complement MSB If new MSB = 0, add 00…01
One's complement	If MSB = 1, complement other bits	If MSB = 1, add 11…11	Complement MSB If new MSB = 1, add 11…11	No change

Figure 1.12: Relationships among bipolar codes

The last code to be considered in this section is *binary-coded decimal (BCD)*, where each base-10 digit (0 to 9) in a decimal number is represented as the corresponding 4-bit straight binary word as shown in Figure 1.13. The minimum digit 0 is represented as 0000, and the digit 9 by 1001. This code is relatively inefficient, since only 10 of the 16 code states for each decade are used. It is, however, a very useful code for interfacing to decimal displays such as in digital voltmeters.

1.1.4 Complementary Codes

Some forms of data converters (for example, early DACs using monolithic NPN quad current switches), require standard codes such as natural binary or BCD, but with all bits represented by their complements. Such codes are called *complementary codes*. All the codes discussed thus far have complementary codes which can be obtained by this method. A *complementary code* should not be confused with a *one's complement* or a *two's complement* code.

In a 4-bit complementary-binary converter, 0 is represented by 1111, half-scale by 0111, and FS−1 LSB by 0000. In practice, the complementary code can usually be obtained by using the complementary output of a register rather than the true output, since both are available.

Base 10 number	Scale	+10V FS	Decade 1	Decade 2	Decade 3	Decade 4
+15	+FS −1LSB = +15/16 FS	9.375	1 0 0 1	0 0 1 1	0 1 1 1	0 1 0 1
+14	+7/8 FS	8.750	1 0 0 0	0 1 1 1	0 1 0 1	0 0 0 0
+13	+13/16 FS	8.125	1 0 0 0	0 0 0 1	0 0 1 0	0 1 0 1
+12	+3/4 FS	7.500	0 1 1 1	0 1 0 1	0 0 0 0	0 0 0 0
+11	+11/16 FS	6.875	0 1 1 0	1 0 0 0	0 1 1 1	0 1 0 1
+10	+5/8 FS	6.250	0 1 1 0	0 0 1 0	0 1 0 1	0 0 0 0
+9	+9/16 FS	5.625	0 1 0 1	0 1 1 0	0 0 1 0	0 1 0 1
+8	+1/2 FS	5.000	0 1 0 1	0 0 0 0	0 0 0 0	0 0 0 0
+7	+7/16 FS	4.375	0 1 0 0	0 0 1 1	0 1 1 1	0 1 0 1
+6	+3/8 FS	3.750	0 0 1 1	0 1 1 1	0 1 0 1	0 0 0 0
+5	+5/16 FS	3.125	0 0 1 1	0 0 0 1	0 0 1 0	0 1 0 1
+4	+1/4 FS	2.500	0 0 1 0	0 1 0 1	0 0 0 0	0 0 0 0
+3	+3/16 FS	1.875	0 0 0 1	1 0 0 0	0 1 1 1	0 1 0 1
+2	+1/8 FS	1.250	0 0 0 1	0 0 1 0	0 1 0 1	0 0 0 0
+1	1LSB = +1/16 FS	0.625	0 0 0 0	0 1 1 0	0 0 1 0	0 1 0 1
0	0	0.000	0 0 0 0	0 0 0 0	0 0 0 0	0 0 0 0

Figure 1.13: Binary-coded decimal (BCD) code

Sometimes the complementary code is useful in inverting the analog output of a DAC. Today many DACs provide differential outputs which allow the polarity inversion to be accomplished without modifying the input code. Similarly, many ADCs provide differential logic inputs which can be used to accomplish the polarity inversion.

1.1.5 DAC and ADC Static Transfer Functions and DC Errors

The most important thing to remember about both DACs and ADCs is that either the input or output is digital, and therefore the signal is quantized. That is, an N-bit word represents one of 2^N possible states, and therefore an N-bit DAC (with a fixed reference) can have only 2^N possible analog outputs, and an N-bit ADC can have only 2^N possible digital outputs. As previously discussed, the analog signals will generally be voltages or currents.

The resolution of data converters may be expressed in several different ways: the weight of the least significant bit (LSB), parts per million of full-scale (ppm FS), millivolts (mV), etc. Different devices (even from the same manufacturer) will be specified differently, so converter users must learn to translate between the different types of specifications if they are to compare devices successfully. The size of the least significant bit for various resolutions is shown in Figure 1.14.

Before we can consider the various architectures used in data converters, it is necessary to consider the performance to be expected, and the specifications which are important. The

Resolution N	2^N	Voltage (10V FS)	ppm FS	% FS	dB FS
2-bit	4	2.5 V	250,000	25	−12
4-bit	16	625 mV	62,500	6.25	−24
6-bit	64	156 mV	15,625	1.56	−36
8-bit	256	39.1 mV	3,906	0.39	−48
10-bit	1,024	9.77 mV (10 mV)	977	0.098	−60
12-bit	4,096	2.44 mV	244	0.024	−72
14-bit	16,384	610 μV	61	0.0061	−84
16-bit	65,536	153 μV	15	0.0015	−96
18-bit	262,144	38 μV	4	0.0004	−108
20-bit	1,048,576	9.54 μV (10 μV)	1	0.0001	−120
22-bit	4,194,304	2.38 μV	0.24	0.000024	−132
24-bit	16,777,216	596 μV*	0.06	0.000006	−144

*600nV is the Johnson Noise in a 10 kHz BW of a 2.2kΩ Resistor @ 25°C

Remember: 10 bits and 10V FS yields an LSB of 10 mV, 1000 ppm, or 0.1%.
all other values may be calculated by powers of 2.

Figure 1.14: Quantization: the size of a least significant bit (LSB)

following sections will consider the definition of errors and specifications used for data converters. This is important in understanding the strengths and weaknesses of different ADC/DAC architectures.

The first applications of data converters were in measurement and control where the exact timing of the conversion was usually unimportant, and the data rate was slow. In such applications, the DC specifications of converters are important, but timing and AC specifications are not. Today many, if not most, converters are used in *sampling* and *reconstruction* systems where AC specifications are critical (and DC ones may not be)—these will be considered in Section 1.3 of this chapter.

Figure 1.15 shows the ideal transfer characteristics for a 3-bit unipolar DAC and a 3-bit unipolar ADC. In a DAC, both the input and the output are quantized, and the graph consists of eight points. While it is reasonable to discuss the line through these points, it is very important to remember that the actual transfer characteristic is *not* a line, but a number of discrete points.

The input to an ADC is analog and is not quantized, but its output is quantized. The transfer characteristic therefore consists of eight horizontal steps. When considering the offset, gain

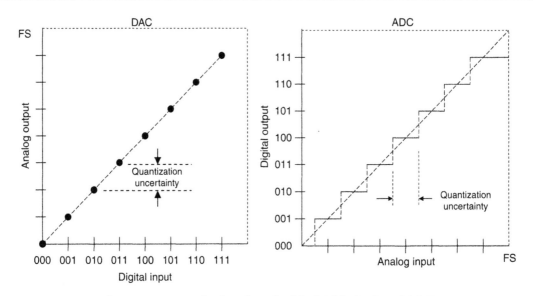

Figure 1.15: Transfer functions for ideal 3-bit DAC and ADC

and linearity of an ADC we consider the line joining the midpoints of these steps—often referred to as the *code centers*.

For both DACs and ADCs, digital full-scale (all "1"s) corresponds to 1 LSB below the analog full-scale (FS). The (ideal) ADC transitions take place at ½ LSB above zero, and thereafter every LSB, until 1½ LSB below analog full-scale. Since the analog input to an ADC can take any value, but the digital output is quantized, there may be a difference of up to ½ LSB between the actual analog input and the exact value of the digital output. This is known as the *quantization error* or *quantization uncertainty* as shown in Figure 1.15. In AC (sampling) applications this quantization error gives rise to *quantization noise* which will be discussed in Section 1.3 of this chapter.

As previously discussed, there are many possible digital coding schemes for data converters: *straight binary*, *offset binary*, *one's complement*, *two's complement*, sign *magnitude*, *Gray code*, *BCD,* and others. This section, being devoted mainly to the *analog* issues surrounding data converters, will use simple *binary* and *offset binary* in its examples and will not consider the merits and disadvantages of these, or any other forms of digital code.

The examples in Figure 1.15 use *unipolar* converters, whose analog port has only a single polarity. These are the simplest type, but *bipolar* converters are generally more useful in real-world applications. There are two types of bipolar converters: the simpler is merely a unipolar converter with an accurate 1 MSB of negative offset (and many converters are arranged

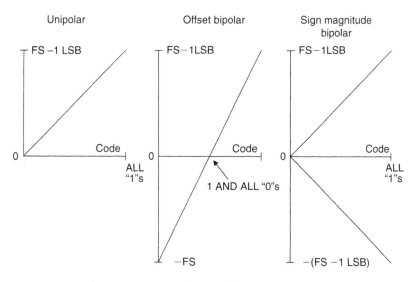

Figure 1.16: Unipolar and bipolar converters

so that this offset may be switched in and out so that they can be used as either unipolar or bipolar converters at will), but the other, known as a *sign-magnitude* converter is more complex, and has N bits of magnitude information and an additional bit which corresponds to the sign of the analog signal.

Sign-magnitude DACs are quite rare, and sign-magnitude ADCs are found mostly in digital voltmeters (DVMs). The unipolar, offset binary, and sign-magnitude representations are shown in Figure 1.16.

The four DC errors in a data converter are *offset error*, *gain error*, and two types of *linearity error (differential and integral)*. Offset and gain errors are analogous to offset and gain errors in amplifiers as shown in Figure 1.17 for a bipolar input range. (Though offset error and zero error, which are identical in amplifiers and unipolar data converters, are not identical in bipolar converters and should be carefully distinguished.)

The transfer characteristics of both DACs and ADCs may be expressed as a straight line given by D = K + GA, where D is the digital code, A is the analog signal, and K and G are constants. In a unipolar converter, the ideal value of K is zero; in an offset bipolar converter it is −1 MSB. The offset error is the amount by which the actual value of K differs from its ideal value.

The gain error is the amount by which G differs from its ideal value, and is generally expressed as the percentage difference between the two, although it may be defined as the gain error contribution (in mV or LSB) to the total error at full scale. These errors can usually be trimmed by the data converter user. Note, however, that amplifier offset is trimmed at zero input, and then the gain is trimmed near to full scale. The trim algorithm for a bipolar data converter is not so straightforward.

The integral linearity error of a converter is also analogous to the linearity error of an amplifier, and is defined as the maximum deviation of the actual transfer characteristic of the converter from a straight line, and is generally expressed as a percentage of full scale (but may be given in LSBs). For an ADC, the most popular convention is to draw the straight line through the mid-points of the codes, or the code centers. There are two common ways of choosing the straight line: *end point* and *best straight line* as shown in Figure 1.18.

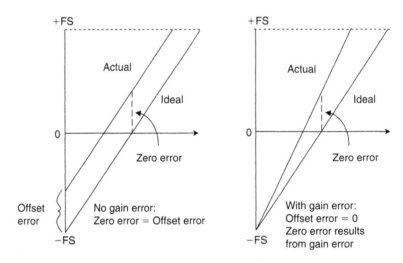

Figure 1.17: Bipolar data converter offset and gain error

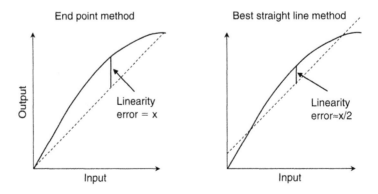

Figure 1.18: Method of measuring integral linearity errors (same converter on both graphs)

In the *end point* system, the deviation is measured from the straight line through the origin and the full-scale point (after gain adjustment). This is the most useful integral linearity measurement for measurement and control applications of data converters (since error budgets depend on deviation from the ideal transfer characteristic, not from some arbitrary "best fit"), and is the one normally adopted by Analog Devices, Inc.

However, the *best straight line* does give a better prediction of distortion in AC applications, and also gives a lower value of "linearity error" on a data sheet. The best fit straight line is drawn through the transfer characteristic of the device using standard curve-fitting techniques, and the maximum deviation is measured from this line. In general, the integral linearity error measured in this way is only 50% of the value measured by end point methods. This makes the method good for producing impressive data sheets, but it is less useful for error budget analysis. For AC applications it is better to specify distortion than DC linearity, so it is rarely necessary to use the best straight line method to define converter linearity.

The other type of converter nonlinearity is *differential nonlinearity* (DNL). This relates to the linearity of the code transitions of the converter. In the ideal case, a change of 1 LSB in digital code corresponds to a change of exactly 1 LSB of analog signal. In a DAC, a change of 1 LSB in digital code produces exactly 1 LSB change of analog output, while in an ADC there should be exactly 1 LSB change of analog input to move from one digital transition to the next. Differential linearity error is defined as the maximum amount of deviation of any quantum (or LSB change) in the entire transfer function from its ideal size of 1 LSB.

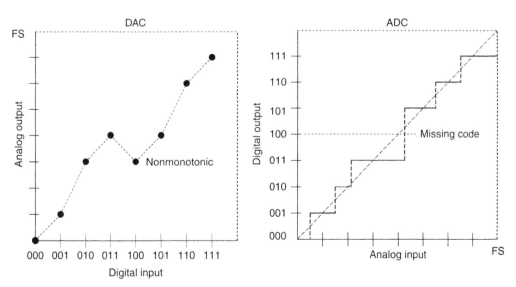

Figure 1.19: Transfer functions for nonideal 3-bit DAC and ADC

Where the change in analog signal corresponding to 1 LSB digital change is more or less than 1 LSB, there is said to be a DNL error. The DNL error of a converter is normally defined as the maximum value of DNL to be found at any transition across the range of the converter. Figure 1.19 shows the nonideal transfer functions for a DAC and an ADC and shows the effects of the DNL error.

The DNL of a DAC is examined more closely in Figure 1.20. If the DNL of a DAC is less than −1 LSB at any transition, the DAC is *nonmonotonic*; i.e., its transfer characteristic contains one or more localized maxima or minima. A DNL greater than +1 LSB does not cause nonmonotonicity, but is still undesirable. In many DAC applications (especially closed-loop systems where nonmonotonicity can change negative feedback to positive feedback), it is critically important that DACs are monotonic. DAC monotonicity is often explicitly specified on data sheets, although if the DNL is guaranteed to be less than 1 LSB (i.e., |DNL| ≤ 1 LSB), the device must be monotonic, even without an explicit guarantee.

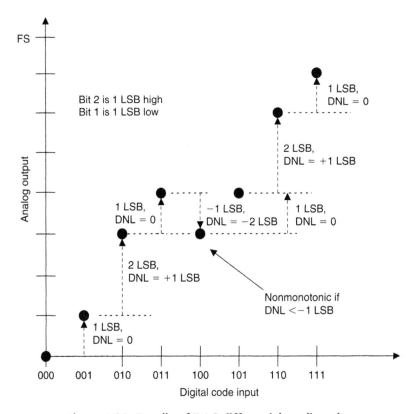

Figure 1.20: Details of DAC differential nonlinearity

In Figure 1.21, the DNL of an ADC is examined more closely on an expanded scale. ADCs can be nonmonotonic, but a more common result of excess DNL in ADCs is *missing codes*. Missing codes in an ADC are as objectionable as nonmonotonicity in a DAC. Again, they result from DNL < -1 LSB.

Not only can ADCs have missing codes, they can also be nonmonotonic as shown in Figure 1.22. As in the case of DACs, this can present major problems—especially in servo applications.

In a DAC, there can be no missing codes—each digital input word will produce a corresponding analog output. However, DACs can be nonmonotonic as previously discussed. In a straight binary DAC, the most likely place a nonmonotonic condition can develop is at midscale between the two codes: 011...11 and 100...00. If a nonmonotonic condition occurs here, it is generally because the DAC is not properly calibrated or trimmed. A successive approximation ADC with an internal nonmonotonic DAC will generally produce missing codes but remain monotonic. However, it is possible for an ADC to be nonmonotonic—again depending on the particular conversion architecture. Figure 1.22 shows the transfer function of an ADC, which is nonmonotonic and has a missing code.

ADCs that use the *subranging* architecture divide the input range into a number of coarse segments, and each coarse segment is further divided into smaller segments—and

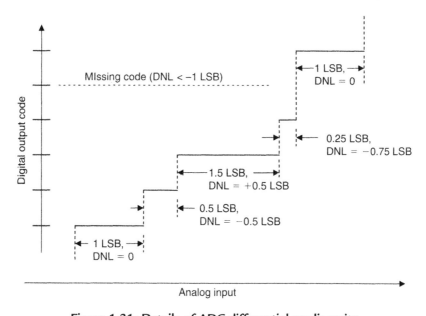

Figure 1.21: Details of ADC differential nonlinearity

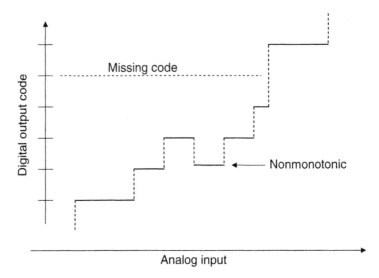

Figure 1.22: Nonmonotonic ADC with missing code

ultimately the final code is derived. An improperly trimmed subranging ADC may exhibit nonmonotonicity, wide codes, or missing codes at the subranging points as shown in Figure 1.23 a, b, and c, respectively. This type of ADC should be trimmed so that drift due to aging or temperature produces wide codes at the sensitive points rather than nonmonotonic or missing codes.

Defining missing codes is more difficult than defining nonmonotonicity. All ADCs suffer from some inherent transition noise as shown in Figure 1.24 (think of it as the flicker between adjacent values of the last digit of a DVM). As resolutions and bandwidths become higher, the range of input over which transition noise occurs may approach, or even exceed, 1 LSB. High resolution wideband ADCs generally have internal noise sources that can be reflected to the input as effective input noise summed with the signal. The effect of this noise, especially if combined with a negative DNL error, may be that there are some (or even all) codes where transition noise is present for the whole range of inputs. Therefore, there are some codes for which there is *no* input that will *guarantee* that code as an output, although there may be a range of inputs that will *sometimes* produce that code.

For low resolution ADCs, it may be reasonable to define *no missing codes* as a combination of transition noise and DNL, which guarantees some level (perhaps 0.2 LSB) of noise-free code for all codes. However, this is impossible to achieve at the very high resolutions achieved by modern sigma-delta ADCs, or even at lower resolutions in wide bandwidth

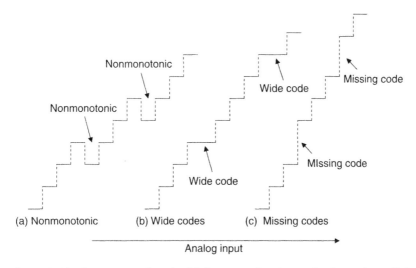

Figure 1.23: Errors associated with improperly trimmed subranging ADC

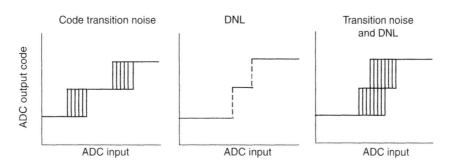

Figure 1.24: Combined effects of code transition noise and DNL

sampling ADCs. In these cases, the manufacturer must define noise levels and resolution in some other way. Which method is used is less important, but the data sheet should contain a clear definition of the method used and the performance to be expected. A complete discussion of effective input noise follows in Section 1.3 of this chapter.

The discussion thus far has dealt with only the most important DC specifications associated with data converters. Other less important specifications require only a definition. For specifications not covered in this section, the reader is referred to Section 1.5 of this chapter for a complete alphabetical listing of data converter specifications along with their definitions.

1.2 Sampling Theory

This section discusses the basics of sampling theory. A block diagram of a typical real-time sampled data system is shown in Figure 1.25. Prior to the actual analog-to-digital conversion, the analog signal usually passes through some sort of signal conditioning circuitry which performs such functions as amplification, attenuation, and filtering. The low-pass/band-pass filter is required to remove unwanted signals outside the bandwidth of interest and prevent aliasing.

The system shown in Figure 1.25 is a real-time system; i.e., the signal to the ADC is continuously sampled at a rate equal to f_s, and the ADC presents a new sample to the DSP at this rate. In order to maintain real-time operation, the DSP must perform all its required computation within the sampling interval, $1/f_s$, and present an output sample to the DAC before arrival of the next sample from the ADC. An example of a typical DSP function would be a digital filter.

In the case of FFT analysis, a block of data is first transferred to the DSP memory. The FFT is calculated at the same time a new block of data is transferred into the memory, in order to maintain real-time operation. The DSP must calculate the FFT during the data transfer interval so it will be ready to process the next block of data.

Note that the DAC is required only if the DSP data must be converted back into an analog signal (as would be the case in a voiceband or audio application, for example). There

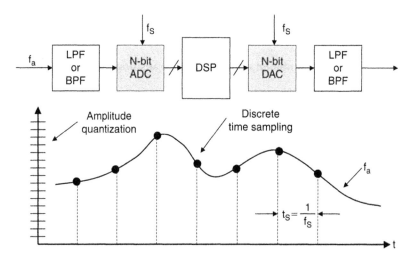

Figure 1.25: Sampled data system

are many applications where the signal remains entirely in digital format after the initial A/D conversion. Similarly, there are applications where the DSP is solely responsible for generating the signal to the DAC. If a DAC is used, it must be followed by an analog anti-imaging filter to remove the image frequencies. Finally, there are slower speed industrial process control systems where sampling rates are much lower—regardless of the system, the fundamentals of sampling theory still apply.

There are two key concepts involved in the actual analog-to-digital and digital-to-analog conversion process: *discrete time sampling* and *finite amplitude resolution due to quantization.* An understanding of these concepts is vital to data converter applications.

1.2.1 The Need for a Sample-and-hold Amplifier (SHA) Function

The generalized block diagram of a sampled data system, shown in Figure 1.25, assumes some type of AC signal at the input. It should be noted that this does not necessarily have to be so, as in the case of modern digital voltmeters (DVMs) or ADCs optimized for DC measurements, but for this discussion assume that the input signal has some upper frequency limit f_a.

Most ADCs today have a built-in sample-and-hold function, thereby allowing them to process AC signals. This type of ADC is referred to as a *sampling ADC.* However many early ADCs, such as Analog Devices' industry-standard AD574, were not of the sampling type, but simply *encoders* as shown in Figure 1.26. If the input signal to a SAR ADC (assuming no SHA function) changes by more than 1 LSB during the conversion time (8 μs in the example),

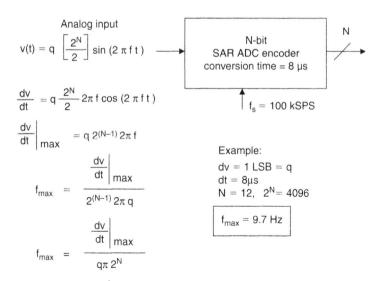

Figure 1.26: Input frequency limitations of nonsampling ADC (encoder)

the output data can have large errors, depending on the location of the code. Most ADC architectures are subject to this type of error—some more, some less—with the possible exception of Flash converters having well-matched comparators.

Assume that the input signal to the encoder is a sinewave with a full-scale amplitude ($q2^N/2$), where q is the weight of 1 LSB.

$$v(t) = q(2^N/2)\sin(2\pi f t). \tag{1.1}$$

Taking the derivative:

$$dv/dt = q2\pi f(2^N/2)\cos(2\pi f t). \tag{1.2}$$

The maximum rate of change is therefore:

$$dv/dt\big|_{max} = q2\pi f(2^N/2). \tag{1.3}$$

Solving for f:

$$f = (dv/dt\big|_{max})/(q\pi 2^N). \tag{1.4}$$

If $N = 12$, and 1 LSB change ($dv = q$) is allowed during the conversion time ($dt = 8\,\mu s$), the equation can be solved for f_{max}, the maximum full-scale signal frequency that can be processed without error:

$$f_{max} = 9.7\,Hz.$$

This implies any input frequency greater than 9.7 Hz is subject to conversion errors, even though a sampling frequency of 100 kSPS is possible with the 8 μs ADC (this allows an extra 2 μs interval for an external SHA to reacquire the signal after coming out of the hold mode).

To process AC signals, a sample-and-hold function is added as shown in Figure 1.27. The ideal SHA is simply a switch driving a hold capacitor followed by a high input impedance buffer. The input impedance of the buffer must be high enough so that the capacitor is discharged by less than 1 LSB during the hold time. The SHA samples the signal in the *sample* mode, and holds the signal constant during the *hold* mode. The timing is adjusted so that the encoder performs the conversion during the hold time. A sampling ADC can therefore process fast signals—the upper frequency limitation is determined by the SHA aperture jitter, bandwidth, distortion, etc., not the encoder. In the example shown, a good

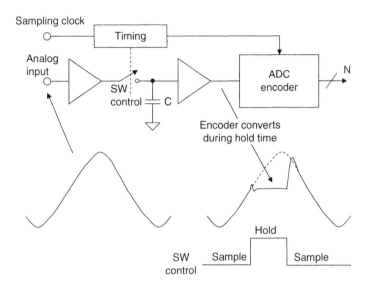

Figure 1.27: Sample-and-hold function required for digitizing AC signals

sample-and-hold could acquire the signal in 2 µs, allowing a sampling frequency of 100 kSPS, and the capability of processing input frequencies up to 50 kHz. A complete discussion of the SHA function including these specifications follows later in this chapter.

It is important to understand a subtle difference between a true *sample-and-hold* amplifier (SHA) and a *track-and-hold* amplifier (T/H, or THA). Strictly speaking, the output of a sample-and-hold is not defined during the sample mode; however, the output of a track-and-hold tracks the signal during the sample or *track* mode. In practice, the function is generally implemented as a track-and-hold, and the terms *track-and-hold* and *sample-and-hold* are often used interchangeably. The waveforms shown in Figure 1.27 are those associated with a track-and-hold.

In order to better understand the types of AC errors an ADC can make without a sample-and-hold function, consider Figure 1.28. The photos show the reconstructed output of an 8-bit ADC (Flash converter) with and without the sample-and-hold function. In an ideal Flash converter the comparators are perfectly matched, and no sample-and-hold is required. In practice, however, there are timing mismatches between the comparators that cause high frequency inputs to exhibit nonlinearities and missing codes as shown in the right-hand photos. The data was taken by driving a DAC with the ADC output. The DAC output is a low frequency aliased sinewave corresponding to the difference between the sampling frequency (20 MSPS) and the ADC input frequency (19.98 MHz). In this case, the alias frequency is 20 kHz. (Aliasing is explained in detail in the next section.)

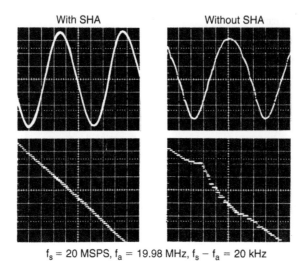

f_s = 20 MSPS, f_a = 19.98 MHz, $f_s - f_a$ = 20 kHz

Figure 1.28: 8-bit, 20 MSPS Flash ADC with and without sample-and-hold

1.2.2 The Nyquist Criteria

A continuous analog signal is sampled at discrete intervals, $t_s = 1/f_s$, which must be carefully chosen to ensure an accurate representation of the original analog signal. It is clear that the more samples taken (faster sampling rates), the more accurate the digital representation; however, if fewer samples are taken (lower sampling rates), a point is reached where critical information about the signal is actually lost. The mathematical basis of sampling was set forth by Harry Nyquist of Bell Telephone Laboratories in two classic papers published in 1924 and 1928, respectively. (See References 1 and 2.) Nyquist's original work was shortly supplemented by R. V. L. Hartley (Reference 3). These papers formed the basis for the PCM work to follow in the 1940s, and in 1948 Claude Shannon wrote his classic paper on communication theory (Reference 4).

Simply stated, Nyquist's criteria require that the sampling frequency be at least twice the highest frequency contained in the signal, or information about the signal will be lost. If the sampling frequency is less than twice the maximum analog signal frequency, a phenomena known as aliasing will occur.

In order to understand the implications of *aliasing* in both the time and frequency domain, first consider the case of a time domain representation of a single tone sinewave sampled as shown in Figure 1.30. In this example, the sampling frequency f_s is not at least $2f_a$, but only slightly more than the analog input frequency f_a—the Nyquist criteria is violated. Notice that the pattern of the actual samples produces an *aliased* sinewave at a lower frequency equal to $f_s - f_a$.

- A signal with a maximum frequency f_a must be sampled at a rate $f_s > 2f_a$ or information about the signal will be lost because of aliasing.

- Aliasing occurs whenever $f_s < 2 f_a$

- The concept of aliasing is widely used in communications applications such as direct IF-to-digital conversion.

- A signal which has frequency components between f_a and f_b must be sampled at a rate $f_s > 2 (f_b - f_a)$ in order to prevent alias components from overlapping the signal frequencies.

Figure 1.29: Nyquist's criteria

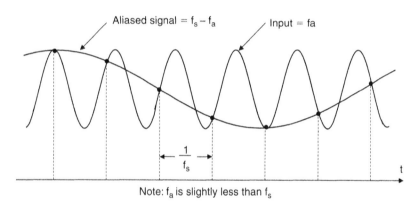

Aliased signal $= f_s - f_a$ Input $=$ fa

$\dfrac{1}{f_s}$

t

Note: f_a is slightly less than f_s

Figure 1.30: Aliasing in the time domain

The corresponding frequency domain representation of this scenario is shown in Figure 1.31b. Now consider the case of a single frequency sinewave of frequency f_a sampled at a frequency f_s by an ideal impulse sampler (see Figure 1.31a). Also assume that $f_s > 2f_a$ as shown. The frequency-domain output of the sampler shows *aliases* or *images* of the original signal around every multiple of f_s; i.e., at frequencies equal to $|\pm Kf_s \pm f_a|$, K $= 1, 2, 3, 4, \ldots$

The *Nyquist* bandwidth is defined to be the frequency spectrum from DC to $f_s/2$. The frequency spectrum is divided into an infinite number of *Nyquist zones*, each having a width equal to $0.5 f_s$ as shown. In practice, the ideal sampler is replaced by an ADC followed by an FFT processor. The FFT processor only provides an output from DC to $f_s/2$, i.e., the signals or aliases that appear in the first Nyquist zone.

Now consider the case of a signal that is outside the first Nyquist zone (Figure 1.31b). The signal frequency is only slightly less than the sampling frequency, corresponding to the condition shown in the time domain representation in Figure 1.30. Notice that even though

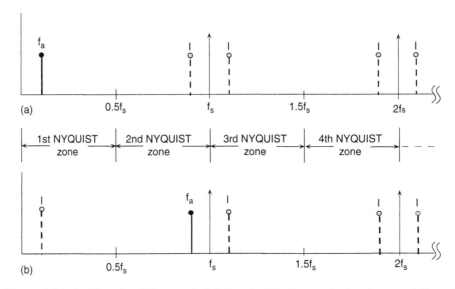

Figure 1.31: Analog signal f_a sampled @ f_s using ideal sampler has images (aliases) at $|\pm Kf_s \pm f_a|$, K = 1, 2, 3, ...

the signal is outside the first Nyquist zone, its image (or *alias*), $f_s - f_a$, falls inside. Returning to Figure 1.31a, it is clear that if an unwanted signal appears at any of the image frequencies of f_a, it will also occur at f_a, thereby producing a spurious frequency component in the first Nyquist zone.

This is similar to the analog mixing process and implies that some filtering ahead of the sampler (or ADC) is required to remove frequency components that are outside the Nyquist bandwidth, but whose aliased components fall inside it. The filter performance will depend on how close the out-of-band signal is to $f_s/2$ and the amount of attenuation required.

1.2.3 Baseband Antialiasing Filters

Baseband sampling implies that the signal to be sampled lies in the first Nyquist zone. It is important to note that with no input filtering at the input of the ideal sampler, *any frequency component (either signal or noise) that falls outside the Nyquist bandwidth in any Nyquist zone will be aliased back into the first Nyquist zone.* For this reason, an antialiasing filter is used in almost all sampling ADC applications to remove these unwanted signals.

Properly specifying the antialiasing filter is important. The first step is to know the characteristics of the signal being sampled. Assume that the highest frequency of interest is f_a. The antialiasing filter passes signals from DC to f_a while attenuating signals above f_a.

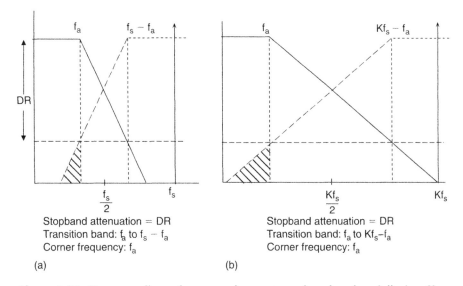

Figure 1.32: Oversampling relaxes requirements on baseband antialiasing filter

Assume that the corner frequency of the filter is chosen to be equal to f_a. The effect of the finite transition from minimum to maximum attenuation on system dynamic range is illustrated in Figure 1.32a.

Assume that the input signal has full-scale components well above the maximum frequency of interest, fa. The diagram shows how full-scale frequency components above $f_s - f_a$ are aliased back into the bandwidth DC to f_a. These aliased components are indistinguishable from actual signals and therefore limit the dynamic range to the value on the diagram which is shown as *DR*.

Some texts recommend specifying the antialiasing filter with respect to the Nyquist frequency, $f_s/2$, but his assumes that the signal bandwidth of interest extends from DC to $f_s/2$ which is rarely the case. In the example shown in Figure 1.32a, the aliased components between f_a and $f_s/2$ are not of interest and do not limit the dynamic range.

The antialiasing filter transition band is therefore determined by the corner frequency f_a, the stopband frequency $f_s - f_a$, and the desired stopband attenuation, DR. The required system dynamic range is chosen based on the requirement for signal fidelity.

Filters become more complex as the transition band becomes sharper, all other things being equal. For instance, a Butterworth filter gives 6 dB attenuation per octave for each filter pole (as do all filters). Achieving 60 dB attenuation in a transition region between 1 MHz and 2 MHz (1 octave) requires a minimum of 10 poles—not a trivial filter, and definitely a design challenge.

Therefore, other filter types are generally more suited to applications where the requirement is for a sharp transition band and in-band flatness coupled with linear phase response. Elliptic filters meet these criteria and are a popular choice. A number of companies specialize in supplying custom analog filters; TTE is an example of such a company (Reference 5).

From this discussion, we can see how the sharpness of the antialiasing transition band can be traded off against the ADC sampling frequency. Choosing a higher sampling rate (oversampling) reduces the requirement on transition band sharpness (hence, the filter complexity) at the expense of using a faster ADC and processing data at a faster rate. This is illustrated in Figure 1.32b which shows the effects of increasing the sampling frequency by a factor of K, while maintaining the same analog corner frequency, f_a, and the same dynamic range, DR, requirement. The wider transition band (f_a to $Kf_s - f_a$) makes this filter easier to design than for the case of Figure 1.32a.

The antialiasing filter design process is started by choosing an initial sampling rate of 2.5 to 4 times f_a. Determine the filter specifications based on the required dynamic range and see if such a filter is realizable within the constraints of the system cost and performance. If not, consider a higher sampling rate which may require using a faster ADC. It should be mentioned that sigma-delta ADCs are inherently highly over-sampled converters, and the resulting relaxation in the analog antialiasing filter requirements is therefore an added benefit of this architecture.

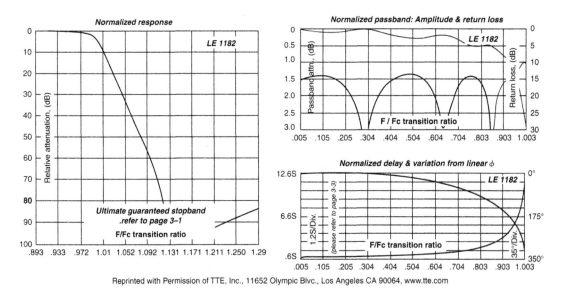

Reprinted with Permission of TTE, Inc., 11652 Olympic Blvd., Los Angeles CA 90064, www.tte.com

Figure 1.33: Characteristics of 11-pole elliptical filter (TTE, Inc., LE1182 Series)

The antialiasing filter requirements can also be relaxed somewhat if it is certain that there will never be a full-scale signal at the stopband frequency $f_s - f_a$. In many applications, it is improbable that full-scale signals will occur at this frequency. If the maximum signal at the frequency $f_s - f_a$ will never exceed X dB below full-scale, then the filter stopband attenuation requirement can be reduced by that same amount. The new requirement for stopband attenuation at $f_s - f_a$ based on this knowledge of the signal is now only DR – X dB. When making this type of assumption, be careful to treat any noise signals that may occur above the maximum signal frequency f_a as unwanted signals that will also alias back into the signal bandwidth.

As an example, the normalized response of the TTE, Inc., LE1182 11-pole elliptic antialiasing filter is shown in Figure 1.33. Notice that this filter is specified to achieve at least 80 dB attenuation between f_c and $1.2 f_c$. The corresponding pass band ripple, return loss, delay, and phase response are also shown in Figure 1.33. This custom filter is available in corner frequencies up to 100 MHz and in a choice of PC board, BNC, or SMA with compatible packages.

1.2.4 Undersampling (Harmonic Sampling, Bandpass Sampling, if Sampling, Direct IF-to-Digital Conversion)

Thus far we have considered the case of baseband sampling, where all the signals of interest lie within the first Nyquist zone. Figure 1.34a shows such a case, where the band of sampled signals is limited to the first Nyquist zone, and images of the original band of frequencies appear in each of the other Nyquist zones.

Consider the case shown in Figure 1.34b, where the sampled signal band lies entirely within the second Nyquist zone. The process of sampling a signal outside the first Nyquist zone is often referred to as *under-sampling*, or *harmonic sampling*. Note that the image which falls in the first Nyquist zone contains all the information in the original signal, with the exception of its original location (the order of the frequency components within the spectrum is reversed, but this is easily corrected by re-ordering the output of the FFT).

Figure 1.34c shows the sampled signal restricted to the third Nyquist zone. Note that the image that falls into the first Nyquist zone has no frequency reversal. In fact, the sampled signal frequencies may lie in *any* unique Nyquist zone, and the image falling into the first Nyquist zone is still an accurate representation (with the exception of the frequency reversal tjat occurs when the signals are located in even Nyquist zones). At this point we can clearly restate the Nyquist criteria:

*A signal must be sampled at a rate equal to or greater than twice its **bandwidth** in order to preserve all the signal information.*

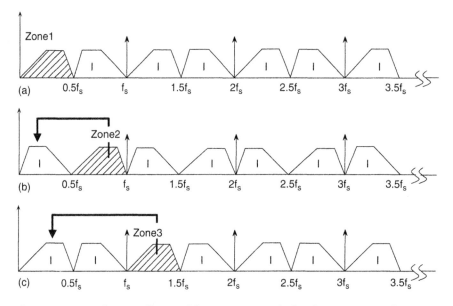

Figure 1.34: Undersampling and frequency translation between Nyquist zones

Notice that there is no mention of the absolute *location* of the band of sampled signals within the frequency spectrum relative to the sampling frequency. The only constraint is that the band of sampled signals be restricted to a *single* Nyquist zone, i.e., the signals must not overlap any multiple of $f_s/2$ (this, in fact, is the primary function of the antialiasing filter).

Sampling signals above the first Nyquist zone has become popular in communications because the process is equivalent to analog demodulation. It is becoming common practice to sample IF signals directly and then use digital techniques to process the signal, thereby eliminating the need for an IF demodulator and filters. Clearly, however, as the IF frequencies become higher, the dynamic performance requirements on the ADC become more critical. The ADC input bandwidth and distortion performance must be adequate at the IF frequency, rather than only baseband. This presents a problem for most ADCs designed to process signals in the first Nyquist zone, therefore an ADC suitable for undersampling applications must maintain dynamic performance into the higher order Nyquist zones.

1.2.5 Antialiasing Filters in Undersampling Applications

Figure 1.35 shows a signal in the second Nyquist zone centered around a carrier frequency, f_c, whose lower and upper frequencies are f_1 and f_2. The antialiasing filter is a bandpass filter. The desired dynamic range is DR, which defines the filter stopband attenuation. The upper transition band is f_2 to $2f_s - f_2$, and the lower is f_1 to $f_s - f_1$. As in the case of base-band

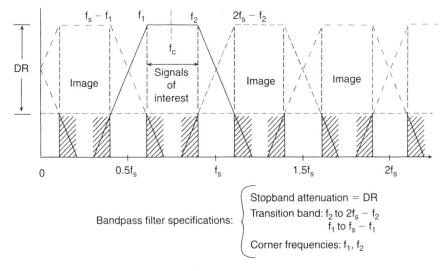

Bandpass filter specifications: $\left\{\begin{array}{l}\text{Stopband attenuation} = \text{DR}\\ \text{Transition band: } f_2 \text{ to } 2f_s - f_2\\ \qquad\qquad\quad f_1 \text{ to } f_s - f_1\\ \text{Corner frequencies: } f_1, f_2\end{array}\right.$

Figure 1.35: Antialiasing filter for undersampling

sampling, the antialiasing filter requirements can be relaxed by proportionally increasing the sampling frequency, but f_c must also be increased so that it is always centered in the second Nyquist zone.

Two key equations can be used to select the sampling frequency, f_s, given the carrier frequency, f_c, and the bandwidth of its signal, Δf. The first is the Nyquist criteria:

$$f_s > 2\Delta f \tag{1.5}$$

The second equation ensures that f_c is placed in the center of a Nyquist zone:

$$f_s = \frac{4f_c}{2NZ - 1} \tag{1.6}$$

where $NZ = 1, 2, 3, 4, \ldots$ and NZ corresponds to the Nyquist zone in which the carrier and its signal fall (see Figure 1.36).

NZ is normally chosen to be as large as possible while still maintaining $f_s > 2\Delta f$. This results in the minimum required sampling rate. If NZ is chosen to be odd, then f_c and its signal will fall in an odd Nyquist zone, and the image frequencies in the first Nyquist zone will not be reversed. Trade-offs can be made between the sampling frequency and the complexity of the antialiasing filter by choosing smaller values of NZ (hence a higher sampling frequency).

As an example, consider a 4 MHz wide signal centered around a carrier frequency of 71 MHz. The minimum required sampling frequency is therefore 8 MSPS. Solving Eq. 1.6 for NZ

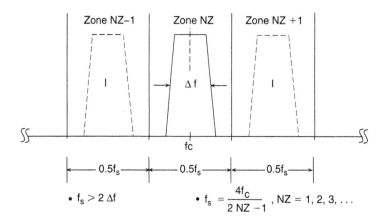

Figure 1.36: Centering an undersampled signal within a Nyquist zone

using $f_c = 71$ MHz and $f_s = 8$ MSPS yields NZ = 18.25. However, NZ must be an integer, so we round 18.25 to the next lowest integer, 18. Solving Eq. 1.6 again for f_s yields $f_s = 8.1143$ MSPS. The final values are therefore $f_s = 8.1143$ MSPS, $f_c = 71$ MHz, and NZ = 18.

Now assume that we desire more margin for the antialiasing filter, and we select f_s to be 10 MSPS. Solving Eq. 1.6 for NZ, using $f_c = 71$ MHz and $f_s = 10$ MSPS yields NZ = 14.7. We round 14.7 to the next lowest integer, giving NZ = 14. Solving Eq. 1.6 again for f_s yields $f_s = 10.519$ MSPS. The final values are therefore $f_s = 10.519$ MSPS, $f_c = 71$ MHz, and NZ = 14.

The above iterative process can also be carried out starting with f_s and adjusting the carrier frequency to yield an integer number for NZ.

1.3 Data Converter AC Errors

This section examines the AC errors associated with data converters. Many of the errors and specifications apply equally to ADCs and DACs, while some are more specific to one or the other. All possible specifications are not discussed here, only the most common ones. Section 1.4 of this chapter contains a comprehensive listing of converter specifications as well as their definitions, including some not discussed in this section.

1.3.1 Theoretical Quantization Noise of an Ideal N-Bit Converter

The only errors (DC or AC) associated with an ideal N-bit data converter are those related to the sampling and quantization processes. The maximum error an ideal converter makes when digitizing a signal is $\pm\frac{1}{2}$ LSB. The transfer function of an ideal N-bit ADC is shown in Figure 1.37. The quantization error for any AC signal that spans more than a few LSBs can

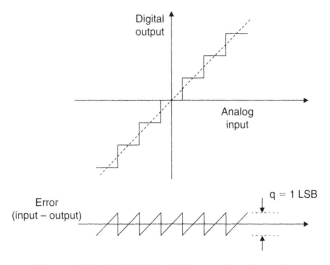

Figure 1.37: Ideal N-bit ADC quantization noise

be approximated by an uncorrelated sawtooth waveform having a peak-to-peak amplitude of q, the weight of an LSB. Although this analysis is not precise, it is accurate enough for most applications. W. R. Bennett of Bell Laboratories analyzed the actual spectrum of quantization noise in his classic 1948 paper (Reference 1). With certain simplifying assumptions, his detailed mathematical analysis simplifies to that of Figure 1.37. Other significant papers on converter noise (References 2–5) followed Bennett's classic publication.

The quantization error as a function of time is shown in Figure 1.38. Again, a simple sawtooth waveform provides a sufficiently accurate model for analysis. The equation of the sawtooth error is given by

$$e(t) = st, -q/2s < t < +q/2s \tag{1.7}$$

The mean-square value of e(t) can be written:

$$\overline{e^2(t)} = \frac{s}{q} \int_{-q/2s}^{+q/2s} (st)^2 \, dt \tag{1.8}$$

Performing the simple integration and simplifying,

$$\overline{e^2(t)} = \frac{q^2}{12} \tag{1.9}$$

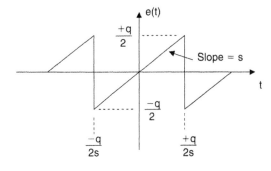

- Error = $\quad e(t) = st, \quad \dfrac{-q}{2s} < t < \dfrac{+q}{2s}$

- Mean-square error = $\quad \overline{e^2(t)} = \dfrac{s}{q} \displaystyle\int_{-q/2s}^{+q/2s} (st)^2 \, dt = \dfrac{q^2}{12}$

- Root-mean-square error = $\sqrt{\overline{e^2(t)}} = \dfrac{q}{\sqrt{12}}$

Figure 1.38: Quantization noise as a function of time

The root-mean-square quantization error is therefore:

$$\text{rms quantization noise} = \sqrt{\overline{r^2(t)}} = \frac{q}{\sqrt{12}} \tag{1.10}$$

As Bennett points out (Reference 1), this noise is approximately Gaussian and spread more or less uniformly over the Nyquist bandwidth DC to $f_s/2$. The underlying assumption here is that the quantization noise is uncorrelated to the input signal. Under certain conditions where the sampling clock and the signal are harmonically related, the quantization noise becomes correlated and the energy is concentrated at the harmonics of the signal—the rms value remains approximately $q/\sqrt{12}$.

The theoretical signal-to-noise ratio can now be calculated assuming a full-scale input sinewave:

$$\text{Input FS Sinewave} = v(t) = \frac{q2^N}{2} \sin(2\pi ft) \tag{1.11}$$

The rms value of the input signal is therefore:

$$\text{rms value of FS input} = \frac{q2^N}{2\sqrt{2}} \tag{1.12}$$

$$\text{FS input} = \quad v(t) = \left[\frac{q2^N}{2}\right] \sin(2\pi \, ft)$$

$$\text{RMS value of FS sinewave} = \frac{q2^N}{2\sqrt{2}}$$

$$\text{RMS value of quantization noise} = \frac{q}{\sqrt{12}}$$

$$\text{SNR} = 20 \log_{10}\left[\frac{\text{RMS value of FS sinewave}}{\text{RMS value of quantization noise}}\right] = 20 \log_{10} 2^N + 20 \log_{10} \sqrt{\frac{3}{2}}$$

$$\boxed{\begin{array}{c} \text{SNR} = 6.02N + 1.76\text{dB} \\ \text{(measured over the Nyquist bandwidth : DC to } f_s/2) \end{array}}$$

Figure 1.39: Theoretical signal-to-quantization noise ratio of an ideal N-bit converter

The rms signal-to-noise ratio for an ideal N-bit converter is therefore:

$$\text{SNR} = 20\log_{10} \frac{\text{rms value of FS input}}{\text{rms value of quantization noise}} \tag{1.13}$$

$$\text{SNR} = 20\log_{10}\left[\frac{q2^N/2\sqrt{2}}{q/\sqrt{12}}\right] = 6.02N + 1.76\,\text{dB, over dc to}\, f_s/2\,\text{bandwidth} \tag{1.14}$$

These relationships are summarized in Figure 1.39.

Bennett's paper shows that although the actual spectrum of the quantization noise is quite complex to analyze—the simplified analysis which leads to Eq. 1.14 is accurate enough for most purposes. However, it is important to emphasize again that the rms quantization noise is measured over the full Nyquist bandwidth, DC to $f_s/2$. In many applications, the actual signal of interest occupies a smaller bandwidth, BW. If digital filtering is used to filter out noise components outside the bandwidth BW, then a correction factor (called *process gain*) must be included in the equation to account for the resulting increase in SNR. The process of sampling a signal at a rate greater than twice its bandwidth is often referred to as *oversampling*. In fact, oversampling in conjunction with quantization noise shaping and digital filtering is a key concept in sigma-delta converters.

$$\text{SNR} = 6.02N + 1.76\,\text{dB} + 10\log_{10}\frac{f_s}{2 \times \text{BW}}, \text{over bandwidth BW} \tag{1.15}$$

The significance of process gain can be seen from the following example. In many digital basestations or other wideband receivers the signal bandwidth is composed of many individual channels, and a single ADC is used to digitize the entire bandwidth. For instance, the analog cellular radio system (AMPS) in the U.S. consists of 416 30-kHz-wide channels, occupying a bandwidth of approximately 12.5 MHz. Assume a 65 MSPS sampling frequency, and that digital filtering is used to separate the individual 30 kHz channels. The process gain due to oversampling is therefore given by:

$$\text{Process Gain} = 10\log_{10}\frac{f_s}{2 \times \text{BW}} = 10\log_{10}\frac{65 \times 10^6}{2 \times 30 \times 10^3} = 30.3\,\text{dB} \qquad (1.16)$$

The process gain is added to the ADC SNR specification to yield the actual SNR in the 30 kHz bandwidth. In the above example, if the ADC SNR specification is 65 dB (DC to $f_s/2$), then it is increased to 95.3 dB in the 30 kHz channel bandwidth (after appropriate digital filtering).

Figure 1.41 shows an application that combines oversampling and undersampling. The signal of interest has a bandwidth BW and is centered around a carrier frequency f_c. The sampling frequency can be much less than f_c and is chosen such that the signal of interest is centered in its Nyquist zone. Analog and digital filtering removes the noise outside the signal bandwidth of interest, and therefore results in process gain per Eq. 1.16.

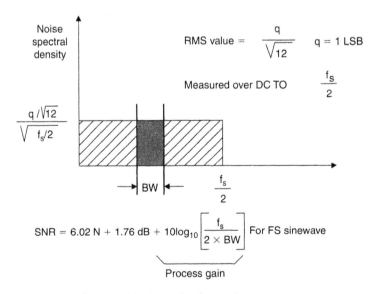

Figure 1.40: Quantization noise spectrum

Although the rms value of the noise is accurately approximated by $q/\sqrt{12}$, its frequency domain content may be highly correlated to the AC input signal. For instance, there is greater correlation for low amplitude periodic signals than for large amplitude random signals. Quite often, the assumption is made that the theoretical quantization noise appears as white noise, spread uniformly over the Nyquist bandwidth DC to $f_s/2$. Unfortunately, this is not true in all cases. In the case of strong correlation, the quantization noise appears concentrated at the various harmonics of the input signal, just where you don't want them. Bennett (Reference 1) has an extensive analysis of the frequency content contained in the quantization noise spectrum in his classic 1948 paper.

In most practical applications, the input to the ADC is a band of frequencies (always summed with some unavoidable system noise), so the quantization noise tends to be random. In spectral analysis applications (or in performing FFTs on ADCs using spectrally pure sinewaves—see Figure 1.42), however, the correlation between the quantization noise and the signal depends upon the ratio of the sampling frequency to the input signal. This is

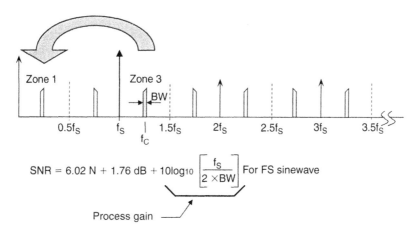

$$\text{SNR} = 6.02\,N + 1.76\,\text{dB} + 10\log_{10}\left[\frac{f_s}{2 \times \text{BW}}\right] \text{ For FS sinewave}$$

Process gain

Figure 1.41: Undersampling and oversampling combined results in process gain

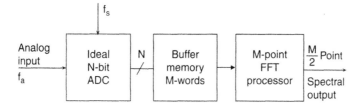

Figure 1.42: Dynamic performance analysis of an ideal N-bit ADC

demonstrated in Figure 1.43, where the output of an ideal 12-bit ADC is analyzed using a 4096-point FFT. In the left-hand FFT plot, the ratio of the sampling frequency to the input frequency was chosen to be exactly 32, and the worst harmonic is about 76 dB below the fundamental. The right hand diagram shows the effects of slightly offsetting the ratio to 4096/127 = 32.25196850394, showing a relatively random noise spectrum, where the SFDR is now about 92 dBc. In both cases, the rms value of all the noise components is approximately $q/\sqrt{12}$, but in the first case, the noise is concentrated at harmonics of the fundamental.

Note that this variation in the apparent harmonic distortion of the ADC is an artifact of the sampling process and the correlation of the quantization error with the input frequency. In a practical ADC application, the quantization error generally appears as random noise because of the random nature of the wideband input signal and the additional fact that there is a usually a small amount of system noise which acts as a dither signal to further randomize the quantization error spectrum.

It is important to understand the above point, because single-tone sinewave FFT testing of ADCs is one of the universally accepted methods of performance evaluation. In order to accurately measure the harmonic distortion of an ADC, steps must be taken to ensure that the test setup truly measures the ADC distortion, not the artifacts due to quantization noise correlation. This is done by properly choosing the frequency ratio and sometimes by injecting a small amount of noise (dither) with the input signal. The exact same precautions apply to measuring DAC distortion with an analog spectrum analyzer.

Figure 1.44 shows the FFT output for an ideal 12-bit ADC. Note that the average value of the noise floor of the FFT is approximately 100 dB below full-scale, but the theoretical SNR of a 12-bit ADC is 74 dB. The FFT noise floor is not the SNR of the ADC, because the FFT acts

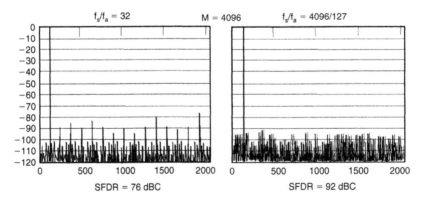

Figure 1.43: Effect of ratio of sampling clock to input frequency on SFDR for ideal 12-bit ADC

like an analog spectrum analyzer with a bandwidth of f_s/M, where M is the number of points in the FFT. The theoretical FFT noise floor is therefore $10\log10(M/2)$ dB below the quantization noise floor due to the processing gain of the FFT. In the case of an ideal 12-bit ADC with an SNR of 74 dB, a 4096-point FFT would result in a processing gain of $10\log10(4096/2) = 33$ dB, thereby resulting in an overall FFT noise floor of $74 + 33 = 107$ dBc. In fact, the FFT noise floor can be reduced even further by going to larger and larger FFTs; just as an analog spectrum analyzer's noise floor can be reduced by narrowing the bandwidth. When testing ADCs using FFTs, it is important to ensure that the FFT size is large enough that the distortion products can be distinguished from the FFT noise floor itself.

1.3.2 Noise in Practical ADCs

A practical sampling ADC (one that has an integral sample-and-hold), regardless of architecture, has a number of noise and distortion sources as shown in Figure 1.45. The wideband analog front-end buffer has wideband noise, nonlinearity, and also finite bandwidth. The SHA introduces further nonlinearity, bandlimiting, and aperture jitter. The actual quantizer portion of the ADC introduces quantization noise, and both integral and differential nonlinearity. In this discussion, assume that sequential outputs of the ADC are loaded into a buffer memory of length M and that the FFT processor provides the spectral output. Also assume that the FFT arithmetic operations themselves introduce no significant errors relative to the ADC. However, when examining the output noise floor, the FFT processing gain (dependent on M) must be considered.

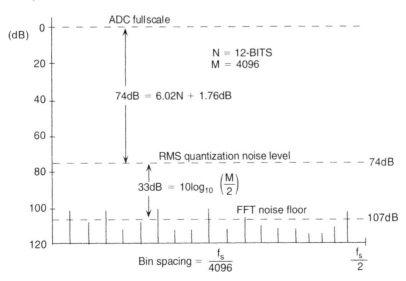

Figure 1.44: Noise floor for an ideal 12-bit ADC using 4096-point FFT

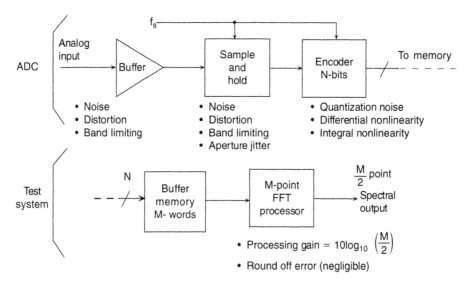

Figure 1.45: ADC model showing noise and distortion sources

1.3.2.1 Equivalent Input Referred Noise

Wideband ADC internal circuits produce a certain amount of rms noise due to resistor noise and "kT/C" noise. This noise is present even for DC input signals, and accounts for the fact that the output of most wideband (or high resolution) ADCs is a distribution of codes, centered around the nominal value of a DC input (Figure 1.46). To measure its value, the input of the ADC is either grounded or connected to a heavily decoupled voltage source, and a large number of output samples are collected and plotted as a histogram (sometimes referred to as a *grounded-input* histogram). Since the noise is approximately Gaussian, the standard deviation of the histogram is easily calculated (Reference 6), corresponding to the effective input rms noise. It is common practice to express this rms noise in terms of LSBs rms, although it can be expressed as an rms voltage referenced to the ADC full-scale input range.

1.3.2.2 Noise-Free (Flicker-Free) Code Resolution

The *noise-free code resolution* of an ADC is the number of bits beyond which it is impossible to distinctly resolve individual codes. The cause is the effective input noise (or input-referred noise) associated with all ADCs and described above. This noise can be expressed as an rms quantity, usually having the units of LSBs *rms*. Multiplying by a factor of 6.6 converts the rms noise into peak-to-peak noise (expressed in LSBs *peak-to-peak*). The total range of an N-bit ADC is 2^N LSBs. The noise-free (or flicker-free) resolution can be calculated using the equation:

$$\text{Noise-Free Code Resolution} = \log_2(2^N / \text{Peak-to-Peak Noise}) \qquad (1.17)$$

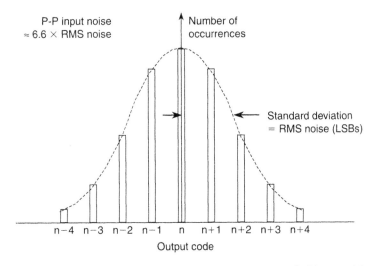

Figure 1.46: Effect of input-referred noise on ADC "grounded input" histogram

The specification is generally associated with high-resolution sigma-delta measurement ADCs, but is applicable to all ADCs.

The ratio of the FS range to the *rms* input noise is sometimes used to calculate resolution. In this case, the term *effective resolution* is used. Note that under identical conditions, effective resolution is larger than noise-free code resolution by $\log_2(6.6)$, or approximately 2.7 bits.

$$\text{Effective Resolution} = \log_2(2^N / \text{RMS Input Noise}) \qquad (1.18)$$

$$\text{Effective Resolution} = \text{Noise-Free Code Resolution} + 2.7\,\text{bits} \qquad (1.19)$$

The calculations are summarized in Figure 1.47.

1.3.3 Dynamic Performance of Data Converters

There are various ways to characterize the AC performance of ADCs. Before the 1970s, there was little standardization with respect to AC specifications, and measurement equipment and techniques were not well understood or available. Over nearly a 30-year period, manufacturers and customers have learned more about measuring the dynamic performance of converters, and the specifications shown in Figure 1.48 represent the most popular ones used today. Practically all the specifications represent the converter's performance in the frequency domain. The FFT is the heart of practically all these measurements.

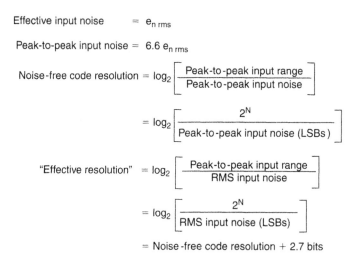

Figure 1.47: Calculating noise-free (flicker-free) code resolution from input-referred noise

- Harmonic distortion
- Worst harmonic
- Total harmonic distortion (THD)
- Total harmonic distortion plus noise (THD + N)
- Signal-to-noise-and-distortion ratio (SINAD, or S/N + D)
- Effective number of bits (ENOB)
- Signal-to-noise ratio (SNR)
- Analog bandwidth (full-power, small-signal)
- Spurious free dynamic range (SFDR)
- Two-tone intermodulation distortion
- Multitone intermodulation distortion
- Noise power ratio (NPR)
- Adjacent channel leakage ratio (ACLR)
- Noise figure
- Settling time, overvoltage recovery time

Figure 1.48: Quantifying data converter dynamic performance

1.3.3.1 Integral and Differential Nonlinearity Distortion Effects

One of the first things to realize when examining the nonlinearities of data converters is that the transfer function of a data converter has artifacts that do not occur in conventional linear devices such as op amps or gain blocks. The overall integral nonlinearity of an ADC is due to the integral nonlinearity of the front-end and SHA as well as the overall integral nonlinearity in the ADC transfer function. However, *differential nonlinearity is due exclusively to the*

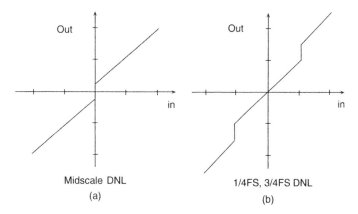

Figure 1.49: Typical ADC/ DAC DNL errors (exaggerated)

encoding process and may vary considerably, dependent on the ADC encoding architecture. Overall integral nonlinearity produces distortion products whose amplitude varies as a function of the input signal amplitude. For instance, second-order intermodulation products increase 2 dB for every 1 dB increase in signal level, and third-order products increase 3 dB for every 1 dB increase in signal level.

The differential nonlinearity in the ADC transfer function produces distortion products which not only depend on the amplitude of the signal but the positioning of the differential nonlinearity errors along the ADC transfer function. Figure 1.49 shows two ADC transfer functions having differential nonlinearity. The left-hand diagram shows an error that occurs at midscale. Therefore, for both large and small signals, the signal crosses through this point producing a distortion product which is relatively independent of the signal amplitude. The right-hand diagram shows another ADC transfer function which has differential nonlinearity errors at 1/4 and 3/4 full-scale. Signals above 1/2 scale peak-to-peak will exercise these codes and produce distortion, while those less than 1/2 scale peak-to-peak will not.

Most high-speed ADCs are designed so that differential nonlinearity is spread across the entire ADC range. Therefore, for signals that are within a few dB of full-scale, the overall integral nonlinearity of the transfer function determines the distortion products. For lower level signals, however, the harmonic content becomes dominated by the differential nonlinearities and does not generally decrease proportionally with decreases in signal amplitude.

1.3.3.2 *Harmonic Distortion, Worst Harmonic, Total Harmonic Distortion (THD), Total Harmonic Distortion Plus Noise (THD + N)*

There are a number of ways to quantify the distortion of an ADC. An FFT analysis can be used to measure the amplitude of the various harmonics of a signal. The harmonics of the input signal can be distinguished from other distortion products by their location in the frequency spectrum. Figure 1.50 shows a 7 MHz input signal sampled at 20 MSPS and the location of the first nine harmonics. Aliased harmonics of f_a fall at frequencies equal to $|\pm Kf_s \pm nf_a|$, where n is the order of the harmonic, and $K=0, 1, 2, 3,\ldots$. The second and third harmonics are generally the only ones specified on a data sheet because they tend to be the largest, although some data sheets may specify the value of the *worst* harmonic.

Harmonic distortion is normally specified in dBc (decibels below *carrier*), although at audio frequencies it may be specified as a percentage. Harmonic distortion is generally specified with an input signal near full-scale (generally 0.5 to 1 dB below full-scale to prevent clipping), but it can be specified at any level. For signals much lower than full-scale, other distortion products due to the DNL of the converter (not direct harmonics) may limit performance.

Total harmonic distortion (THD) is the ratio of the rms value of the fundamental signal to the mean value of the root-sum-square of its harmonics (generally, only the first five are significant). THD of an ADC is also generally specified with the input signal close to full-scale, although it can be specified at any level.

Total harmonic distortion plus noise (THD + N) is the ratio of the rms value of the fundamental signal to the mean value of the root-sum-square of its harmonics plus all noise components (excluding DC). The bandwidth over which the noise is measured must be specified. In the case of an FFT, the bandwidth is DC to $f_s/2$. (If the bandwidth of the measurement is DC to $f_s/2$, THD+N is equal to SINAD—see below).

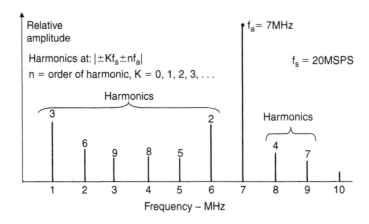

Figure 1.50: Location of distortion products: input signal = 7 MHz, sampling rate = 20 MSPS

1.3.3.3 Signal-to-Noise-and-Distortion Ratio (SINAD), Signal-to-Noise Ratio (SNR), and Effective Number of Bits (ENOB)

SINAD and SNR deserve careful attention, because there is still some variation between ADC manufacturers as to their precise meaning. Signal-to-Noise-and Distortion (SINAD, or S/(N + D) is the ratio of the rms signal amplitude to the mean value of the root-sum-square (rss) of all other spectral components, *including harmonics*, but excluding DC (Figure 1.50). SINAD is a good indication of the overall dynamic performance of an ADC as a function of input frequency because it includes all components which make up noise (including thermal noise) and distortion. It is often plotted for various input amplitudes. SINAD is equal to THD + N if the bandwidth for the noise measurement is the same. A typical plot for the AD9226 12-bit, 65 MSPS ADC is shown in Figure 1.52.

The SINAD plot shows where the AC performance of the ADC degrades due to high-frequency distortion and is usually plotted for frequencies well above the Nyquist frequency so that performance in undersampling applications can be evaluated. SINAD is often converted to *effective-number-of-bits* (ENOB) using the relationship for the theoretical SNR of an ideal N-bit ADC: SNR=6.02N + 1.76 dB. The equation is solved for N, and the value of SINAD is substituted for SNR:

$$\text{ENOB} = \frac{\text{SINAD} - 1.76\,\text{dB}}{6.02} \tag{1.20}$$

Signal-to-noise ratio (SNR, or SNR-*without-harmonics*) is calculated the same as SINAD except that the signal harmonics are excluded from the calculation, leaving only the noise

- SINAD (Signal-to-noise-and-distortion ratio):
  - The ratio of the rms signal amplitude to the mean value of the root-sum-squares (RSS) of all other spectral components, including harmonics, but excluding dc
- ENOB (effective number of bits):

$$\text{ENOB} = \frac{\text{SINAD} - 1.76\,\text{dB}}{6.02}$$

- SNR (Signal-to-noise ratio, or signal-to-noise ratio without harmonics:
  - The ratio of the rms signal amplitude to the mean value of the root-sum-squares (RSS) of all other spectral components, excluding the first five harmonics and dc

Figure 1.51: SINAD, ENOB, and SNR

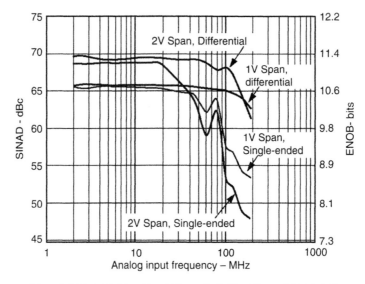

Figure 1.52: AD9226 12-bit, 65 MSPS ADC SINAD and ENOB for various input full-scale spans (range)

terms. In practice, it is only necessary to exclude the first five harmonics since they dominate. The SNR plot will degrade at high frequencies, but not as rapidly as SINAD because of the exclusion of the harmonic terms.

Many current ADC data sheets somewhat loosely refer to SINAD as SNR, so the engineer must be careful when interpreting these specifications.

1.3.3.4 Analog Bandwidth

The analog bandwidth of an ADC is that frequency at which the spectral output of the *fundamental* swept frequency (as determined by the FFT analysis) is reduced by 3 dB. It may be specified for either a small signal (SSBW—*small signal bandwidth*), or a full-scale signal (FPBW—*full power bandwidth*), so there can be a wide variation in specifications between manufacturers.

Like an amplifier, the analog bandwidth specification of a converter does not imply that the ADC maintains good distortion performance up to its bandwidth frequency. In fact, the SINAD (or ENOB) of most ADCs will begin to degrade considerably before the input frequency approaches the actual 3 dB bandwidth frequency. Figure 1.53 shows ENOB and full-scale frequency response of an ADC with a FPBW of 1 MHz, however, the ENOB begins to drop rapidly above 100 kHz.

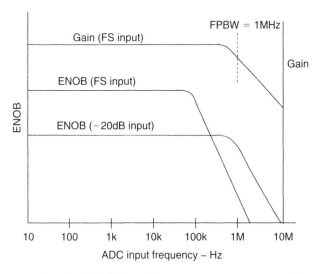

Figure 1.53: ADC gain (bandwidth) and ENOB versus frequency shows importance
of ENOB specification

1.3.3.5 Spurious Free Dynamic Range (SFDR)

Probably the most significant specification for an ADC used in a communications application is its *spurious free dynamic range* (SFDR). SFDR of an ADC is defined as the ratio of the rms signal amplitude to the rms value of the *peak spurious spectral content* measured over the bandwidth of interest. Unless otherwise stated, the bandwidth is assumed to be the Nyquist bandwidth DC to $f_s/2$.

Occasionally the frequency spectrum is divided into an *in-band* region (containing the signals of interest) and an *out-of-band* region (signals here are filtered out digitally). In this case there may be an *in-band* SFDR specification and an *out-of-band* SFDR specification, respectively.

SFDR is generally plotted as a function of signal amplitude and may be expressed relative to the signal amplitude (dBc) or the ADC full-scale (dBFS) as shown in Figure 1.54.

For a signal near full-scale, the peak spectral spur is generally determined by one of the first few harmonics of the fundamental. However, as the signal falls several dB below full-scale, other spurs generally occur which are not direct harmonics of the input signal. This is because of the differential nonlinearity of the ADC transfer function as discussed earlier. Therefore, SFDR considers *all* sources of distortion, regardless of their origin.

The AD6645 is a 14-bit, 80 MSPS wideband ADC designed for communications applications where high SFDR is important. The single-tone SFDR for a 69.1 MHz input and a sampling

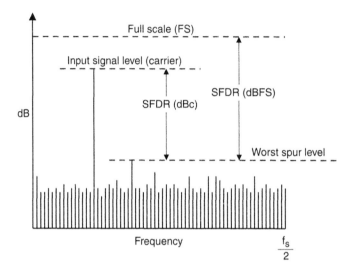

Figure 1.54: Spurious free dynamic range (SFDR)

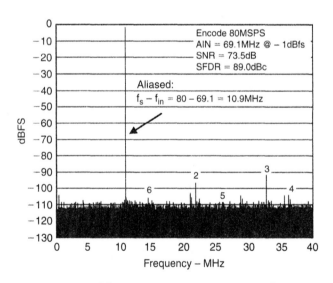

Figure 1.55: AD6645 14-bit, 80/105 MSPS ADC SFDR for 69.1 MHz input

frequency of 80 MSPS is shown in Figure 1.55. Note that a minimum of 89 dBc SFDR is obtained over the entire first Nyquist zone (DC to 40 MHz).

SFDR as a function of signal amplitude is shown in Figure 1.56 for the AD6645. Notice that over the entire range of signal amplitudes, the SFDR is greater than 90 dBFS. The abrupt

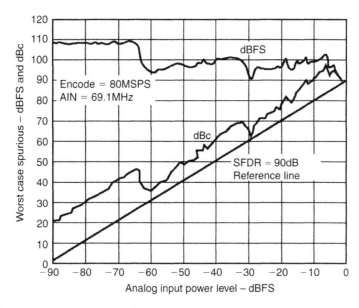

Figure 1.56: AD6645 14-bit, 80/105 MSPS ADC SFDR versus input power level for 69.1 MHz input

changes in the SFDR plot are due to the differential nonlinearities in the ADC transfer function. The nonlinearities correspond to those shown in Figure 1.49B, and are offset from midscale such that input signals less than about 65 dBFS do not exercise any of the points of increased DNL. It should be noted that the SFDR can be improved by injecting a small out-of-band dither signal—at the expense of a slight degradation in SNR.

SFDR is generally much greater than the ADCs theoretical N-bit SNR (6.02N + 1.76 dB). For example, the AD6645 is a 14-bit ADC with an SFDR of 90 dBc and a typical SNR of 73.5 dB (the theoretical SNR for 14 bits is 86 dB). This is because there is a fundamental distinction between noise and distortion measurements. The process gain of the FFT (33 dB for a 4096-point FFT) allows frequency spurs well below the noise floor to be observed. Adding extra resolution to an ADC may serve to increase its SNR but may or may not increase its SFDR.

1.3.3.6 Two-Tone Intermodulation Distortion (IMD)

Two-tone IMD is measured by applying two spectrally pure sinewaves to the ADC at frequencies f_1 and f_2, usually relatively close together. The amplitude of each tone is set slightly more than 6 dB below full scale so that the ADC does not clip when the two tones add in-phase. The location of the second- and third-order products are shown in Figure 1.57. Notice that the second-order products fall at frequencies that can be removed by digital filters.

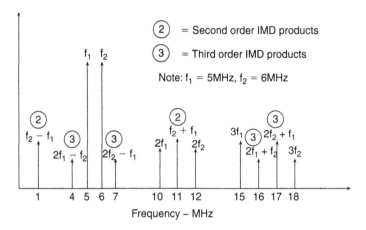

Figure 1.57: Second and third-order intermodulation products for f_1 = 5 MHz, f_2 = 6 MHz

However, the third-order products, $2f_2 - f_1$ and $2f_1 - f_2$, are close to the original signals and more difficult to filter. Unless otherwise specified, two-tone IMD refers to these third-order products. The value of the IMD product is expressed in dBc relative to the value of *either* of the two original tones, and not to their sum.

Note, however, that if the two tones are close to $f_s/4$, the aliased third harmonics of the fundamentals can make the identification of the actual $2f_2 - f_1$ and $2f_1 - f_2$ products difficult. This is because the third harmonic of $f_s/4$ is $3f_s/4$, and the alias occurs at $f_s - 3f_s/4 = f_s/4$. Similarly, if the two tones are close to $f_s/3$, the aliased second harmonics may interfere with the measurement. The same reasoning applies here; the second harmonic of $f_s/3$ is $2f_s/3$, and its alias occurs at $f_s - 2f_s/3 = f_s/3$.

1.3.3.7 Second- and Third-Order Intercept Points, 1 dB Compression Point

Third-order IMD products are especially troublesome in multichannel communications systems where the channel separation is constant across the frequency band. Third-order IMD products can mask out small signals in the presence of larger ones.

In amplifiers, it is common practice to specify the third-order IMD products in terms of the *third-order* intercept point, as shown by Figure 1.58. Two spectrally pure tones are applied to the system. The output signal power in a single tone (in dBm) as well as the relative amplitude of the third-order products (referenced to a single tone) are plotted as a function of input signal power. The fundamental is shown by the slope = 1 curve in the diagram. If the system nonlinearity is approximated by a power series expansion, it can be shown that second-order IMD amplitudes increase 2 dB for every 1 dB of signal increase, as represented by *slope* = 2 curve in the diagram.

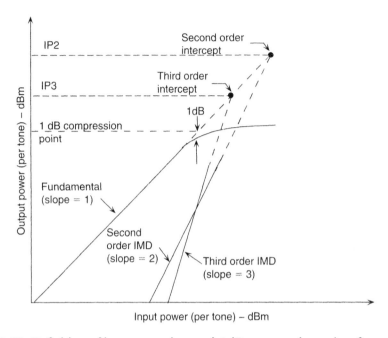

Figure 1.58: Definition of intercept points and 1 dB compression points for amplifiers

Similarly, the third-order IMD amplitudes increase 3 dB for every 1 dB of signal increase, as indicated by the *slope* = 3 plotted line. With a low level two-tone input signal, and two data points, one can draw the second- and third-order IMD lines as they are shown in Figure 1.58 (using the principle that a point and a slope define a straight line).

Once the input reaches a certain level however, the output signal begins to soft-limit, or compress. A parameter of interest here is the 1 dB *compression point*. This is the point where the output signal is compressed 1 dB from an ideal input/output transfer function. This is shown in Figure 1.58 within the region where the ideal slope = 1 line becomes dotted, and the actual response exhibits compression (solid).

Nevertheless, both the second- and third-order intercept lines may be extended, to intersect the (dotted) extension of the ideal output signal line. These intersections are called the *second-* and *third-order intercept points,* respectively, or IP2 and IP3. These power level values are usually referenced to the output power of the device delivered to a matched load (usually, but not necessarily 50 Ω) expressed in dBm.

It should be noted that IP2, IP3, and the 1 dB compression point are all a function of frequency and, as one would expect, the distortion is worse at higher frequencies.

For a given frequency, knowing the third-order intercept point allows calculation of the approximate level of the third-order IMD products as a function of output signal level.

The concept of *second- and third-order intercept points* is not valid for an ADC, because the distortion products do not vary in a predictable manner (as a function of signal amplitude). The ADC does not gradually begin to compress signals approaching full scale (there is no 1 dB compression point); it acts as a *hard limiter* as soon as the signal exceeds the ADC input range, thereby suddenly producing extreme amounts of distortion because of clipping. On the other hand, for signals much below full scale, the distortion floor remains relatively constant and is independent of signal level. This is shown graphically in Figure 1.59.

The IMD curve in Figure 1.59 is divided into three regions. For low level input signals, the IMD products remain relatively constant regardless of signal level. This implies that as the input signal increases 1 dB, the ratio of the signal to the IMD level will also increase 1 dB. When the input signal is within a few dB of the ADC full-scale range, the IMD may start to increase (but it might not in a very well-designed ADC). The exact level at which this occurs is dependent on the particular ADC under consideration—some ADCs may not exhibit significant increases in the IMD products over their full input range, however, most will. As

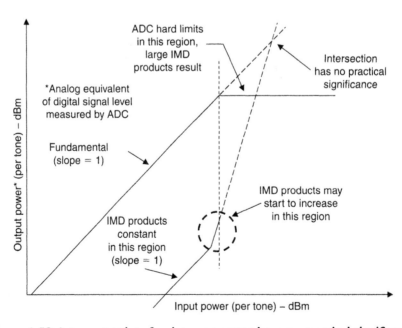

Figure 1.59: Intercept points for data converters have no practical significance

the input signal continues to increase beyond full scale, the ADC should function to act as an ideal limiter, and the IMD products become very large.

For these reasons, the second and third order IMD intercept points are not specified for ADCs. It should be noted that essentially the same arguments apply to DACs. In either case, the single- or multitone SFDR specification is the most accepted way to measure data converter distortion.

1.3.3.8 Multitone Spurious Free Dynamic Range

Two-tone and multitone SFDR is often measured in communications applications. The larger number of tones more closely simulates the wideband frequency spectrum of cellular telephone systems such as AMPS or GSM. Figure 1.60 shows the two-tone intermodulation performance of the AD6645 14-bit, 80/105 MSPS ADC. The input tones are at 55.25 MHz and 56.25 MHz and are located in the second Nyquist Zone.

The aliased tones therefore occur at 23.75 MHz and 24.75 MHz in the first Nyquist Zone. High SFDR increases the receiver's ability to capture small signals in the presence of large ones, and prevents the small signals from being masked by the intermodulation products of the larger ones. Figure 1.61 shows the AD6645 two-tone SFDR as a function of input signal amplitude for the same input frequencies.

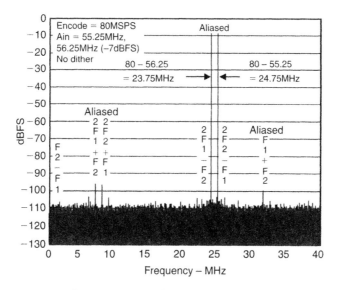

Figure 1.60: Two-tone SFDR for AD6645 14-bit, 80/105 MSPS ADC, input tones: 55.25 MHz and 56.25 MHz

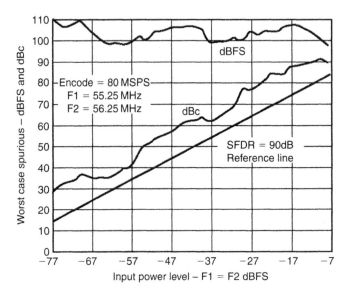

Figure 1.61: Two-tone SFDR versus input amplitude for AD6645 14-bit, 80/105 MSPS ADC

1.3.3.9 Wideband CDMA (WCDMA) Adjacent Channel Power Ratio (ACPR) and Adjacent Channel Leakage Ratio (ADLR)

A wideband CDMA channel has a bandwidth of approximately 3.84 MHz, and channel spacing is 5 MHz. The ratio in dBc between the measured power within a channel relative to its adjacent channel is defined as the *adjacent channel power ratio* (ACPR).

The ratio in dBc between the measured power within the channel bandwidth relative to the noise level in an adjacent empty carrier channel is defined as *adjacent channel leakage ratio* (ACLR).

Figure 1.62 shows a single wideband CDMA channel centered at 140 MHz sampled at a frequency of 76.8 MSPS using the AD6645. This is a good example of undersampling (direct IF-to-digital conversion). The signal lies within the fourth Nyquist zone: $3f_s/2$ to $2f_s$ (115.2 MHz to 153.6 MHz). The aliased signal within the first Nyquist zone is therefore centered at $2f_s - f_a = 153.6 - 140 = 13.6$ MHz. The diagram also shows the location of the aliased harmonics. For example, the second harmonic of the input signal occurs at $2 \times 140 = 280$ MHz, and the aliased component occurs at $4f_s - 2f_a = 4 \times 76.8 - 280 = 307.2 - 280 = 27.2$ MHz.

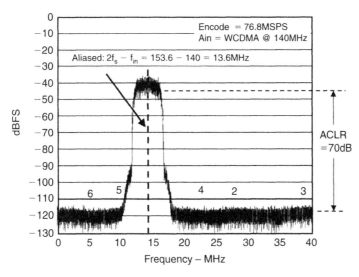

Figure 1.62: Wideband CDMA (WCDMA) adjacent channel leakage ratio (ACLR)

1.3.3.10 Noise Power Ratio (NPR)

Noise power ratio has been used extensively to measure the transmission characteristics of frequency division multiple access (FDMA) communications links (Reference 7). In a typical FDMA system, 4kHz wide voice channels are "stacked" in frequency bins for transmission over coaxial, microwave, or satellite equipment. At the receiving end, the FDMA data is demultiplexed and returned to 4kHz individual base-band channels. In an FDMA system having more than approximately 100 channels, the FDMA signal can be approximated by Gaussian noise with the appropriate bandwidth. An individual 4kHz channel can be measured for "quietness" using a narrow-band notch (band-stop) filter and a specially tuned receiver which measures the noise power inside the 4kHz notch (Figure 1.63).

Noise Power Ratio (NPR) measurements are straightforward. With the notch filter out, the rms noise power of the signal inside the notch is measured by the narrowband receiver. The notch filter is then switched in, and the residual noise inside the slot is measured. The ratio of these two readings expressed in dB is the NPR. Several slot frequencies across the noise bandwidth (low, midband, and high) are tested to characterize the system adequately. NPR measurements on ADCs are made in a similar manner except the analog receiver is replaced by a buffer memory and an FFT processor.

The NPR is plotted as a function of rms noise level referred to the peak range of the system. For very low noise loading level, the undesired noise (in nondigital systems) is primarily

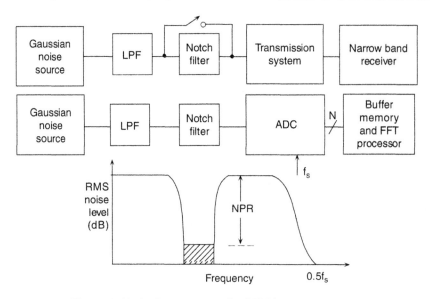

Figure 1.63: Noise power ratio (NPR) measurements

thermal noise and is independent of the input noise level. Over this region of the curve, a 1 dB increase in noise loading level causes a 1 dB increase in NPR. As the noise loading level is increased, the amplifiers in the system begin to overload, creating intermodulation products that cause the noise floor of the system to increase. As the input noise increases further, the effects of "overload" noise predominate, and the NPR is dramatically reduced. FDMA systems are usually operated at a noise loading level a few dB below the point of maximum NPR.

In a digital system containing an ADC, the noise within the slot is primarily quantization noise when low levels of noise input are applied. The NPR curve is linear in this region. As the noise level increases, there is a one-for-one correspondence between the noise level and the NPR. At some level, however, "clipping" noise caused by the hard-limiting action of the ADC begins to dominate. A theoretical curve for 10-, 11-, and 12-bit ADCs is shown in Figure 1.64 (References 8 and 21).

Figure 1.65 shows the maximum theoretical NPR and the noise loading level at which the maximum value occurs for 8- to 16-bit ADCs. The ADC input range is 2 V_O peak-to-peak. The rms noise level is σ, and the noise-loading factor k (crest factor) is defined as V_O/σ, the peak-to-rms ratio (k is expressed either as numerical ratio or in dB).

In multichannel high frequency communication systems, where there is little or no phase correlation between channels, NPR can also be used to simulate the distortion caused by a

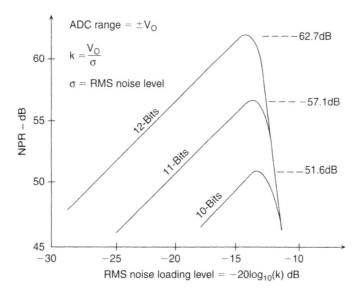

Figure 1.64: Theoretical NPR for 10-, 11-, 12-bit ADCs

Bits	k optimum	k(dB)	Max NPR (dB)
8	3.92	11.87	40.60
9	4.22	12.50	46.05
10	4.50	13.06	51.56
11	4.76	13.55	57.12
12	5.01	14.00	62.71
13	5.26	14.41	68.35
14	5.49	14.79	74.01
15	5.72	15.15	79.70
16	5.94	15.47	85.40

ADC range $= \pm V_o$
$k = V_O/\sigma$
$\sigma = $ RMS noise level

Figure 1.65: Theoretical maximum NPR for 8- to 16-bit ADCs

large number of individual channels, similar to an FDMA system. A notch filter is placed between the noise source and the ADC, and an FFT output is used in place of the analog receiver. The width of the notch filter is set for several MHz as shown in Figure 1.66 for the AD9430 12-bit 170/210 MSPS ADC. The notch is centered at 19 MHz, and the NPR is the "depth" of the notch. An ideal ADC will only generate quantization noise inside the

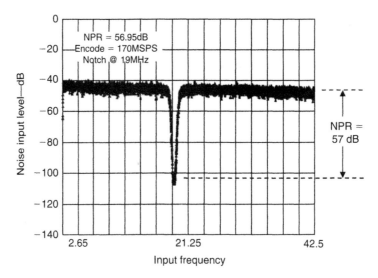

Figure 1.66: AD9430 12-bit, 170/210 MSPS ADC NPR measures 57 dB (62.7 dB theoretical)

notch; however, a practical one has additional noise components due to additional noise and intermodulation distortion caused by ADC imperfections. Notice that the NPR is about 57 dB compared to 62.7 dB theoretical.

1.3.3.11 Noise Factor (F) and Noise Figure (NF)

Noise figure (NF) is a popular specification among RF system designers. It is used to characterize RF amplifiers, mixers, etc., and widely used as a tool in radio receiver design. Many excellent textbooks on communications and receiver design treat noise figure extensively (see Reference 9, for example)—it is not the purpose here to discuss the topic in much detail, but only how it applies to data converters.

Since many wideband operational amplifiers and ADCs are now being used in RF applications, the inevitable day has come where the noise figure of these devices becomes important. As discussed in Reference 10, in order to determine the noise figure of an op amp correctly, one must not only know op amp voltage and current noise, but the exact circuit conditions—closed-loop gain, gain-setting resistor values, source resistance, bandwidth, etc. Calculating the noise figure for an ADC is even more of a challenge as will be seen.

Figure 1.67 shows the basic model for defining the noise figure of an ADC. The *noise factor*, F, is simply defined as the ratio of the total effective input noise power of the ADC to the amount of that noise power caused by the source resistance alone. Because the impedance is

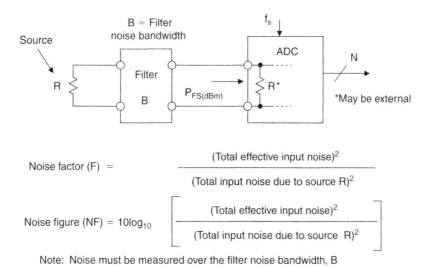

Noise factor (F) = $\dfrac{\text{(Total effective input noise)}^2}{\text{(Total input noise due to source R)}^2}$

Noise figure (NF) = $10\log_{10}\left[\dfrac{\text{(Total effective input noise)}^2}{\text{(Total input noise due to source R)}^2}\right]$

Note: Noise must be measured over the filter noise bandwidth, B

Figure 1.67: Noise figure for ADCs: use with caution

matched, the square of the voltage noise can be used instead of noise power. The *noise figure*, NF, is simply the noise factor expressed in dB, NF = $10\log_{10}F$.

This model assumes the input to the ADC comes from a source having a resistance, R, and that the input is band-limited to $f_s/2$ with a filter having a noise bandwidth equal to $f_s/2$. It is also possible to further band-limit the input signal resulting in oversampling and process gain, and this condition will be discussed shortly.

It is also assumed that the input impedance to the ADC is equal to the source resistance. Many ADCs have a high input impedance, so this termination resistance may be external to the ADC or used in parallel with the internal resistance to produce an equivalent termination resistance equal to R. The full-scale input power is the power of a sinewave whose peak-to-peak amplitude fills the entire ADC input range. The full-scale input sinewave given by the following equation has a peak-to-peak amplitude of $2V_O$ corresponding to the peak-to-peak input range of the ADC:

$$v(t) = V_O \sin 2\pi ft \qquad (1.21)$$

The full-scale power in this sinewave is given by:

$$P_{FS} = \frac{(V_O/\sqrt{2})^2}{R} = \frac{V_O^{\,2}}{2R} \qquad (1.22)$$

It is customary to express this power in dBm (referenced to 1 mW) as follows:

$$P_{FS(dBm)} = 10 \log_{10} \left[\frac{P_{FS}}{1mW} \right] \tag{1.23}$$

The *noise bandwidth* of a nonideal brick wall filter is defined as the bandwidth of an ideal brick wall filter which will pass the same noise power as the nonideal filter. Therefore, the noise bandwidth of a filter is always greater than the 3 dB bandwidth of the filter by a factor which depends upon the sharpness of the cutoff region of the filter. Figure 1.68 shows the relationship between the noise bandwidth and the 3 dB bandwidth for Butterworth filters up to five poles. Note that for two poles, the noise bandwidth and 3 dB bandwidth are within 11% of each other, and beyond that the two quantities are essentially equal.

The first step in the NF calculation is to calculate the effective input noise of the ADC from its SNR. The SNR of the ADC is given for a variety of input frequencies, so be sure and use the value corresponding to the input frequency of interest. Also, make sure that the harmonics are not included in the SNR number—some ADC data sheets may confuse SINAD with SNR. Once the SNR is known, the equivalent input rms voltage noise can be calculated starting from the equation:

$$SNR = 20 \log_{10} \left[\frac{V_{FS\ RMS}}{V_{NOISE\ RMS}} \right] \tag{1.24}$$

Solving for $V_{NOISE\ RMS}$:

$$V_{NOISE\ RMS} = V_{FS\ RMS} \times 10^{-SNR/20} \tag{1.25}$$

Number of poles	Noise BW/3dB BW
1	1.57
2	1.11
3	1.05
4	1.03
5	1.02

**Figure 1.68: Relationship between noise bandwidth and 3 dB bandwidth
for Butterworth filter**

This is the total effective input rms noise voltage at the carrier frequency measured over the Nyquist bandwidth, DC to $f_s/2$. Note that this noise includes the source resistance noise. These results are summarized in Figure 1.69.

The next step is to actually calculate the noise figure. In Figure 1.70 notice that the amount of the input voltage noise due to the source resistance is the voltage noise of the source resistance $\sqrt{(4kTBR)}$ divided by two, or $\sqrt{(kTBR)}$ because of the 2:1 attenuator formed by the ADC input termination resistor. The expression for the noise factor F can be written:

$$F = \frac{V_{NOISE\ RMS}^{2}}{kTRB} = \left[\frac{V_{FS\ RMS}^{2}}{R}\right]\left[\frac{1}{kT}\right]\left[10^{-SNR/10}\right]\left[\frac{1}{B}\right] \tag{1.26}$$

The noise figure is obtained by converting F into dB and simplifying:

$$NF = 10_{10}\log F = P_{FS(dBm)} + 174\,dBm - SNR - 10_{10}\log B, \tag{1.27}$$

where SNR is in dB, B in Hz, $T = 300\,K$, $k = 1.38 \times 10^{-23}\,J/K$.

Oversampling and filtering can be used to decrease the noise figure as a result of the process gain as has been previously discussed. In this case, the signal bandwidth B is less than $f_s/2$. Figure 1.71 shows the correction factor which results in the following equation:

$$NF = 10_{10}\log F = P_{FS(dBm)} + 174\,dBm - SNR - 10\log_{10}\left[f_s/2B\right] - 10\log_{10} B. \tag{1.28}$$

- Start with the SNR of the ADC measured at the carrier frequency (Note: this SNR value does not include the harmonics of the fundamental and is measured over the Nyquist bandwidth, dc to $f_s/2$)

$$SNR = 20\log_{10} \frac{V_{FS\text{-}RMS}}{V_{NOISE\text{-}RMS}}$$

$$V_{NOISE\text{-}RMS} = V_{FS\text{-}RMS}\, 10^{-SNR/20}$$

- This is the total ADC effective input noise at the carrier frequency measured over the Nyquist bandwidth, dc to $f_s/2$

Figure 1.69: Calculating ADC total effective input noise from SNR

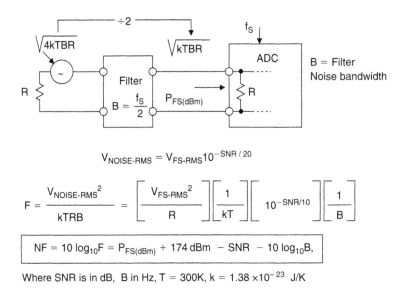

$$V_{NOISE-RMS} = V_{FS-RMS}10^{-SNR/20}$$

$$F = \frac{V_{NOISE-RMS}^2}{kTRB} = \left[\frac{V_{FS-RMS}^2}{R}\right]\left[\frac{1}{kT}\right]\left[10^{-SNR/10}\right]\left[\frac{1}{B}\right]$$

$$NF = 10\log_{10}F = P_{FS(dBm)} + 174\,dBm - SNR - 10\log_{10}B,$$

Where SNR is in dB, B in Hz, T = 300K, k = 1.38 ×10⁻²³ J/K

Figure 1.70: ADC noise figure in terms of SNR, sampling rate, and input power

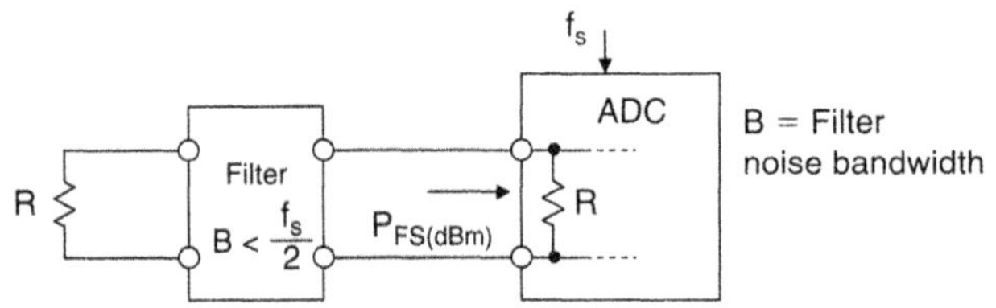

$$NF = P_{FS(dBm)} + 174dBm - SNR - \underbrace{10\log_{10}\left[\frac{f_s/2}{B}\right]}_{\substack{\text{Measured} \\ \text{DC to } f_s/2}} \underbrace{}_{\substack{\text{Process} \\ \text{gain}}} - 10\log_{10}B,$$

where SNR is in dB, B in Hz, T = 300K, k = 1.38 × 10⁻²³ J/K

Figure 1.71: Effect of oversampling and process gain on ADC noise figure

Figure 1.72 shows an example NF calculation for the AD6645 14-bit, 80 MSPS ADC. A 52.3 Ω resistor is added in parallel with the AD6645 input impedance of 1 kΩ to make the net input impedance 50 Ω. The ADC is operating under Nyquist conditions, and the SNR of 74 dB is the starting point for the calculations using Eq. 1.28 above. A noise figure of 34.8 dB is obtained.

Figure 1.73 shows how using an RF transformer with voltage gain can improve the noise figure. Figure 1.73A shows a 1:1 turns ratio, and the noise figure (from Figure 1.72) is 34.8.

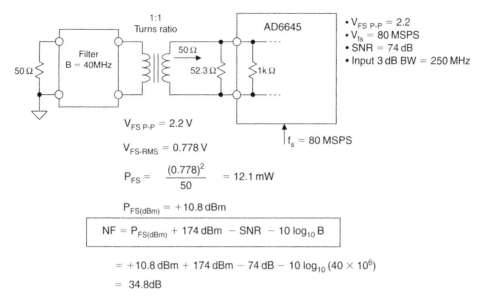

$$V_{FS\ P-P} = 2.2\,V$$

$$V_{FS-RMS} = 0.778\,V$$

$$P_{FS} = \frac{(0.778)^2}{50} = 12.1\,mW$$

$$P_{FS(dBm)} = +10.8\,dBm$$

$$NF = P_{FS(dBm)} + 174\,dBm - SNR - 10\,\log_{10} B$$

$$= +10.8\,dBm + 174\,dBm - 74\,dB - 10\,\log_{10}(40 \times 10^6)$$

$$= 34.8\,dB$$

Figure 1.72: Example calculation of noise figure under Nyquist conditions for AD6645

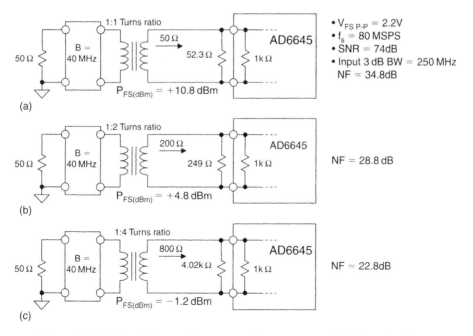

Figure 1.73: Using RF transformers to improve overall ADC noise figure

Figure 1.73B shows a transformer with a 1:2 turns ratio. The 249 Ω resistor in parallel with the AD6645 internal resistance results in a net input impedance of 200 Ω. The noise figure is improved by 6 dB because of the "noise-free" voltage gain of the transformer. Figure 1.73C shows a transformer with a 1:4 turns ratio. The AD6645 input is paralleled with a 4.02 kΩ resistor to make the net input impedance 800 Ω. The noise figure is improved by another 6 dB. Transformers with higher turns ratios are not generally practical because of bandwidth and distortion limitations.

Even with the 1:4 turns ratio transformer, the overall noise figure for the AD6645 was still 22.8 dB, still relatively high by RF standards. The solution is to provide low noise high gain stages ahead of the ADC. Figure 1.74 shows how the Friis equation is used to calculate the noise factor for cascaded gain stages. Notice that high gain in the first stage reduces the contribution of the noise factor of the second stage—the noise factor of the first stage dominates the overall noise factor.

Figure 1.75 shows the effects of a high-gain (25 dB) low-noise (NF = 4 dB) stage placed in front of a relatively high NF stage (30 dB)—the noise figure of the second stage is typical of high performance ADCs. The overall noise figure is 7.53 dB, only 3.53 dB higher than the first stage noise figure of 4 dB.

In summary, applying the noise figure concept to characterize wideband ADCs must be done with extreme caution to prevent misleading results. Simply trying to minimize the noise figure using the equations can actually increase circuit noise.

For instance, NF decreases with increasing source resistance according to the calculations, but increased source resistance increases circuit noise. Also, NF decreases with increasing

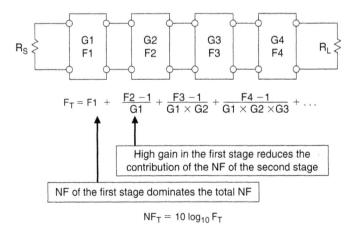

Figure 1.74: Cascaded noise figure using the Friis equation

ADC input bandwidth if there is no input filtering. This is also contradictory, because widening the bandwidth increases noise. In both these cases, the circuit noise increases, and the NF decreases. The reason NF decreases is that the source noise makes up a larger component of the total noise (which remains relatively constant because the ADC noise is much greater than the source noise); therefore, according to the calculation, NF decreases, but actual circuit noise increases.

It is true that on a standalone basis ADCs have relatively high noise figures compared to other RF parts such as LNAs or mixers. In the system the ADC should be preceded with low noise gain blocks as shown in the example of Figure 1.75. Noise figure considerations for ADCs are summarized in Figure 1.76.

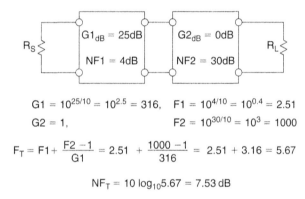

$$G1 = 10^{25/10} = 10^{2.5} = 316, \quad F1 = 10^{4/10} = 10^{0.4} = 2.51$$
$$G2 = 1, \qquad\qquad\qquad F2 = 10^{30/10} = 10^{3} = 1000$$

$$F_T = F1 + \frac{F2 - 1}{G1} = 2.51 + \frac{1000 - 1}{316} = 2.51 + 3.16 = 5.67$$

$$NF_T = 10 \log_{10} 5.67 = 7.53 \text{ dB}$$

- The first stage dominates the overall NF
- It should have the highest gain possible with the lowest NF possible

Figure 1.75: Example of two-stage cascaded network

- NF decreases with increasing source resistance.
- NF decreases with increasing ADC input bandwidth if there is no input filtering.
- In both cases, the circuit noise increases, and the NF decreases.
- The reason NF decreases is that the source noise makes up a larger component of the total noise (which remains relatively constant because the ADC noise is much greater than the source noise).
- In practice, input filtering is used to limit the input noise bandwidth and reduce overall system noise.
- ADCs have relatively high NF compared to other RF parts. In the system the ADC should be preceded with low-noise gain blocks.
- Exercise caution when using NF.

Figure 1.76: Noise figure considerations for ADCs: summary and caution

1.3.3.12 Aperture Time, Aperture Delay Time, and Aperture Jitter

Perhaps the most misunderstood and misused ADC and sample-and-hold (or track-and-hold) specifications are those that include the word *aperture*. The most essential dynamic property of a SHA is its ability to disconnect quickly the hold capacitor from the input buffer amplifier as shown in Figure 1.77. The short (but non-zero) interval required for this action is called *aperture time (or sampling aperture)*, t_a. The actual value of the voltage held at the end of this interval is a function of both the input signal slew rate and the errors introduced by the switching operation itself. Figure 1.77 shows what happens when the hold command is applied with an input signal of two arbitrary slopes labeled as 1 and 2. For clarity, the sample-to-hold pedestal and switching transients are ignored. The value that is finally held is a delayed version of the input signal, averaged over the aperture time of the switch as shown in Figure 1.77. The first-order model assumes that the final value of the voltage on the hold capacitor is approximately equal to the average value of the signal applied to the switch over the interval during which the switch changes from a low to high impedance (t_a).

The model shows that the finite time required for the switch to open (t_a) is equivalent to introducing a small delay (t_e) in the sampling clock driving the SHA. This delay is constant and may be either positive or negative. The diagram shows that the same value of t_e works for the two signals, even though the slopes are different. This delay is called *effective aperture delay time, aperture delay time, or simply aperture delay,* t_e. In an ADC, the aperture delay

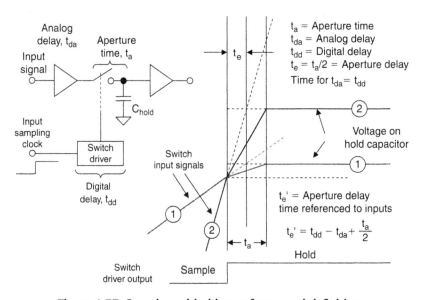

Figure 1.77: Sample-and-hold waveforms and definitions

time is referenced to the input of the converter, and the effects of the analog propagation delay through the input buffer, t_{da} and the digital delay through the switch driver, t_{dd}, must be considered. Referenced to the ADC inputs, aperture time, $t_e{}'$, is defined as the time difference between the analog propagation delay of the front-end buffer, t_{da}, and the switch driver digital delay, t_{dd}, plus one-half the aperture time, $t_a/2$.

The effective aperture delay time is usually positive, but may be negative if the sum of one-half the aperture time, $t_a/2$, and the switch driver digital delay, t_{dd}, is less than the propagation delay through the input buffer, t_{da}. The aperture delay specification thus establishes when the input signal is actually sampled with respect to the sampling clock edge.

Aperture delay time can be measured by applying a bipolar sinewave signal to the ADC and adjusting the synchronous sampling clock delay such that the output of the ADC is midscale (corresponding to the zero-crossing of the sinewave). The relative delay between the input sampling clock edge and the actual zero-crossing of the input sinewave is the aperture delay time (see Figure 1.78).

Aperture delay produces no errors (assuming it is relatively short with respect to the hold time), but acts as a fixed delay in either the sampling clock input or the analog input (depending on its sign). However, in simultaneous sampling applications or in direct I/Q demodulation where two or more ADCs must be well matched, variations in the aperture delay between converters can produce errors on fast slewing signals. In these applications, the aperture delay mismatches must be removed by properly adjusting the phases of the individual sampling clocks to the various ADCs.

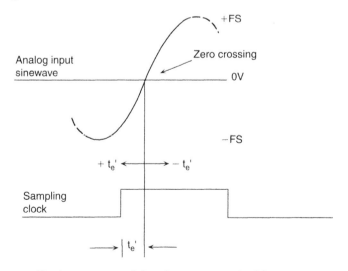

Figure 1.78: Effective aperture delay time measured with respect to ADC input

If, however, there is *sample-to-sample* variation in aperture delay (*aperture jitter*), a corresponding voltage error is produced as shown in Figure 1.79. This sample-to-sample variation in the instant the switch opens is called *aperture uncertainty,* or *aperture jitter* and is usually measured in rms picoseconds. The amplitude of the associated output error is related to the rate-of-change of the analog input. For any given value of aperture jitter, the aperture jitter error increases as the input dv/dt increases. The effects of phase jitter on the external sampling clock (or the analog input for that matter) produce exactly the same type of error.

The effects of aperture and sampling clock jitter on an ideal ADC's SNR can be predicted by the following simple analysis. Assume an input signal given by

$$v(t) = V_O \sin 2\pi ft \qquad (1.29)$$

The rate of change of this signal is given by:

$$dv/dt = 2\pi f V_O \cos 2\pi ft \qquad (1.30)$$

The rms value of dv/dt can be obtained by dividing the amplitude, $2\pi f V_O$, by $\sqrt{2}$:

$$dv/dt\big|_{rms} = 2\pi f V_O / \sqrt{2} \qquad (1.31)$$

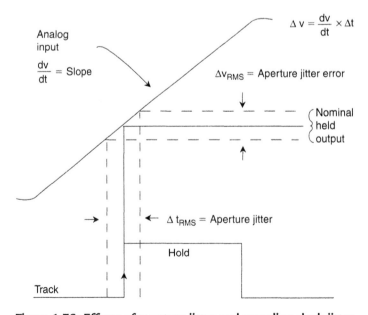

Figure 1.79: Effects of aperture jitter and sampling clock jitter

Now let Δv_{rms} = the rms voltage error and Δt = the rms aperture jitter t_j, and substitute:

$$\Delta v_{rms}/t_j = 2\pi f V_O/\sqrt{2} \qquad (1.32)$$

Solving for Δv_{rms}:

$$\Delta v_{rms} = 2\pi f V_O t_j/\sqrt{2} \qquad (1.33)$$

The rms value of the full-scale input sinewave is $V_o/\sqrt{2}$, therefore the rms signal to rms noise ratio is given by:

$$SNR = 20\log_{10}\left[\frac{V_O/\sqrt{2}}{\Delta v_{rms}}\right] = 20\log_{10}\left[\frac{V_O/\sqrt{2}}{2\pi f \, V_O t_j/\sqrt{2}}\right] = 20\log_{10}\left[\frac{1}{2\pi f \, t_j}\right] \qquad (1.34)$$

This equation assumes an infinite-resolution ADC where aperture jitter is the only factor in determining the SNR. This equation is plotted in Figure 1.80 and shows the serious

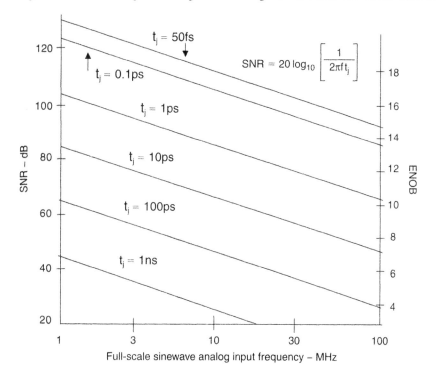

Figure 1.80: Theoretical SNR and ENOB due to jitter versus full-scale sinewave input frequency

effects of aperture and sampling clock jitter on SNR, especially at higher input/output frequencies. Therefore, extreme care must be taken to minimize phase noise in the sampling/ reconstruction clock of any sampled data system.

This care must extend to all aspects of the clock signal: the oscillator itself (for example, a 555 timer is absolutely inadequate, but even a quartz crystal oscillator can give problems if it uses an active device that shares a chip with noisy logic); the transmission path (these clocks are very vulnerable to interference of all sorts), and phase noise introduced in the ADC or DAC. As discussed, a very common source of phase noise in converter circuitry is aperture jitter in the integral sample-and-hold (SHA) circuitry; however, the total rms jitter will be composed of a number of components—the actual SHA aperture jitter often being the least of them.

1.3.3.13 A Simple Equation for the Total SNR of an ADC

A relatively simple equation for the ADC SNR in terms of sampling clock and aperture jitter, DNL, effective input noise, and the number of bits of resolution is shown in Figure 1.81. The equation combines the various error terms on an rss basis. The average DNL error, ε, is computed from histogram data. This equation is used in Figure 1.82 to predict the SNR performance of the AD6645 14-bit, 80 MSPS ADC as a function of sampling clock and aperture jitter.

Before the 1980s, most sampling ADCs were generally built up from a separate SHA and ADC. Interface design was difficult, and a key parameter was aperture jitter in the

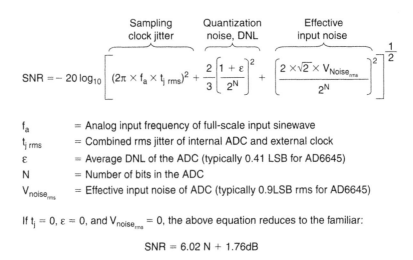

$$SNR = -20\log_{10}\left[(2\pi \times f_a \times t_{j\,rms})^2 + \frac{2}{3}\left[\frac{1+\varepsilon}{2^N}\right]^2 + \left[\frac{2 \times \sqrt{2} \times V_{Noise_{rms}}}{2^N}\right]^2 \right]^{\frac{1}{2}}$$

f_a = Analog input frequency of full-scale input sinewave

$t_{j\,rms}$ = Combined rms jitter of internal ADC and external clock

ε = Average DNL of the ADC (typically 0.41 LSB for AD6645)

N = Number of bits in the ADC

$V_{noise_{rms}}$ = Effective input noise of ADC (typically 0.9LSB rms for AD6645)

If $t_j = 0$, $\varepsilon = 0$, and $V_{noise_{rms}} = 0$, the above equation reduces to the familiar:

$$SNR = 6.02\,N + 1.76dB$$

Figure 1.81: Relationship between SNR, sampling clock jitter, quantization noise, DNL, and input noise

SHA. Today, almost all sampled data systems use *sampling* ADCs that contain an integral SHA. The aperture jitter of the SHA may not be specified as such, but this is not a cause of concern if the SNR or ENOB is clearly specified, since a guarantee of a specific SNR is an implicit guarantee of an adequate aperture jitter specification. However, the use of an additional high-performance SHA will sometimes improve the high frequency ENOB of even the best sampling ADC by presenting "DC" to the ADC, and may be more cost effective than replacing the ADC with a more expensive one.

1.3.3.14 ADC Transient Response and Overvoltage Recovery

Most high-speed ADCs designed for communications applications are specified primarily in the frequency domain. However, in general-purpose data acquisition applications the transient response (or settling time) of the ADC is important. The *transient response* of an ADC is the time required for the ADC to settle to rated accuracy (usually 1 LSB) after the application of a full-scale step input. The typical response of a general-purpose 12-bit, 10 MSPS ADC is shown in Figure 1.83, showing a 1 LSB settling time of less than 40 ns. The settling time specification is critical in the typical data acquisition system application where the ADC is being driven by an analog multiplexer as shown in Figure 1.84. The multiplexer output can deliver a full-scale sample-to-sample change to the ADC input. If both the multiplexer and the ADC have not settled to the required accuracy, channel-to-channel crosstalk will result, even though only DC or low frequency signals are present on the multiplexer inputs.

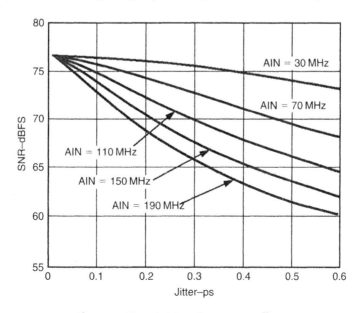

Figure 1.82: AD6645 SNR versus jitter

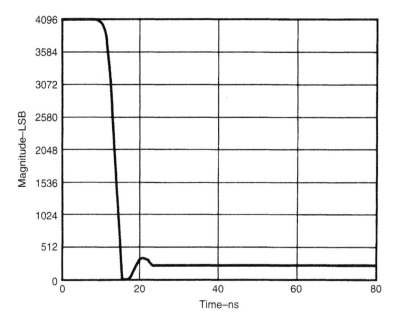

Figure 1.83: ADC transient response (settling time)

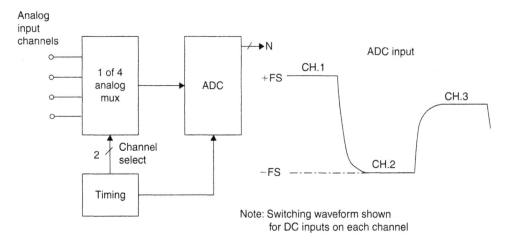

Note: Switching waveform shown
for DC inputs on each channel

Figure 1.84: Settling time is critical in multiplexed applications

Most ADCs have settling times which are less than $1/f_{s\,max}$, even if not specified. However sigma-delta ADCs have a built-in digital filter that can take several output sample intervals to settle. This should be kept in mind when using sigma-delta ADCs in multiplexed applications.

Resolution, # of bits	LSB (%FS)	# of time constants
6	1.563	4.16
8	0.391	5.55
10	0.0977	6.93
12	0.0244	8.32
14	0.0061	9.70
16	0.00153	11.09
18	0.00038	12.48
20	0.000095	13.86
22	0.000024	15.25

Figure 1.85: Settling time as function of time constant for various resolutions

The importance of settling time in multiplexed systems can be seen in Figure 1.85, where the ADC input is modeled as a single-pole filter having a corresponding time constant, $\tau = RC$. The required number of time constants to settle to a given accuracy (1 LSB) is shown. A simple example will illustrate the point.

Assume a multiplexed 16-bit data acquisition system uses an ADC with a sampling frequency $f_s = 100\,kSPS$. The ADC must settle to 16-bit accuracy for a full-scale step function input in less than $1/f_s = 10\,\mu s$. The chart shows that 11.09 time constants are required to settle to 16-bit accuracy. The input filter time constant must therefore be less than $\tau = 10\,\mu s/11.09 = 900\,ns$. The corresponding rise time $t_r = 2.2\tau = 1.98\,\mu s$. The required ADC full power input bandwidth can now be calculated from $BW = 0.35/t_r = 177\,kHz$. This neglects the settling time of the multiplexer and second-order settling time effects in the ADC.

Overvoltage recovery time is defined as that amount of time required for an ADC to achieve a specified accuracy, measured from the time the overvoltage signal re-enters the converter's range, as shown in Figure 1.86. This specification is usually given for a signal that is some stated percentage outside the ADC's input range. Needless to say, the ADC should act as an ideal limiter for out-of-range signals and should produce either the positive full-scale code or the negative full-scale code during the overvoltage condition. Some converters provide over- and underrange flags to allow gain-adjustment circuits to be activated. Care should always be taken to avoid overvoltage signals that will damage an ADC input.

1.3.3.15 ADC Sparkle Codes, Metastable States, and Bit Error Rate (BER)

A primary concern in the design of many digital communications systems using ADCs is the bit error rate (BER). Unfortunately, ADCs contribute to the BER in ways that are not predictable by simple analysis. This section describes the mechanisms within the ADCs that can contribute to the error rate, ways to minimize the problem, and methods for measuring the BER.

Random noise, regardless of the source, creates a finite probability of errors (deviations from the expected output). Before describing the error code sources, however, it is important to define what constitutes an ADC error code. Noise generated prior to or inside the ADC can be analyzed in the traditional manner. Therefore, an ADC error code is any deviation from the expected output that is not attributable to the equivalent input noise of the ADC. Figure 1.87 illustrates an exaggerated output of a low amplitude sinewave applied to an ADC that has error codes. Note that the noise of the ADC creates some uncertainty in the output. These anomalies are not considered error codes, but are simply the result of ordinary noise and quantization. The large errors are more significant and are not expected. These errors are random and so infrequent that an SNR test of the ADC will rarely detect them. These types of errors plagued a few of the early ADCs for video applications, and were given the name *sparkle codes*

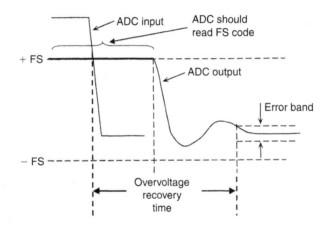

Figure 1.86: Overvoltage recovery time

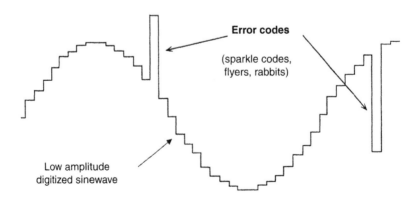

Figure 1.87: Exaggerated output of ADC showing error codes

because of their appearance on a TV screen as small white dots or "sparkles" under certain test conditions. These errors have also been called *rabbits* or *flyers*. In digital communications applications, this type of error increases the overall system bit error rate (BER).

In order to understand the causes of the error codes, we will first consider the case of a simple Flash converter. The comparators in a Flash converter are latched comparators usually arranged in a master-slave configuration. If the input signal is in the center of the threshold of a particular comparator, that comparator will balance, and its output will take a longer period of time to reach a valid logic level after the application of the latch strobe than the outputs of its neighboring comparators which are being overdriven. This phenomenon is known as *metastability* and occurs when a balanced comparator cannot reach a valid logic level in the time allowed for decoding. If simple binary decoding logic is used to decode the thermometer code, a metastable comparator output may result in a large output code error. Consider the case of a simple 3-bit Flash converter shown in Figure 1.88. Assume that the input signal

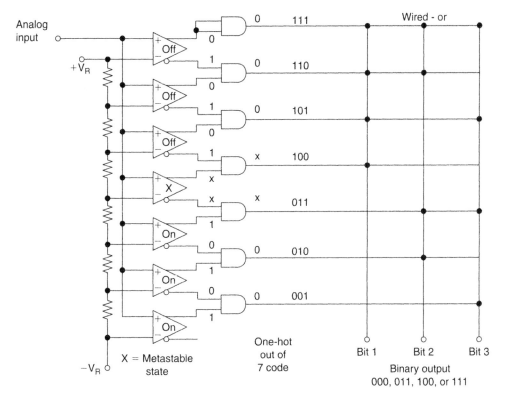

Figure 1.88: Metastable comparator output states may cause error codes in data converters

is exactly at the threshold of Comparator 4 and random noise is causing the comparator to toggle between a "1" and a "0" output each time a latch strobe is applied. The corresponding binary output should be interpreted as either 011 or 100. If, however, the comparator output is in a metastable state, the simple binary decoding logic shown may produce binary codes 000, 011, 100, or 111. The codes 000 and 111 represent a one-half scale departure from the expected codes.

The probability of errors due to metastability increases as the sampling rate increases because less time is available for a metastable comparator to settle.

Various measures have been taken in Flash converter designs to minimize the metastable state problem. Decoding schemes described in References 12 to 15 minimize the magnitude of these errors. Optimizing comparator designs for regenerative gain and small time constants is another way to reduce these problems.

Metastable state errors may also appear in successive approximation and subranging ADCs that make use of comparators as building blocks. The same concepts apply, although the magnitudes and locations of the errors may be different.

The test system shown in Figure 1.89 may be used to test for BER in an ADC. The analog input to the ADC is provided by a high stability low noise sinewave generator. The analog input level is set slightly greater than full-scale, and the frequency such that there is always slightly less than 1 LSB change between samples as shown in Figure 1.90.

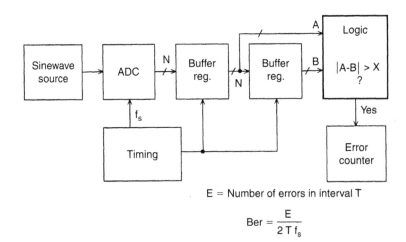

$$E = \text{Number of errors in interval T}$$

$$Ber = \frac{E}{2\,T\,f_s}$$

Figure 1.89: ADC bit error rate test setup

The test set uses series latches to acquire successive codes A and B. A logic circuit determines the absolute difference between A and B. This difference is then compared to the error limit, chosen to allow for expected random noise spikes and ADC quantization errors. Errors that cause the difference to be larger than the limit will increment the counters. The number of errors, E, are counted over a period of time, T. The error rate is then calculated as BER = $E/2Tf_s$. The factor of 2 in the denominator is required because the hardware records a second error when the output returns to the correct code after making the initial error. The error counter is therefore incremented twice for each error. It should be noted that the same function can be accomplished in software if the ADC outputs are stored in a memory and analyzed by a computer program.

The input frequency must be carefully chosen such that at least one sample is taken per code.

Assume a full-scale input sinewave having an amplitude of $2^N/2$:

$$v(t) = \frac{2^N}{2} \sin 2\pi ft \qquad (1.35)$$

The maximum rate of change of this signal is

$$\left. \frac{dv}{dt} \right|_{max} \leq 2^N \pi f \qquad (1.36)$$

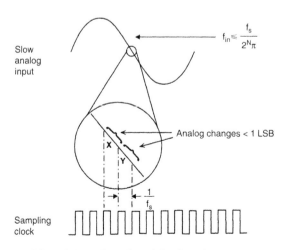

Figure 1.90: ADC analog signal for low frequency BER test

Letting dv = 1 LSB, dt = 1/f_s, and solving for the input frequency:

$$f_{in} \leq \frac{f_s}{2^N \pi} \qquad (1.37)$$

Choosing an input frequency less than this value will ensure that there is at least one sample per code. The same test can be conducted at high frequencies by applying an input frequency slightly offset from f_s/2 as shown in Figure 1.91. This causes the ADC to slew full-scale between conversions. Every other conversion is compared, and the "beat" frequency is chosen such that there is slightly less than 1 LSB change between alternate samples. The equation for calculating the proper frequency for the high frequency BER test is derived as follows.

Assume an input full-scale sinewave of amplitude 2^N/2 whose frequency is slightly less than f_s/2 by a frequency equal to Δf.

$$v(t) = \frac{2^N}{2} \sin\left[2\pi\left(\frac{f_s}{2} - \Delta f\right)t\right] \qquad (1.38)$$

The maximum rate of change of this signal is:

$$\left.\frac{dv}{dt}\right|_{max} \leq 2^N \pi\left(\frac{f_s}{2} - \Delta f\right) \qquad (1.39)$$

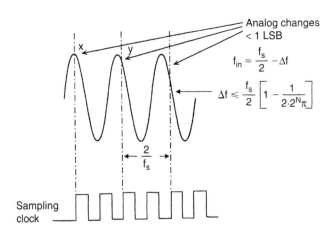

Figure 1.91: ADC analog input for high frequency BER Test

Letting dv = 1 LSB and dt = $2/f_s$, and solving for the input frequency Δf:

$$\Delta f \leq \frac{f_s}{2}\left(1 - \frac{1}{2 \times 2^N \pi}\right) \quad (1.40)$$

Establishing the BER of a well-behaved ADC is a difficult, time-consuming task; a single unit can sometimes be tested for days without an error. For example, tests on a typical 8-bit Flash converter operating at a sampling rate of 75 MSPS yield a BER of approximately 3.7×10^{-12} (1 error per hour) with an error limit of 4 LSBs. Meaningful tests for longer periods of time require special attention to EMI/RFI effects (possibly requiring a shielded screen room), isolated power supplies, isolation from soldering irons with mechanical thermostats, isolation from other bench equipment, etc. Figure 1.92 shows the average time between errors as a function of BER for a sampling frequency of 75 MSPS. This illustrates the difficulty in measuring low BER because the long measurement times increase the probability of power supply transients, noise, etc., causing an error.

1.3.4 DAC Dynamic Performance

The AC specifications most likely to be important with DACs are settling time, glitch impulse area, distortion, and Spurious Free Dynamic Range (SFDR).

1.3.4.1 DAC Settling Time

The input to output settling time of a DAC is the time from a change of digital code (t = 0) to when the output comes within *and remains within* some error band as shown in Figure 1.93. With amplifiers, it is hard to make comparisons of settling time, since their specified error bands may differ from amplifier to amplifier, but with DACs the error band will almost invariably be

Bit error rate (BER)	Average time between errors
1×10^{-8}	1.3 seconds
1×10^{-9}	13.3 seconds
1×10^{-10}	2.2 minutes
1×10^{-11}	22 minutes
1×10^{-12}	3.7 hours
1×10^{-13}	1.5 days
1×10^{-14}	15 days

Figure 1.92: Average time between errors versus BER when sampling at 75 MSPS

specified as ± 1 or $\pm \frac{1}{2}$ LSB. Note that in some cases, the output settling time may be of more interest, in which case it is referenced to the time the output first leaves the error band.

The input to output settling time of a DAC is made up of four different periods: the *switching time* or *dead time* (during which the digital switching, but not the output, is changing), the *slewing time* (during which the rate of change of output is limited by the slew rate of the DAC output), the *recovery time* (when the DAC is recovering from its fast slew and may overshoot), and the *linear settling time* (when the DAC output approaches its final value in an exponential or near-exponential manner). If the slew time is short compared to the other three (as is usually the case with current output DACs), the settling time will largely be independent of the output step size. On the other hand, if the slew time is a significant part of the total, the larger the step, the longer the settling time.

Settling time is especially important in video display applications. For example a standard 1024×768 display updated at a 60 Hz refresh rate must have a pixel rate of $1024 \times 768 \times 60\,\text{Hz} = 47.2\,\text{MHz}$ with no overhead. Allowing 35% overhead time increases the pixel frequency to 64 MHz corresponding to a pixel duration of $1/(64 \times 10^6) = 15.6\,\text{ns}$. In order to accurately reproduce a single fully-white pixel located between two black pixels, the DAC settling time should be less than the pixel duration time of 15.6 ns.

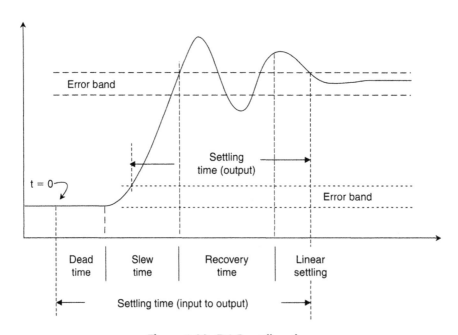

Figure 1.93: DAC settling time

Higher resolution displays require even faster pixel rates. For example, a 2048×2048 display requires a pixel rate of approximately 330 MHz at a 60 Hz refresh rate.

1.3.4.2 Glitch Impulse Area

Ideally, when a DAC output changes it should move from one value to its new one monotonically. In practice, the output is likely to overshoot, undershoot, or both (Figure 1.94). This uncontrolled movement of the DAC output during a transition is known as a *glitch*. It can arise from two mechanisms: capacitive coupling of digital transitions to the analog output, and the effects of some switches in the DAC operating more quickly than others and producing temporary spurious outputs.

Capacitive coupling frequently produces roughly equal positive and negative spikes (sometimes called a *doublet* glitch) which more or less cancel in the longer term. The glitch produced by switch timing differences is generally unipolar, much larger, and of greater concern.

Glitches can be characterized by measuring the *glitch impulse area*, sometimes inaccurately called *glitch energy*. The term glitch energy is a misnomer, since the unit for glitch impulse area is volt-seconds (or more probably μV-sec or pV-sec. The peak *glitch area* is the area of the largest of the positive or negative glitch areas. The glitch impulse area is the net area under the voltage-versus-time curve and can be estimated by approximating the waveforms by triangles, computing the areas, and subtracting the negative area from the positive area as shown in Figure 1.95.

The midscale glitch produced by the transition between the codes 0111…111 and 1000…000 is usually the worst glitch because all switches are changing states. Glitches at other code transition points (such as 1/4 and 3/4 full scale) are generally less. Figure 1.96 shows the midscale glitch for a fast low glitch DAC. The peak and net glitch areas are estimated using triangles as described above. Settling time is measured from the time the waveform leaves the

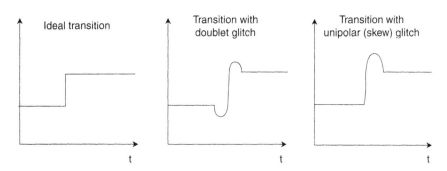

Figure 1.94: DAC transitions (showing glitch)

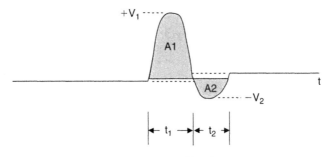

- Peak glitch impulse area = A1 $\approx \dfrac{V_1 \times t_1}{2}$

- Net glitch impulse area = A1 − A2 $\approx \dfrac{V_1 \times t_1}{2} - \dfrac{V_2 \times t_2}{2}$

Figure 1.95: Calculating net glitch impulse area

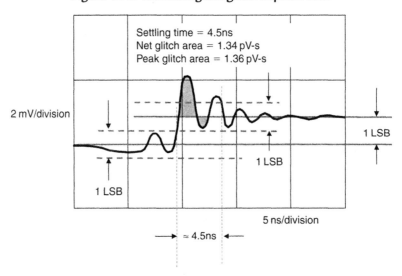

Figure 1.96: DAC midscale glitch shows 1.34 pV-s net impulse area and settling time of 4.5 ns

initial 1 LSB error band until it enters and remains within the final 1 LSB error band. The step size between the transition regions is also 1 LSB.

1.3.4.3 DAC SFDR and SNR

DAC settling time is important in applications such as RGB raster scan video display drivers, but frequency-domain specifications such as SFDR are generally more important in communications.

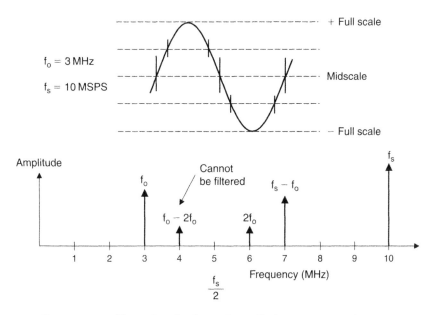

Figure 1.97: Effect of code-dependent glitches on spectral output

If we consider the spectrum of a waveform reconstructed by a DAC from digital data, we find that in addition to the expected spectrum (which will contain one or more frequencies, depending on the nature of the reconstructed waveform), there will also be noise and distortion products. Distortion may be specified in terms of harmonic distortion, Spurious Free Dynamic Range (SFDR), intermodulation distortion, or all of the above. Harmonic distortion is defined as the ratio of harmonics to fundamental when a (theoretically) pure sine wave is reconstructed, and is the most common specification. Spurious free dynamic range is the ratio of the worst spur (usually, but not necessarily always a harmonic of the fundamental) to the fundamental.

Code-dependent glitches will produce both out-of-band and in-band harmonics when the DAC is reconstructing a digitally generated sinewave as in a Direct Digital Synthesis (DDS) system. The midscale glitch occurs twice during a single cycle of a reconstructed sinewave (at each midscale crossing), and will therefore produce a second harmonic of the sinewave, as shown in Figure 1.97. Note that the higher order harmonics of the sinewave, which alias back into the Nyquist bandwidth (DC to $f_s/2$), cannot be filtered.

It is difficult to predict the harmonic distortion or SFDR from the glitch area specification alone. Other factors, such as the overall linearity of the DAC, also contribute to distortion as shown in Figure 1.98. In addition, certain ratios between the DAC output frequency and the

- Resolution
- Integral nonlinearity
- Differential nonlinearity
- Code-dependent glitches
- Ratio of clock frequency to output frequency (even in an ideal DAC)
- Mathematical analysis is difficult

Figure 1.98: Contributors to DDS DAC distortion

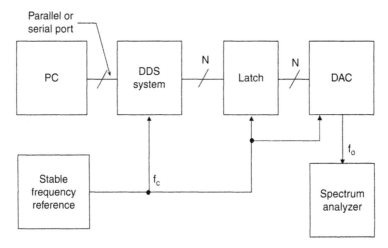

Figure 1.99: Test setup for measuring DAC SFDR

sampling clock cause the quantization noise to concentrate at harmonics of the fundamental thereby increasing the distortion at these points.

It is therefore customary to test reconstruction DACs in the frequency domain (using a spectrum analyzer) at various clock rates and output frequencies as shown in Figure 1.99. Typical SFDR for the 16-bit AD9777 Transmit TxDAC is shown in Figure 1.100. The clock rate is 160 MSPS, and the output frequency is swept to 50 MHz. As in the case of ADCs, quantization noise will appear as increased harmonic distortion if the ratio between the clock frequency and the DAC output frequency is an integer number. These ratios should be avoided when making the SFDR measurements.

There is nearly an infinite combination of possible clock and output frequencies for a low distortion DAC, and SFDR is generally specified for a limited number of selected combinations. For this reason, Analog Devices offers fast turnaround on customer-specified test vectors for the Transmit TxDAC family. A test vector is a combination of amplitudes,

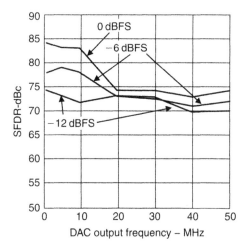

Figure 1.100: AD9777 16-bit TxDAC SFDR, data update rate = 160 MSPS

output frequencies, and update rates specified directly by the customer for SFDR data on a particular DAC.

1.3.4.4 Measuring DAC SNR with an Analog Spectrum Analyzer

Analog spectrum analyzers are used to measure the distortion and SFDR of high performance DACs. Care must be taken that the front end of the analyzer is not overdriven by the fundamental signal. If overdrive is a problem, a band-stop filter can be used to filter out the fundamental signal so the spurious components can be observed.

Spectrum analyzers can also be used to measure the SNR of a DAC provided attention is given to bandwidth considerations. SNR of an ADC is normally defined as the signal-to-noise ratio measured over the Nyquist bandwidth DC to $f_s/2$. However, spectrum analyzers have a resolution bandwidth less than $f_s/2$—this therefore lowers the analyzer noise floor by the process gain equal to $10 \log_{10}[f_s/(2 \times BW)]$, where BW is the resolution noise bandwidth of the analyzer (Figure 1.101).

It is important that the noise bandwidth (not the 3-dB bandwidth) be used in the calculation; however, from Figure 1.68 the error is small assuming that the analyzer narrowband filter is at least two poles. The ratio of the noise bandwidth to the 3-dB bandwidth of a one-pole Butterworth filter is 1.57 (causing an error of 1.96 dB in the process gain calculation). For a two-pole Butterworth filter, the ratio is 1.11 (causing an error of 0.45 dB in the process gain calculation).

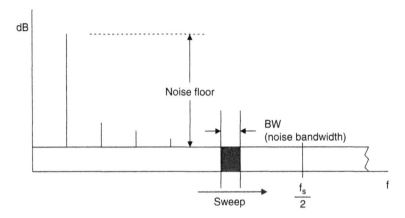

Figure 1.101: Measuring DAC SNR with an Analog spectrum analyzer

1.3.4.5 DAC Output Spectrum and sin (x)/x Frequency Roll-off

The output of a reconstruction DAC can be represented as a series of rectangular pulses whose width is equal to the reciprocal of the clock rate as shown in Figure 1.102. Note that the reconstructed signal amplitude is down 3.92 dB at the Nyquist frequency, $f_c/2$. An inverse sin(x)/x filter can be used to compensate for this effect in most cases. The images of the fundamental signal occur as a result of the sampling function and are also attenuated by the sin(x)/x function.

1.3.4.6 Oversampling Interpolating DACs

In ADC-based systems, oversampling can ease the requirements on the antialiasing filter. In a DAC-based system (such as DDS), the concept of interpolation can be used in a similar manner. This concept is common in digital audio CD players, where the basic update rate of the data from the CD is 44.1 kSPS. Early CD players used traditional binary DACs and inserted "Zeros" into the parallel data, thereby increasing the effective update rate to 4 times, 8 times, or 16 times the fundamental throughput rate. The 4×, 8×, or 16× data stream is passed through a digital interpolation filter that generates the extra data points. The high oversampling rate moves the image frequencies higher, thereby allowing a less complex filter with a wider transition band. The sigma-delta 1-bit DAC architecture uses a much higher oversampling rate and represents the ultimate extension of this concept and has become popular in modern CD players.

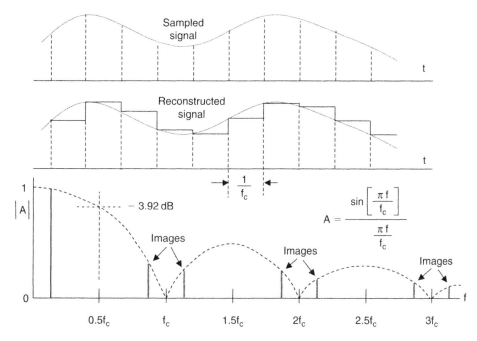

Figure 1.102: DAC sin (x)/x roll-off (amplitude normalized)

The same concept of oversampling can be applied to high speed DACs used in communications applications, relaxing the requirements on the output filter as well as increasing the SNR due to process gain.

Assume a traditional DAC is driven at an input word rate of 30 MSPS (Figure 1.103A). Assume the DAC output frequency is 10 MHz. The image frequency component at $30 - 10 = 20$ MHz must be attenuated by the analog antialiasing filter, and the transition band of the filter is 10 to 20 MHz. Assume that the image frequency must be attenuated by 60 dB. The filter must therefore go from a pass band of 10 MHz to 60 dB stopband attenuation over the transition band lying between 10 and 20 MHz (one octave). A filter gives 6 dB attenuation per octave for each pole. Therefore, a minimum of 10 poles is required to provide the desired attenuation. Filters become even more complex as the transition band becomes narrower.

Assume that the DAC update rate is increased to 60 MSPS and insert a "zero" between each original data sample. The parallel data stream is now 60 MSPS, but must now be determined the value of the zero-value data points. This is done by passing the 60 MSPS data stream with the added zeros through a digital interpolation filter that computes the additional data points. The response of the digital filter relative to the 2 times oversampling frequency is shown in

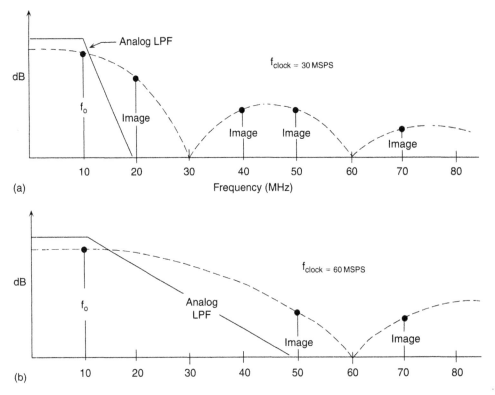

Figure 1.103: Analog filter requirements for f$_o$=10 MHz: (A) f$_c$=30 MSPS, and (B) f$_c$=60 MSPS

Figure 1.103B. The analog antialiasing filter transition zone is now 10 to 50 MHz (the first image occurs at $2f_c - f_o = 60 - 10 = 50$ MHz). This transition zone is a little greater than two octaves, implying that a 5- or 6-pole filter is sufficient.

The AD9773/AD9775/AD9777 (12-/14-/16-bit) series of Transmit DACs (TxDAC) are selectable 2×, 4×, or 8× oversampling interpolating dual DACs, and a simplified block diagram is shown in Figure 1.104. These devices are designed to handle 12-/14-/16-bit input word rates up to 160 MSPS. The output word rate is 400 MSPS maximum. For an output frequency of 50 MHz, an input update rate of 160 MHz, and an oversampling ratio of 2×, the image frequency occurs at 320 MHz − 50 MHz = 270 MHz. The transition band for the analog filter is therefore 50 MHz to 270 MHz. Without 2× oversampling, the image frequency occurs at 160 MHz − 50 MHz = 110 MHz, and the filter transition band is 50 MHz to 110 MHz.

Notice also that an oversampling interpolating DAC allows both a lower frequency input clock and input data rate, which are much less likely to generate noise within the system.

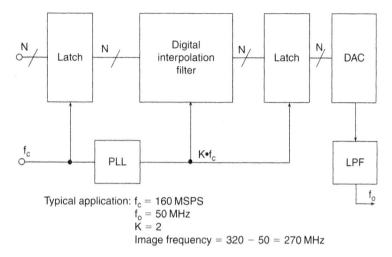

Typical application: f_c = 160 MSPS
f_o = 50 MHz
K = 2
Image frequency = 320 − 50 = 270 MHz

Figure 1.104: Oversampling interpolating TxDAC simplified block diagram

1.4 General Data Converter Specifications

1.4.1 Overall Considerations

Data converters, as we have observed, have a digital port and an analog port and, like all integrated circuits, they require power supplies and will draw current from those supplies. Data converter specifications will therefore include the usual specifications common to any integrated circuit, including supply voltage and supply current, logic interfaces, power on and standby timing, package and thermal issues and ESD. We shall not consider these at any length, but there are some issues that may require a little consideration.

An overriding piece of advice here is *read the data sheet*. There is no excuse for being unaware of the specifications of a device for which one owns a data sheet—and it is often possible to deduce extra information that is not printed on it by understanding the issues and conventions involved in preparing it.

Traditional precision analog integrated circuits (which include amplifiers, converters, and other devices) were designed for operation from supplies of ±15 V, and many (but not all—it is important to check with the data sheet) would operate within specification over quite a wide range of supply voltages. Today the processes used for many, but by no means all, modern converters have low breakdown voltages and absolute maximum ratings of only a few volts. Converters built with these processes may only work to specification over a narrow range of supply voltages.

It is therefore important when selecting a data converter to check both the absolute maximum supply voltage(s) and the range of voltages where correct operation can be expected. Some low-voltage devices work equally well with both 5 V and 3.3 V supplies, others are sold in 5 V and 3.3 V versions with different suffixes on their part numbers—with these it is important to use the correct one.

Absolute maximum ratings are ratings that can never be exceeded without grave risk of damage to the device concerned—they are not safe operating limits, but they are conservative. Integrated circuit manufacturers try to set absolute maximum ratings so that every device they manufacture will survive brief exposures to absolute maximum conditions. As a result many devices will, in fact, appear to operate safely and continuously outside the permitted limits. Good engineers do not take advantage of this for three reasons: (1) components are not tested outside their absolute maximum limits so, although they may be operating, they may not be operating at their specified accuracy. Also the damage done by incorrect operation may not be immediately fatal, but may cause low levels of disruption which, in turn, may (2) shorten the device's life, or (3) may affect its subsequent accuracy even when it is operated within specification again. None of these effects is at all desirable and absolute maximum ratings should always be respected.

The supply current in a data converter specification is usually the no-load current—i.e., the current consumption when the data converter output is driving a high impedance or open-circuit load. CMOS logic, and to a lesser extent some other types, have current consumption that is proportional to clock speed so a CMOS data converter current may be defined at a specific clock frequency and will be higher if the clock runs faster. Current consumption will also be higher when the output (or the reference output if there is one, or both) is loaded. There may be another figure for "standby" current—the current that flows when the data converter is connected to a power supply but is internally shut down into a non-operational low power state to conserve current.

When power is first applied to some data converters they may take several tens, hundreds, or even thousands of microseconds for their reference and amplifier circuitry to stabilize and, although this is less common, some may even take a long time to "wake up" from a power saving standby mode. It is therefore important to ensure that data converters that have such delays are not used in applications where full functionality is required within a short time of power-up or wake-up.

All integrated circuits are vulnerable to electrostatic discharge (ESD), but precision analog circuits are, on the whole, more vulnerable than some other types. This is because the technologies available for minimizing such damage also tend to degrade the performance of precision circuitry, and there is a necessary compromise between robustness and performance. It is always a good idea to ensure that when handling amplifiers, converters and other vulnerable circuits the necessary steps are taken to avoid ESD.

Specifications of packages, operating temperature ranges, and similar issues, although important, do not need further discussion here.

1.4.2 Logic Interface Issues

As it is important to read and understand power supply specifications, so it is equally important to read and understand logic specifications. In the past most integrated circuit logic circuitry (with the exception of emitter-coupled logic or ECL) operated from 5 V supplies and had compatible logic levels—with a few exceptions 5 V logic would interface with other 5 V logic. Today, with the advent of low voltage logic operating with supplies of 3.3 V, 2.7 V, or even less, it is important to ensure that logic interfaces are compatible. There are several issues which must be considered—absolute maximum ratings, worst-case logic levels, and timing. The logic inputs of integrated circuits generally have absolute maximum ratings, as do most other inputs, of 300 mV outside the power supply. Note that these are instantaneous ratings. If an IC has such a rating and is currently operating from a +5 V supply, the logic inputs may be between −0.3 V and +5.3 V—but if the supply is not present, that input must be between +0.3 V and −0.3 V, not the −0.3 V to +5.3 V which are the limits once the power is applied—ICs cannot predict the future.

The reason for the rating of 0.3 V is to ensure that no parasitic diode on the IC is ever turned on by a voltage outside the IC's absolute maximum rating. It is quite common to protect an input from such overvoltage with a Schottky diode clamp. At low temperatures the clamp voltage of a Schottky diode may be a little more than 0.3 V, and so the IC may see voltages just outside its absolute maximum rating. Although, strictly speaking, this subjects the IC to stresses outside its absolute maximum ratings and so is forbidden, this is an acceptable exception to the general rule provided the Schottky diode is at a temperature similar to the IC it is protecting (say within ±10°C).

Some low voltage devices, however, have inputs with absolute maximum ratings that are substantially greater than their supply voltage. This allows such circuits to be driven by higher voltage logic without additional interface or clamp circuitry. But it is important to read the data sheets and ensure that both logic levels and absolute maximum voltages are compatible for all combinations of high and low supplies.

This is the general rule when interfacing different low voltage logic circuitry—it is always necessary to check that at the lowest value of its power supply (a) the Logic 1 output from the driving circuit applied to its worst-case load is greater than the specified minimum Logic 1 input for the receiving circuit, and (b) with its output sinking maximum allowed current, the logic 0 output is less than the specified Logic 0 input of the receiver. If the logic specifications of the chosen devices do not meet these criteria it will be necessary to select different devices, use different power supplies, or use additional interface circuitry to ensure that the required levels are available. Note that additional interface circuitry introduces extra delays in timing.

It is not sufficient to build an experimental setup and test it. In general, logic thresholds are generously specified and usually logic circuits will work correctly well outside their specified limits—but it is not possible to rely on this in a production design. At some point a batch of devices near the limit on low output swing will be required to drive some devices needing slightly more drive than usual—and will be unable to do so.

1.4.3 Data Converter Logic: Timing and Other Issues

It is not the purpose of this brief section to discuss logic architectures, so we shall not define the many different data converter logic interface operations and their timing specifications except to note that data converter logic interfaces may be more complex than expected—*read the data sheet*. Do not expect that because there is a pin with the same name on memory and interface chips it will behave in exactly the same way in a data converter. Also, some data converters reset to a known state on power-up but many more do not.

It is very necessary to consider general timing issues. The new low voltage processes used for many modern data converters have a number of desirable features. One that is often overlooked by users (but not by converter designers) is their higher logic speed. DACs built on older processes frequently had logic that was orders of magnitude slower than the microprocessors with which they interfaced, and it was sometimes necessary to use separate buffers, or multiple WAIT instructions, to make the two compatible. Today it is much more common for the write times of DACs to be compatible with those of the fast logic with which they interface.

Nevertheless, not all DACs are speed compatible with all logic interfaces, and it is still important to ensure that minimum data setup times and write pulsewidths are observed. Again, experiments will often show that devices work with faster signals than their specification requires—but at the limits of temperature or supply voltage some may not, and interfaces should be designed on the basis of specified rather than measured timing.

1.5 Defining the Specifications

The following list, in alphabetical order, should prove helpful regarding specifications and their definitions. The original source for these definitions was provided by Dan Sheingold from Chapter 11 in his classic book *Analog-to-Digital Conversion Handbook, Third Edition*, Prentice-Hall, 1986.

Accuracy, Absolute. Absolute accuracy error of a *DAC* is the difference between actual analog output and the output that is expected when a given digital code is applied to the converter. Error is usually commensurate with resolution, i.e., less 1/2 LSB of full-scale, for example. However, accuracy may be much better than resolution in some applications; for example, a 4-bit DAC having only 16 discrete digitally chosen levels would have a resolution of 1/16, but might have an accuracy to within 0.01 % of each ideal value.

Absolute accuracy error of an *ADC* at a given output code is the difference between the actual and the theoretical analog input voltages required to produce that code. Since the code can be produced by any analog voltage in a finite band (see *Quantizing Uncertainty*), the "input required to produce that code" is usually defined as the midpoint of the band of inputs that will produce that code. For example, if 5 V, ± 1.2 mV, will theoretically produce a 12-bit half-scale code of 1000 0000 0000, then a converter for which any voltage from 4.997 V to 4.999 V will produce that code will have absolute error of $(1/2)(4.997 + 4.999) - 5$ V $= +2$ mV.

Sources of error include gain (calibration) error, zero error, linearity errors, and noise. Absolute accuracy measurements should be made under a set of standard conditions with sources and meters traceable to an internationally accepted standard.

Accuracy, Logarithmic DACs. The difference (measured in dB) between the actual transfer function and the ideal transfer function, as measured after calibration of gain error at 0 dB.

Accuracy, Relative. Relative accuracy error, expressed in %, ppm, or fractions of 1 LSB, is the deviation of the analog value at any code (relative to the full analog range of the device transfer characteristic) from its theoretical value (relative to the same range), after the full-scale range (FSR) has been calibrated (see *Full-Scale Range*).

Since the discrete analog values that correspond to the digital values ideally lie on a straight line, the specified worst-case relative accuracy error of a linear ADC or DAC can be interpreted as a measure of end-point nonlinearity (see *Linearity*).

The "discrete points" of a DAC transfer characteristic are measured by the actual analog outputs. The "discrete points" of an ADC transfer characteristic are the midpoints of the quantization bands at each code (see *Accuracy, Absolute*).

Acquisition Time. The acquisition time of a track-and-hold circuit for a step change is the time required by the output to reach its final value, within a specified error band, after the track command has been given. Included are switch delay time, the slewing interval, and settling time for a specified output voltage change.

Adjacent Channel Power Ratio (ACPR). The ratio in dBc between the measured power within a channel relative to its adjacent channel. See *Adjacent Channel Leakage Ratio* (ACLR).

Adjacent Channel Leakage Ratio (ACLR). The ratio in dBc between the measured power within the carrier bandwidth relative to the noise level in an adjacent empty carrier channel. Both ACPR and ACLR are Wideband CDMA (WCDMA) specifications. The channel bandwidth for WCDMA is approximately 3.84 MHz with 5 MHz spacing between channels.

Aliasing. A signal within a bandwidth f_a must be sampled at a rate $f_s > 2f_a$ in order to avoid the loss of information. If $f_s < 2f_a$, a phenomenon called *aliasing*, inherent in the spectrum of the sampled signal, will cause a frequency equal to $f_s - f_a$, called an *alias*, to appear in the Nyquist bandwidth, DC to $f_s/2$. For example, if $f_s = 4$ kSPS and $f_a = 3$ kHz, a 1 kHz alias will appear. Note also that for $f_a = 1$ kHz (within the DC to $f_s/2$ bandwidth), an alias will occur at 3 kHz (outside the DC to $f_s/2$ bandwidth). Since noise is also aliased, it is essential to provide low-pass (or band-pass) filtering prior to the sampling stage to prevent out-of-band noise on the input signal from being aliased into the signal range and thereby degrading the SNR.

Analog Bandwidth. For an ADC, the analog input frequency at which the spectral power of the fundamental frequency (as determined by the FFT analysis) is reduced by 3 dB. This can be specified as full power bandwidth, or small signal bandwidth. (See also *Bandwidth, Full Linear* and *Bandwidth, Full Power.*)

Analog Bandwidth, 0.1 dB. For an ADC, the analog input frequency at which the spectral power of the fundamental frequency (as determined by the FFT analysis) is reduced by 0.1 dB. This is a popular video specification. (See also *Bandwidth, Full Linear* and *Bandwidth, Full Power.*)

Aperture Time (classic definition). Aperture time in a sample-and-hold is defined as the time required for the internal switch to switch from the closed position (zero resistance) to the fully open position (infinite resistance). A first-order analysis that neglects nonlinear effects assumes that the input signal is averaged over this time interval to produce the final output signal. The analysis shows that this does not introduce an error as long as the switch opens in a repeatable fashion, and as long as the aperture time is reasonably short with respect to the hold time. There exists an effective sampling point in time that will cause an ideal sample-and-hold to produce the same held voltage. The difference between this effective sampling point and the actual sampling point is defined as effective aperture delay time.

Aperture Delay Time, or *Effective Aperture Delay Time.* In a sample-and-hold or track-and-hold, there exists an effective sampling point in time that will cause an ideal sample-and-hold to produce the same held voltage. The difference between this effective sampling point and the actual sampling point is defined as the aperture delay time or effective aperture delay time. In a sampling ADC, aperture delay time can be measured by sampling the zero crossing of a sinewave with a sampling clock locked to the sinewave. The phase of the sampling clock is adjusted until the output of the ADC is 100…00. The time difference between the leading edge of the sampling clock and the zero crossing of the sinewave—referenced to the analog input—is the effective aperture delay time. A dual trace oscilloscope can be used to make the measurement.

Aperture Uncertainty (or Aperture Jitter). The sample-to-sample variation in the sampling point because of jitter. Aperture jitter is expressed as an rms quantity and produces a corresponding rms voltage error in the sample-and-hold output. In an ADC it is caused by internal noise and jitter in the sampling clock path from the sampling clock input pin to the internal switch. Jitter in the external sampling clock produces the same type of error.

Automatic Zero. To achieve zero stability in many integrating-type converters, a time interval is provided during each conversion cycle to allow the circuitry to compensate for drift errors. The drift error in such converters is substantially zero. A similar function exists in many high resolution sigma-delta ADCs.

Bandwidth, Full-Linear. The full-linear bandwidth of an ADC is the input frequency at which the slew-rate limit of the sample-and-hold amplifier is reached. Up to this point, the amplitude of the reconstructed fundamental signal will have been attenuated by less than 0.1 dB. Beyond this frequency, distortion of the sampled input signal increases significantly.

Bandwidth, Full-Power (FPBW). The full-power bandwidth is that input frequency at which the amplitude of the reconstructed fundamental signal (measured using FFTs) is reduced by 3 dB for a full-scale input. In order to be meaningful, the FPBW must be examined in conjunction with the signal-to-noise ratio (SNR), signal-to-noise-plus-distortion ratio (SINAD), effective number of bits (ENOB), and harmonic distortion in order to ascertain the true dynamic performance of the ADC at the FPBW frequency.

Bandwidth, Analog Input Small-Signal. Analog input bandwidth is measured similarly to FPBW at a reduced analog input amplitude. This specification is similar to the small signal bandwidth of an op amp. The amplitude of the input signal at which the small signal bandwidth is measured should be specified on the data sheet.

Bandwidth, Effective Resolution (ERB). Some ADC manufacturers define the frequency at which SINAD drops 3 dB as the *effective resolution bandwidth (ERB)*. This is the same frequency at which the ENOB drops ½ bit. This specification is a misnomer, however, since bandwidth normally is associated with signal amplitude.

Bias Current. The zero-signal DC current required from the signal source by the inputs of many semiconductor circuits. The voltage developed across the source resistance by bias current constitutes an (often negligible) offset error. When an instrumentation amplifier performs measurements of a source that is remote from the amplifier's power-supply, there *must* be a return path for bias currents. If it does not already exist and is not provided, those currents will charge stray capacitances, causing the output to drift uncontrollably or to saturate. Therefore, when amplifying outputs of "floating" sources, such as transformers, insulated thermocouples, and AC-coupled circuits, there must be a high impedance DC leakage path from each input to common, or to the driven-guard terminal (if present). If a DC return path is impracticable, an *isolator* must be used.

Bipolar Mode. (See *Offset.*)

Bipolar Offset. (See *Offset.*)

Bus. A bus is a parallel path of binary information signals—usually 4, 8, 16, 32, or 64-bits wide. Three common types of information usually found on buses are data, addresses, and control signals. Three-state output switches (inactive, high, and low) permit many sources— such as ADCs—to be connected to a bus, while only one is active at any time.

Byte. A byte is a binary digital word, usually 8 bits wide. A byte is often part of a longer word that must be placed on an 8-bit bus in two stages. The byte containing the MSB is called the *high byte*; that containing the LSB is called the *low byte*. A 4-bit byte is called a *nibble* on an 8-bit or greater bus.

Channel-to-Channel Isolation. In multiple DACs, the proportion of analog input signal from one DAC's reference input that appears at the output of the other DAC, expressed logarithmically in dB. See also *crosstalk*.

Charge Transfer, Charge Injection (or *Offset Step*). The principal component of *sample-to-hold offset* (or *pedestal*) is the small charge transferred to the storage capacitor via interelectrode capacitance of the switch and stray capacitance when switching to the *hold* mode. The offset step is directly proportional to this charge:

Offset error=Incremental Charge/Capacitance=$\Delta Q/C$

It can be reduced somewhat by lightly coupling an appropriate polarity version of the *hold* signal to the capacitor for first-order cancellation. The error can also be reduced by increasing the capacitance, but this increases *acquisition time*.

Code Width. This is a fundamental quantity for ADC specifications. In an ADC where the code transition noise is a fraction of an LSB, it is defined as the range of analog input values for which a given digital output code will occur. The nominal value of a code width (for all but the first and last codes) is the voltage equivalent of 1 least significant bit (LSB) of the full-scale range, or 2.44 mV out of 10 V for a 12-bit ADC. Because the full-scale range is fixed, the presence of excessively wide codes implies the existence of narrow and perhaps even missing codes. Code transition noise can make the measurement of code width difficult or impossible. In wide bandwidth and high resolution ADCs additional noise modulates the effective code width and appears as input-referred noise. Many ADCs have input-referred noise that spans several code widths, and histogram techniques must be used to accurately measure differential linearity.

Common-Mode Range. Common-mode rejection usually varies with the magnitude of the range through which the input signal can swing, determined by the sum of the common-mode and the differential voltage. *Common-mode range* is that range of *total* input voltage over which specified common-mode rejection is maintained. For example, if the common-mode signal is ± 5 V and the differential signal is ± 5 V, the common-mode range is ± 10 V.

Common-Mode Rejection (CMR). A measure of the change in output voltage when both inputs are changed by equal amounts of AC and/or DC voltage. Common-mode rejection is usually expressed either as a ratio (e.g., CMRR = 1,000,000:1) or in decibels: CMR = $20\log_{10}$CMRR; if CMRR = 10^6, CMR = 120 dB. A CMRR of 10^6 means that 1 V of common mode is processed by the device as though it were a differential signal of 1 μV at the input.

CMR is usually specified for a full range common-mode voltage change (CMV), at a given frequency, and a specified imbalance of source impedance (e.g., 1 kΩ source unbalance, at 60 Hz). In amplifiers, the common-mode rejection ratio is defined as the ratio of the signal gain, G, to the common-mode gain (the ratio of common-mode signal appearing at the output to the CMV at the input.

Common-Mode Voltage (CMV). A voltage that appears in common at both input terminals of a differential-input device, with respect to its output reference (usually "ground"). For inputs, V_1 and V_2, with respect to ground, CMV = $\frac{1}{2}(V_1 + V_2)$. An ideal differential-input device would ignore CMV. *Common-mode error (CME)* is any error at the output due to the common-mode input voltage. The errors due to supply voltage variation, an internal common-mode effect, are specified separately.

Compliance-Voltage Range. For a current-output DAC, the maximum range of (output) terminal voltage for which the device will maintain the specified current-output characteristics.

Conversion Complete. An ADC digital output signal which indicates the end of conversion. When this signal is in the opposite state, the ADC is considered to be "busy." Also called *end-of-conversion (EOC)*, *data ready*, or *status* in some converters.

Conversion Time and *Conversion Rate.* For an ADC without a sample-and-hold, the time required for a complete measurement is called *conversion time.* For most converters (assuming no significant additional systemic delays), conversion time is essentially identical with the inverse of *conversion rate.* For simple sampling ADCs, however, the conversion rate is the inverse of the conversion time plus the sample-and-hold's acquisition time. However, in many high speed converters, because of pipelining, new conversions are initiated before the results of prior conversions have been determined; thus, there can one, two, three, or more clock cycles of conversion delay (plus a fixed delay in some cases). Once a train of conversions has been initiated, as in signal-processing applications, the conversion rate can therefore be much faster than the conversion time would imply.

Crosstalk. Leakage of signals, usually via capacitance between circuits or channels of a multichannel system or device, such as a multiplexer, multiple input ADC, or multiple DAC. Crosstalk is usually determined by the impedance parameters of the physical circuit, and actual values are frequency-dependent. See also *channel-to-channel isolation.*

Multiple DACs have a *digital crosstalk* specification: the spike (sometimes called a *glitch*) impulse appearing at the output of one converter due to a change in the digital input code of another of the converters. It is specified in nanovolt- or picovolt-seconds and measured at $V_{REF} = 0\,V$.

Data Ready. (See *Conversion Complete.*)

Deglitcher (See *Glitch.*) A device that removes or reduces the effects of time-skew in D/A conversion. A deglitcher normally employs a track-and-hold circuit, often specifically designed as part of the DAC. When the DAC is updated, the deglitcher holds the output of the DAC's output amplifier constant at the previous value until the switches reach equilibrium, then acquires and tracks the new value.

DAC Glitch. A glitch is a switching transient appearing in the output during a code transition. The worst-case DAC glitch generally occurs when the DAC is switched between the 011...111 and 100...000 codes. The net area under the glitch is referred to as *glitch impulse area* and is measured in millivolt-nanoseconds, nanovolt-seconds, or picovolt-seconds. Sometimes the term *glitch energy* is used to describe the net area under the glitch—this terminology is incorrect because the unit of measurement is not energy.

Differential Analog Input Resistance, Differential Analog Input Capacitance, and *Differential Analog Input Impedance.* The real and complex impedances measured at each analog input port of an ADC. The resistance is measured statically and the capacitance and differential input impedances are measured with a network analyzer.

Differential Analog Input Voltage Range. The peak-to-peak differential voltage that must be applied to the converter to generate a full-scale response. Peak differential voltage is computed by observing the voltage on a single pin and subtracting the voltage from the other pin, which is 180 degrees out of phase. Peak-to-peak differential is computed by rotating the inputs phase 180 degrees and taking the peak measurement again. The difference is then computed between both peak measurements.

Differential Gain (ΔG). A video specification that measures the variation in the amplitude (in percent) of a small amplitude color subcarrier signal as it is swept across the video range from black to white.

Differential Phase ($\Delta \phi$). A video specification that measures the phase variation (in degrees) of a small amplitude color subcarrier signal as it is swept across the video range from black to white.

Droop Rate. When a sample-and-hold circuit using a capacitor for storage is in *hold,* it will not hold the information forever. Droop rate is the rate at which the output voltage changes (by increasing or decreasing), and hence gives up information. The change of output occurs as a result of leakage or bias currents flowing through the storage capacitor. The polarity of change depends on the sources of leakage within a given device. In integrated circuits with external capacitors, it is usually specified as a *(droop* or *drift)* current, in ICs having internal capacitors, a rate of change. Note: dv/dt (volts/second)=I/C (picoamperes/picofarads).

Dual-Slope Converter. An integrating ADC in which the unknown signal is converted to a proportional time interval, which is then measured digitally. This is done by integrating the unknown for a predetermined length of time. A reference input is then switched to the integrator, which integrates "down" from the level determined by the unknown until the starting level is reached. The time for the second integration process, as determined by the counter, is proportional to the average of the unknown signal level over the predetermined integrating period. The counter provides the digital readout.

*Effective Input Noise. (*See *Input-Referred Noise.)*

Effective Number of Bits (ENOB). With a sinewave input, Signal-to-Noise-and-Distortion (SINAD) can be expressed in terms of the number of bits. Rewriting the theoretical SNR formula for an ideal N-bit ADC and solving for N:

N=(SNR−1.76dB)/6.02

The actual ADC SINAD is measured using FFT techniques, and ENOB is calculated from:

ENOB=(SINAD−1.76dB)/6.02

Effective Resolution. (See Noise-Free Code Resolution.)

Encode Command. (See Encode, Sampling Clock.)

Encode (Sampling Clock) Pulsewidth/Duty Cycle. Pulsewidth high is the minimum amount of time that the ENCODE pulse should be left in Logic 1 state to achieve rated performance; pulsewidth low is the minimum time ENCODE or pulse should be left in low state. See timing implications of changing the width in the text of high speed ADC data sheets. At a given clock rate, these specs define an acceptable ENCODE duty cycle.

Feedthrough. Undesirable signal coupling around switches or other devices that are supposed to be turned off or provide isolation, *e.g., feedthrough error* in a sample-and-hold, multiplexer, or multiplying DAC. Feedthrough is variously specified in percent, dB, parts per million, fractions of 1 LSB, or fractions of 1 V, with a given set of inputs, at a specified frequency.

In a multiplying *DAC,* feedthrough error is caused by capacitive coupling from an AC V_{REF} to the output, with all switches off. In a *sample-and-hold, feedthrough* is the fraction of the input signal variation or AC input waveform that appears at the output in *hold.* It is caused by stray capacitive coupling from the input to the storage capacitor, principally across the open switch.

Flash Converter. A converter in which all the bit choices are made at the same time. It requires $2^N - 1$ voltage-divider taps and comparators and a comparable amount of priority encoding logic. A scheme that gives extremely fast conversion, it requires large numbers of nearly identical components, hence it is well suited to integrated-circuit form for resolutions up to eight bits. Several Flash converters are often used in multistage *subranging converters,* to provide high resolution at somewhat slower speed than pure Flash conversion.

Four-Quadrant. In a multiplying DAC, "four quadrant" refers to the fact that both the reference signal and the number represented by the digital input may be of either positive or negative polarity. Such a DAC can be thought of as a gain control for AC signals ("reference" input) with a range of positive and negative digitally controlled gains. A four-quadrant multiplier is expected to obey the rules of multiplication for algebraic sign.

Frequency-to-Voltage Conversion (FVC). The input of an FVC device is an AC waveform—usually a train of pulses (in the context of conversion); the output is an analog voltage, proportional to the number of pulses occurring in a given time. FVC is usually performed by a voltage-to-frequency converter in a feedback loop. Important specifications, in addition to the accuracy specs typical of VFCs (see *Voltage-to-Frequency conversion*), include *output ripple* (for specified input frequencies), *threshold* (for recognition that another cycle has been initiated, and for versatility in interfacing several types of sensors directly), *hysteresis,* to provide a degree of insensitivity to noise superimposed on a slowly varying input waveform, and *dynamic response* (important in motor control).

Full-Scale Input Power (ADC). Expressed in dBm (power level referenced to 1 mW). Computed using the following equation, where $V_{\text{Full Scale rms}}$ is in volts, and Z_{input} is in Ω.

$$\text{Power}_{\text{Full Scale}} = 10 \log_{10} \left[\dfrac{\dfrac{V^2_{\text{Full Scale rms}}}{|Z|_{\text{Input}}}}{0.001} \right].$$

Full-Scale Range (FSR). For binary ADCs and DACs, that magnitude of voltage, current, or—in a multiplying DAC—gain, of which the MSB is specified to be exactly one-half or for which any bit or combination of bits is tested against its (their) prescribed ideal ratio(s). FSR is independent of resolution; the value of the LSB (voltage, current, or gain) is 2^{-N} FSR. There are several other terms, with differing meanings, that are often used in the context of discussions or operations involving full-scale range. They are:

Full-scale—similar to full-scale range, but pertaining to a single polarity. Thus, full-scale for a unipolar device is twice the prescribed value of the MSB and has the same polarity. For a bipolar device, *positive or negative full-scale* is that positive or negative value, of which the next bit after the polarity bit is tested to be one-half.

Span—the scalar voltage or current range corresponding to FSR.

All-1s—*All bits on,* the condition used, in conjunction with *all-zeros,* for gain adjustment of an ADC or DAC, in accordance with the manufacturer's instructions. Its magnitude, for a binary device, is $(1-2^{-N})$ FSR. *All-1s* is a *positive-true* definition of a specific magnitude relationship; for complementary coding the "all-1s" code will actually be all zeros. To avoid confusion, all-1s should never be called *full-scale*; FSR and FS are independent of the number of bits, all-1s isn't.

All-0s—*All bits off,* the condition used in offset (and gain) adjustment of a DAC or ADC, according to the manufacturer's instructions. All-0s corresponds to zero output in a unipolar DAC and negative full-scale in an offset bipolar DAC with positive output reference. In a sign-magnitude device, all-0s refers to all bits after the sign bit. Analogous to "all-1s," "all-0s" is a *positive-true* definition of the *all-bits-off* condition; in a complementary-coded device, it is expressed by all ones. To avoid confusion, all-0s should not be called "zero" unless it accurately corresponds to true analog zero output from a DAC.

Gain. The "gain" of a converter is that analog scale factor setting that establishes the nominal conversion relationship, e.g., 10 V full-scale. In a multiplying DAC or ratiometric ADC, it is indeed a gain. In a device with fixed internal reference, it is expressed as the full-scale magnitude of the output parameter (e.g., 10 V or 2 mA). In a fixed-reference converter, where the use of the internal reference is optional, the converter gain and the reference may be specified separately. Gain and zero adjustment are discussed under *zero*.

Glitch. Transients associated with code changes generally stem from several sources. Some are spikes, known as digital-to-analog feedthrough, or charge transfer, coupled from the digital signal (clock or data) to the analog output, defined with zero reference. These spikes are generally fast, fairly uniform, code-independent, and hence filterable. However, there is a more insidious form of transient, code-dependent, and difficult to filter, known as the "glitch."

If the output of a counter is applied to the input of a DAC to develop a "staircase" voltage, the number of bits involved in a code change between two adjacent codes establish "major" and "minor" transitions. The most major transition is at ½-scale, when the DAC switches all bits, i.e., from 011...111 to 100...000. If, for digital inputs having no skew, the switches are faster to switch *off* than *on,* this means that, for a short time, the DAC will seek zero output, and then return to the required 1 LSB above the previous reading. This large transient spike is commonly known as a "glitch." The better matched the input transitions and the switching times, the faster the switches, the smaller will be the area of the glitch. Because the size of the glitch is not proportional to the signal change, linear filtering may be unsuccessful and may, in fact, make matters worse. *(See also Deglitcher.)*

The severity of a glitch is specified by *glitch impulse area,* the product of its duration and its average magnitude, i.e., the net area under the curve. This product will be recognized as the physical quantity, *impulse* (electromotive *force*$\times\Delta$*time*)*;* however, it has also been incorrectly termed "glitch energy" and "glitch charge." Glitch impulse area is usually expressed, for fast converters, in units of pV-s or mV-ns.

The glitch can be minimized through the use of fast, nonsaturating logic, such as ECL, LVDS, matched latches, and nonsaturating CMOS switches.

Glitch Charge, Glitch Energy, Glitch Impulse, Glitch Impulse Area. (See *Glitch.*)

Harmonic Distortion, 2nd. The ratio of the rms signal amplitude to the rms value of the second harmonic component, reported in dBc.

Harmonic Distortion, 3rd. The ratio of the rms signal amplitude to the rms value of the third harmonic component, reported in dBc.

Harmonic Distortion, Total (THD). The ratio of the rms signal amplitude to the rms sum of all harmonics (neglecting noise components). In most cases, only the first five harmonics are included in the measurement because the rest have negligible contribution to the result. The THD can be derived from the FFT of the ADC's output spectrum. For harmonics that are above the Nyquist frequency, the aliased component is used.

Harmonic Distortion, Total, Plus Noise (THD+N). Total harmonic distortion plus noise (THD+N) is the ratio of the rms signal amplitude to the rms sum of all harmonics and noise components. THD+N can be derived from the FFT of the ADC's output spectrum and is a popular specification for audio applications.

Impedance, Input. The dynamic load of an ADC presented to its input source. In unbuffered CMOS switched-capacitor ADCs, the presence of current transients at the converter's clock frequency mandates that the converter be driven from a low impedance (at the frequencies contained in the transients) in order to accurately convert. For buffered-input ADCs, the input impedance is generally represented by a resistive and capacitive component.

Input-Referred Noise (Effective Input Noise). Input-referred noise can be viewed as the net effect of all internal ADC noise sources referred to the input. It is generally expressed in *LSBs rms*, but can also be expressed as a voltage. It can be converted to a peak-to-peak value by multiplying by the factor 6.6. The peak-to-peak input-referred noise can then be used to calculate the *noise-free code resolution*. (See *Noise-Free Code Resolution*).

Intermodulation Distortion (IMD). With inputs consisting of sinewaves at two frequencies, f_1 and f_2, any device with nonlinearities will create distortion products of order $(m+n)$, at sum and difference frequencies of $mf_1 \pm nf_2$, where m, n=0, 1, 2, 3, ... Intermodulation terms are those for which m or n is not equal to zero. For example, the second-order terms are (f_1+f_2) and (f_2-f_1), and the third-order terms are $(2f_1+f_2)$, $(2f_1-f_2)$, (f_1+2f_2), and (f_1-2f_2). The IMD products are expressed as the dB ratio of the rms sum of the distortion terms to the rms sum of the measured input signals.

Latency. (See *Pipelining.*)

Leakage Current, Output. Current that appears at the output terminal of a DAC with all bits "off." For a converter with two complementary outputs (for example, many fast CMOS DACs), output leakage current is the current measured at OUT 1, with all digital inputs *low*— *and* the current measured at OUT 2, with all digital inputs *high*.

Least-Significant Bit (LSB). In a system in which a numerical magnitude is represented by a series of binary (i.e., two-valued) digits, the *least-significant bit* is that digit (or "bit") that carries the smallest value, or weight. For example, in the natural binary number 1101 (decimal 13, or $(1\times2^3)+(1\times2^2)+(0\times2^1)+(1\times2^0)$), the rightmost digit is the LSB. Its analog weight, in relation to full-scale (see *Full-Scale Range*), is 2^{-N}, where N is the number of binary digits. It represents the smallest analog change that can be resolved by an n-bit converter.

In data converter nomenclature, the LSB is bit N; in bus nomenclature (integer binary), it is Data Bit 0.

Left-Justified Data. When a 12-bit word is placed on an 8-bit bus in two bytes, the high byte contains the 4 or 8 most-significant bits. If 8, the word is said to be left justified; if 4 (plus filled-in leading sign bits), the word is said to be right justified.

*Linearity. (*See also *Nonlinearity.)* Linearity error of a converter *(*also, *integral nonlinearity*— see *Linearity, Differential),* expressed in % or parts per million of full-scale range, or (sub)multiples of 1 LSB, is a deviation of the analog values, in a plot of the measured conversion relationship, from a straight line. The straight line can be either a "best straight line," determined empirically by manipulation of the gain and/or offset to equalize maximum positive and negative deviations of the actual transfer characteristic from this straight line; or, it can be a straight line passing through the end points of the transfer characteristic after they have been calibrated, sometimes referred to as "end- point" linearity. "End-point" nonlinearity is similar to relative accuracy error *(*see *Accuracy, Relative).* It provides an easier method for users to calibrate a device, and it is a more conservative way to specify linearity.

For multiplying DACs, the *analog* linearity error, at a specified analog gain (digital code), is defined in the same way as for analog multipliers, i.e., by deviation from a "best straight line" through the plot of the analog output-input response.

Linearity, Differential. In a DAC, any two adjacent digital codes should result in measured output values that are exactly 1 LSB apart (2^{-N} of full-scale for an N-bit converter). Any positive or negative deviation of the measured "step" from the ideal difference is called *differential nonlinearity,* expressed in (sub)multiples of 1 LSB. It is an important specification, because a differential linearity error more negative than -1 LSB can lead to nonmonotonic response in a DAC and missed codes in an ADC using that DAC.

Similarly, in an ADC, midpoints between code transitions should be 1 LSB apart. Differential nonlinearity is the deviation between the actual difference between midpoints and 1 LSB, for adjacent codes. If this deviation is equal to or more negative than -1 LSB, a code will be missed (See *Missing Codes.)*

Often, instead of a maximum differential nonlinearity specification, there will be a simple specification of "monotonicity" or "no missing codes," which implies that the differential nonlinearity cannot be more negative than -1 for any adjacent pair of codes. However, the differential linearity error may still be more positive than $+1$ LSB.

Linearity, Integral. (See Linearity.) While *differential linearity* deals with errors in step size, *integral linearity* has to do with deviations of the overall shape of the conversion response. Even converters that are not subject to differential linearity errors. (e.g., integrating types) have integral linearity (sometimes just "linearity") errors.

Maximum Conversion Rate. The maximum sampling (encode) rate at which parametric testing is performed.

Minimum Conversion (Sampling) Rate. The encode rate at which the SNR of the lowest analog signal frequency drops by no more than 3 dB below the guaranteed limit.

Missing Codes. An ADC is said to have missing codes when a transition from one quantum of the analog range to the adjacent one does not result in the adjacent digital code, but in a code removed by one or more counts. Missing codes can be caused by large negative differential linearity errors, noise, or changing inputs during conversion. A converter's proclivity towards missing codes is also a function of the architecture and temperature.

Monotonicity. An DAC is said to be *monotonic* if its output either increases or remains constant as the digital input increases, with the result that the output will always be a single-valued function of the input. The condition "monotonic" requires that the derivative of the transfer function never change sign. Monotonic behavior requires that the differential nonlinearity be more positive than -1 LSB. The same basic definition applies to an ADC— the digital output code either increases or remains constant as the digital input increases. In practice, however, noise will cause the ADC output code to oscillate between two code transitions over a small range of analog input. Input-referred noise can make this effect worse, so histogram techniques are often used to measure ADC monotonicity in these situations.

Most Significant Bit (MSB). In a system in which a numerical magnitude is represented by a series of binary (i.e., two-valued) digits, the *most-significant bit* is that digit (or "bit") that carries the greatest value or weight. For example, in the natural binary number 1101 (decimal 13, or $(1\times2^3)+(1\times2^2)+(0\times2^1)+(1\times2^0)$), the leftmost "1" is the MSB, with a weight of ½ nominal peak-to-peak full-scale (full-scale range). In bipolar devices, the sign bit is the MSB.

In converter nomenclature, the MSB is bit 1; in bus nomenclature, it is Data Bit $(N-1)$.

Multiplying DAC. A multiplying DAC differs from the conventional fixed-reference DAC in being designed to operate with varying (or AC) reference signals. The output signal of such a DAC is proportional to the product of the "reference" (i.e., analog input) voltage and the fractional equivalent of the digital input number. (See also *Four-Quadrant.*)

Multitone Spurious Free Dynamic Range (SFDR). The ratio of the rms value of an input tone to the rms value of the peak spurious component. The peak spurious component may or may not be an intermodulation distortion (IMD) product. May be reported in dBc (dB relative to the carrieror in dBFS (dB relative to full-scale). The amplitudes of the individual tones are equal and chosen such that the ADC is not overdriven when they add in-phase.

Noise-Free (Flicker-Free) Code Resolution. The noise-free code resolution of an ADC is the number of bits beyond which it is impossible to distinctly resolve individual codes. The cause is the effective input noise (or input-referred noise) associated with all ADCs. This noise can be expressed as an rms quantity, usually having the units of *LSBs rms*. Multiplying by a factor of 6.6 converts the rms noise into peak-to-peak noise (expressed in *LSBs peak-to-peak*). The total range of an N-bit ADC is 2^N. The noise-free (or flicker-free) resolution can be calculated using the equation:

Noise-Free Code Resolution$=\log_2(2^N/\text{Peak-to-Peak Noise})$

The specification is generally associated with high-resolution sigma-delta measurement ADCs, but is applicable to all ADCs.

The ratio of the FS range to the *rms* input noise is sometimes used to calculate resolution. In this case, the term *effective resolution* is used. Note that effective resolution is larger than noise-free code resolution by $\log_2(6.6)$, or approximately 2.7 bits.

Effective Resolution$=\log_2 (2^N /\text{RMS Input Noise})$.

Noise, Peak and RMS. Internally generated random noise is not a major factor in DACs, except at extreme resolutions and dynamic ranges. Random noise is characterized by rms specifications for a given bandwidth, or as a spectral density (current or voltage per root hertz); if the distribution is Gaussian, the probability of peak-to-peak values exceeding $6.6\times$ the rms value is less than 0.1%.

Of much greater importance in DACs is interference, in the form of high amplitude, low energy (hence low rms) spikes appearing at a DAC's output, caused by coupling of digital signals in a surprising variety of ways; they include coupling via stray capacitance, via power supplies, via inadequate ground systems, via feedthrough, and by glitch-generation (see *Glitch*). Their presence underscores the necessity for maximum application of the designer's art, including layout, shielding, guarding, grounding, bypassing, and deglitching.

Noise in ADCs in effect narrows the region between transitions. Sources of noise include the input sample-and-hold, resistor noise, "KT/C" noise, the reference, the analog signal itself, and pickup in infinite variety.

Noise Power Ratio (NPR). In this measurement, wideband Gaussian noise (bandwidth $< f_s/2$) is applied to an ADC through a narrowband notch filter. The notch filter removes all noise within its bandwidth. The output of the ADC is examined with a large FFT. The ratio of the rms noise level to the rms noise level inside the notch (due to quantization noise, thermal noise, and intermodulation distortion) is defined as the *noise power ratio (NPR)*. The rms noise level at the input to the ADC is generally adjusted to give the best NPR value.

No Missing Codes Resolution. (See *Resolution, No Missing Codes.*)

Nonlinearity (or "*gain nonlinearity*") The deviation from a straight line on the plot of output versus input. The magnitude of linearity error is the maximum deviation from a "best straight line," with the output swinging through its full-scale range. Nonlinearity is usually specified in percent of full-scale output range.

Normal Mode. For an amplifier used in instrumentation, the *normal-mode* signal is the actual difference signal being measured. This signal often has noise associated with it. Signal conditioning systems and digital panel instruments usually contain input filtering to remove high frequency and line frequency noise components. *Normal-mode rejection* (NMR), is a logarithmic measure of the attenuation of normal-mode noise components at specified frequencies in dB.

Offset, Bipolar. For the great majority of bipolar converters (e.g., ± 10 V output), negative currents are not actually generated to correspond to negative numbers; instead, a unipolar DAC is used, and the output is offset by half full-scale (1 MSB). For best results, this offset voltage or current is derived from the same reference supply that determines the gain of the converter.

Because of nonlinearity, a device with perfectly calibrated end points may have offset error at analog zero.

Offset Step. (See *Pedestal.*)

Output Propagation Delay. For an ADC having a single-ended sampling (or ENCODE) clock input, the delay between the 50% point of the sampling clock and the time when all output data bits are within valid logic levels. For an ADC having differential sampling clock inputs, the delay is measured with respect to the zero crossing of the differential sampling clock signal.

Output Voltage Tolerance. For a reference, the maximum deviation from the normal output voltage at 25°C and specified input voltage, as measured by a device traceable to a recognized fundamental voltage standard.

Overload. An input voltage exceeding the ADC's full-scale input range producing an overload condition.

Overvoltage Recovery Time. Overvoltage recovery time is defined as the amount of time required for an ADC to achieve a specified accuracy after an overvoltage (usually 50% greater than full-scale range), measured from the time the overvoltage signal reenters the converter's range. The ADC should act as an ideal limiter for out-of-range signals, producing a positive or negative full-scale code during the overvoltage condition. Some ADCs provide over- and underrange flags to allow gain-adjustment circuits to be activated.

Overrange, Overvoltage. An input signal that exceeds the full-scale input range of an ADC, but is less than an overload.

Pedestal, or *Sample-to-Hold Offset Step.* In sample/track-and-hold amplifiers, a shift in level between the last value in *sample* and the value settled-to in *hold;* in devices having fixed internal capacitors, it includes *charge transfer,* or *offset step.* However, for devices that may use external capacitors, it is often defined as the residual step error after the *charge transfer* is accounted for and/or cancelled. Since it is unpredictable in magnitude and may be a function of the signal, it is also known as *offset nonlinearity.*

Pipelining. A pipelined converter is a multistage converter capable of accepting a new signal before it has completed the conversion of one or more previous ones. A new signal arrives while others are still "in the pipeline." This is a technique used where a fast conversion rate is desired and the latency of individual conversions is relatively unimportant.

Power-Supply Rejection Ratio (PSRR). The ratio of a change in DC power supply voltage to the resulting change in the specified device error, expressed in percentage, parts per million, or fractions of 1 LSB. It may also be expressed logarithmically, in dB, $PSR = 20 \log_{10} (PSRR)$.

Quad-Slope Converter. This is an integrating analog-to-digital converter that goes through two cycles of *dual-slope* conversion, once with zero input and once with the analog input being measured. The errors determined during the first cycle are subtracted digitally from the result in the second cycle. The scheme can result in high-accuracy conversion.

Quantizing Uncertainty (or "Quantization Error"). The analog continuum is partitioned into 2^N discrete ranges for N-bit conversion and processing. All analog values within a given quantum are represented by the same digital code, usually assigned to the nominal midrange value. There is, therefore, an inherent quantization uncertainty of $\pm\frac{1}{2}$ LSB, in addition to the actual conversion errors. In integrating ADCs, this "error" is often expressed as "±1 count." Depending on the system context, it may be interpreted as a truncation (round-off) error or as noise.

Ratiometric. The output of an ADC is a digital number proportional to the *ratio* of (some measure of) the input to a reference voltage. Most requirements for conversions call for an absolute measurement, i.e., against a fixed reference; but this presumes that the signal applied to the converter is either reference-independent or in some way derived from another fixed reference. However, real references are not truly fixed; the references for both the converter and the signal source vary with time, temperature, loading, etc. Therefore, if the converter is used with signal sources that also rely on references (for example, strain-gage bridges, RTDs, thermistors), it makes sense to replace this multiplicity of references by a single system reference. In this case, reference-caused errors will tend to cancel out. This can be done by using the converter's internal reference (if it has one) as the system reference. Another way is to use a separate external system reference, which also becomes the reference for a *ratiometric* converter. For instance, if a bridge is excited with the same voltage used for the ADC reference, ratiometric operation is achieved, and the ADC output code is not a function of the reference. This is because the bridge output signal is proportional to the same voltage which defines the ADC input range.

Resolution. An N-bit binary converter has N digital data inputs (DAC) or N digital data outputs (ADC). A converter that satisfies this criterion is said to have a *resolution* of N bits.

Resolution, No Missing Codes. The *no missing code resolution* of an ADC is the maximum number of bits of resolution beyond which the ADC will have missing codes. For instance, if an 18-bit ADC has a no missing code resolution of 16 bits, there will be no missing codes if only the 16 MSBs are utilized. Codes may be missed at the 17- and 18-bit level. The smallest output change that can be resolved by a linear DAC is 2^{-N} of the full-scale span. Thus, for example, the resolution of an 8-bit DAC would be 2^{-8}, or 1/256. On the other hand, a nonlinear device, such as the AD7111 LOGDAC™, can ideally achieve a dynamic range of 89.625 dB, or 30,000:1, in 0.375 dB steps, using only 8 bits of digital resolution.

Right-Justified Data. When a 12-bit word is placed on an 8-bit bus in two stages, the high byte contains the 4 or 8 most-significant bits. If 8, the word is said to be left justified; if 4 (plus filled-in leading sign bits), the word is said to be right justified.

Sample-to-Hold Offset.(See *Pedestal.*)

Sampling ADC. A sampling ADC includes a sample-and-hold function that acquires the input value at a given instant and holds it throughout the conversion time (or until the converter is ready for the next sample point). Flash ADCs and sigma-delta ADCs are inherently sampling devices.

Sampling Clock. (See *Encode Command*).

Sampling Frequency. The rate at which an ADC converts an analog input signal into digital outputs, not to be confused with *conversion time*.

Serial Output. A bit-serial output consists of a series of bits clocked out on a single line. There must be some means of identifying the beginning and ends of words; this can be accomplished via an additional clock line, by using synchronized clocks, and/or by providing a consistent identifying signature for the beginning of a word. Byte-serial consists of a series of bytes transmitted in sequence on a bus. (See *Byte.*)

Settling Time—ADC. The time required, following an analog input step change (usually full-scale), for the digital output of the ADC to reach and remain within a given fraction (usually ±½ LSB).

Settling Time—DAC. The time required, following a prescribed data change, for the output of a DAC to reach and remain within an error band (usually ±½ LSB) of the final value. Typical prescribed changes are full-scale, 1 MSB, and 1 LSB at a major carry. Settling time of current-output DACs is quite fast. The major share of settling time of a voltage-output DAC is usually contributed by the settling time of the output op-amp. DAC settling time can also be defined with respect to the output. Output settling time is the time measured from the point the output signal leaves an error band referenced to the initial output value until the time the signal enters and remains within the error band referenced to the final output value.

Signal-to-Noise-and-Distortion Ratio (SINAD). The ratio of the rms signal amplitude (set 1 dB below full-scale to prevent overdrive) to the rms value of the sum of all other spectral components, including harmonics but excluding DC.

Signal-to-Noise Ratio (without Harmonics). The ratio of the rms signal amplitude (set at 1 dB below full-scale to prevent overdrive) to the rms value of the sum of all other spectral components, excluding the first five harmonics and DC. Technically, all harmonics should be excluded, but in practice, only the first five are generally significant.

Single-Slope Conversion. In the single-slope converter, a reference voltage is integrated until the output of the integrator is equal to the input voltage. The time period required for the integrator to go from zero to the level of the input is proportional to the magnitude of the input voltage and is measured by an internal clock. Measurement accuracy is sensitive to clock speed and integrating capacitance, as well as the reference accuracy.

Slew(ing) Rate. A limitation in the rate of change of output voltage, usually imposed by some basic circuit consideration, such as limited current to charge a capacitor. The output slewing speed of a voltage-output DAC is usually limited by the slew rate of the amplifier used at its output.

Spurious-Free Dynamic Range (SFDR). The ratio of the rms signal amplitude to the rms value of the peak spurious spectral component. The peak spurious component may or may not be a harmonic. May be reported in dBc (i.e., degrades as signal level is lowered) or dBFS (related back to converter full-scale).

Stability. In a well-designed, intelligently applied converter, *dynamic stability* is not an important question. The term stability usually applies to the insensitivity of the converter's characteristics to time, temperature, etc. All measurements of stability are difficult and time consuming, but stability versus temperature is sufficiently critical in most applications to warrant universal inclusion in tables of specifications (see *Temperature Coefficient).*

Staircase. A voltage or current, increasing in equal increments as a function of time and having the appearance of a staircase (in a time plot); it is generated by applying a pulse train to a counter, and the output of the counter to the input of a DAC.

Subranging ADCs. In this type of converter, a fast converter produces the most-significant portion of the output word. This portion is stored in a holding register and also converted back to analog with a fast, high-accuracy DAC. The analog result is subtracted from the input, and the resulting residue is amplified, converted to digital at high speed, and combined with the results of the earlier conversion to form the output word. In *digitally corrected subranging* (DCS) ADCs, the two conversions are combined in a manner that corrects for the error of the LSB of the most significant bits. For example, using 8-bit and 5-bit conversion, plus this technique and a great deal of video-speed converter expertise, a full-accuracy high speed 12-bit ADC can be built. Many pipelined subranging ADCs use more than two stages with error correction between each stage.

Successive Approximation. Successive approximation is a method of conversion by comparing an unknown against a group of weighted references. The operation of a successive-approximation ADC is generally similar to the orderly weighing of an unknown quantity on a precision balance, using a set of weights, such as 1 gram ½ gram, ¼ gram, etc. The weights are tried in order, starting with the largest. Any weight that tips the scale is removed. At the end of the process, the sum of the weights remaining on the scale will be within 1 LSB of the actual weight ($\pm$½ LSB, if the scale is properly biased—see *Zero).* The successive approximation ADC is often called a SAR ADC, because the logic block that controls the conversion process is known as a successive approximation register (SAR).

Switching Time. In a DAC, the switching time is the time taken for an analog switch to change to a new state from the previous one. It includes propagation delay time, and rise time from 10% to 90%, but does not include settling time.

Temperature Coefficient. In general, temperature instabilities are expressed as %/°C, ppm/°C, fractions of 1 LSB per degree C, or as a change in a parameter over a specified temperature range. Measurements are usually made at room temperature (25°C) and at the extremes of the specified range, and the temperature coefficient (tempco, TC) is defined as the change in the parameter, divided by the corresponding temperature change. Parameters of interest include gain, linearity, offset (bipolar), and zero.

a. *Gain Tempco:* Two factors principally affect converter gain stability with temperature. In fixed-reference converters, the reference voltage will vary with temperature. The reference circuitry and switches (and comparator in ADCs) will also contribute to the overall gain TC.

b. *Linearity Tempco:* Sensitivity of linearity (integral and/or differential linearity) to temperature, in % FSR/°C or ppm FSR/°C, over the specified range. Monotonic behavior in DACs is achieved if the differential nonlinearity is less than 1 LSB at any temperature in the range of interest. The *differential nonlinearity temperature coefficient* may be expressed as a ratio, as a maximum change over a temperature range, and/or implied by a statement that the device is monotonic over the specified temperature range. To avoid missing codes in noiseless ADCs, it is sufficient that the differential nonlinearity error be less than -1 LSB at any temperature in the range of interest. The differential nonlinearity temperature coefficient is often implied by the statement that there are no missed codes when operating within a specified temperature range. In DACs, the differential nonlinearity TC is often implied by the statement that the DAC is monotonic over a specified temperature range.

c. *Zero TC (unipolar converters):* The temperature stability of a unipolar fixed-reference DAC, measured in % FSR/°C or ppm FSR/°C, is principally affected by current leakage (current-output DAC), and offset voltage and bias current of the output op amp (voltage-output DAC). The zero stability of an ADC is dependent on the zero stability of the DAC or integrator and/or the input buffer and the comparator. It is typically expressed in μV/°C or in percent or ppm of full-scale range (FSR) per degree C.

d. *Offset Tempco:* The temperature coefficient of the all-DAC-switches-off (minus full-scale) point of a bipolar converter (in % FSR/°C or ppm FSR/°C) depends on three major factors—the tempco of the reference source, the voltage zero-stability of the output amplifier, and the tracking capability of the bipolar-offset resistors and the gain resistors. In an ADC, the corresponding tempco of the negative full-scale point depends on similar quantities—the tempco of the reference source, the voltage stability of the input buffer and the sample-and-hold, and the tracking capabilities of the bipolar offset resistors and the gain-setting resistors.

Thermal Tail. The slow drift of an amplifier having a thermally induced offset due to self-heating as it settles to a final electrical equilibrium value corresponding to internal thermal equilibrium.

Total Unadjusted Error. A comprehensive specification on some devices which includes full-scale error, relative-accuracy and zero-code errors, under a specified set of conditions.

Transient Response. (See *Settling Time.*)

Two-Tone SFDR. The ratio of the rms value of either input tone to the rms value of the peak spurious component. The peak spurious component may or may not be an intermodulation distortion (IMD) product. May be reported in dBc (i.e., degrades as signal level is lowered) or in dBFS (always related back to converter full-scale).

Worst Other Spur. The ratio of the rms signal amplitude to the rms value of the worst spurious component (excluding the second and third harmonic) reported in dBc.

References

1.1 Coding and Quantization

1. Cattermole KW. *Principles of Pulse Code Modulation*. New York NY: American Elsevier Publishing Company, Inc.; 1969 ISBN 444-19747-8. (An excellent tutorial and historical discussion of data conversion theory and practice, oriented towards PCM, but covers practically all aspects. This one is a must for anyone serious about data conversion.)

2. Gray F. Pulse Code Communication. U.S. Patent 2,632,058, filed November 13, 1947, issued March 17, 1953. (Detailed patent on the Gray code and its application to electron beam coders.)

3. Sears RW. Electron Beam Deflection Tube for Pulse Code Modulation. *Bell System Technical Journal* January 1948;**27**:44–57. (Describes an electron-beam deflection tube 7-bit,100 kSPS Flash converter for early experimental PCM work.)

4. Edson JO, Henning HH. Broadband Codecs for an Experimental 224Mb/s PCM Terminal. *Bell System Technical Journal* November 1965;**44**:1887–940. (Summarizes experiments on ADCs based on the electron tube coder as well as a bit-per-stage Gray code 9-bit solid-state ADC. The electron beam coder was 9 bits at 12 MSPS, and represented the fastest of its type.)

5. Sheingold D. *Analog-Digital Conversion Handbook*. 3rd Edition. Analog Devices and Prentice-Hall; 1986 ISBN-0-13-032848-0. (The defining and classic book on data conversion.)

1.2 Sampling Theory

6. Nyquist H. Certain Factors Affecting Telegraph Speed. *Bell System Technical Journal* April 1924;**3**:324–46.

7. Nyquist H. Certain Topics in Telegraph Transmission Theory. *A.I.E.E. Transactions* April 1928;**47**:617–44.

8. Hartley RVL. Transmission of Information. *Bell System Technical Journal* July 1928;**7**:535–63.

9. Shannon CE. A Mathematical Theory of Communication. *Bell System Technical Journal* July 1948;**27**:379–423 and October 1948, pp. 623–656.

10. TTE, Inc., 11652 Olympic Blvd., Los Angeles, CA 90064, www.tte.com

1.3 Data Converter AC Errors

11. Bennett WR. Spectra of Quantized Signals. *Bell System Technical Journal* July 1948;**27**:446–71.

12. Oliver BM, Pierce JR, Shannon CE. The Philosophy of PCM. *Proceedings IRE* November 1948;**36**:1324–31.

13. Bennett WR. Noise in PCM Systems. *Bell Labs Record* December 1948;**26**:495–99.

14. Black HS, Edson JO. Pulse Code Modulation. *AIEE Transactions* 1947;**66**:895–99.

15. Black HS. Pulse Code Modulation. *Bell Labs Record* July 1947;**25**:265–69.

16. Ruscak S, Singer L. Using Histogram Techniques to Measure A/D Converter Noise. *Analog Dialogue* 1995;**29-2**.

17. Tant MJ. *The White Noise Book*: Marconi Instruments; July 1974.

18. Gray GA, Zeoli GW. Quantization and Saturation Noise due to A/D Conversion. *IEEE Trans. Aerospace and Electronic Systems* January 1971:222–23.

19. McClaning K, Vito T. *Radio Receiver Design*. Noble Publishing; 2000. ISBN 1-88-4932-07-X.

20. Jung WG, editor. *Op Amp Applications*: Analog Devices, Inc.; 2002.ISBN 0-916550-26-5, pp. 6.144–6.152.

21. Brannon B. *Aperture Uncertainty and ADC System Performance* Application Note AN-501: Analog Devices, Inc.; January 1998 (available for download at www.analog.com).

22. Mangelsdorf CW. A 400 MHz Input Flash Converter with Error Correction. *IEEE Journal of Solid-State Circuits* February 1990;**25**(1):184–91.

23. Woodward CE. A Monolithic Voltage-Comparator Array for A/D Converters. *IEEE Journal of Solid State Circuits* December 1975;**SC-10**(6):392–99.

24. Akazawa Y, et al. A 400MSPS 8 Bit Flash A/D Converter. *1987 ISSCC Digest of Technical Papers*, pp. 98–99.

25. Matsuzawa A, et al. An 8b 600 MHz Flash A/D Converter with Multistage Duplex-gray Coding. *Symposium VLSI Circuits, Digest of Technical Papers*, May 1991, pp. 113–114.

26. Waltman R, Duff D. Reducing Error Rates in Systems Using ADCs. *Electronics Engineer* April 1993:98–104.

27. Cattermole KW. *Principles of Pulse Code Modulation*. New York NY: American Elsevier Publishing Company, Inc.; 1969 ISBN 444-19747-8. (*An excellent tutorial and historical discussion of data conversion theory and practice, oriented towards PCM, but covers practically all aspects. This one is a must for anyone serious about data conversion.*).

28. Witte RA. Distortion Measurements Using a Spectrum Analyzer. *RF Design* September 1992:75–84.

29. Kester W. Confused About Amplifier Distortion Specs? *Analog Dialogue* 1993;**27-1**:27–29.

30. Sheingold D, editor. *Analog-to-Digital Conversion Handbook*. Third Edition: Prentice-Hall; 1986.

31. Irons FH. The Noise Power Ratio—Theory and ADC Testing. *IEEE Transactions on Instrumentation and Measurement* June 2000;**49**(3):659–65.

Bibliography

Baker, R. J. (2002). *CMOS Circuit Design Volumes I and II*. John Wiley-IEEE Computer Society ISBN 0-4712-7256-6.

Bruck, D. B. (1974). *Data Conversion Handbook*. Hybrid Systems Corporation.

Candy, J. C., & Temes, G. C. (1992). *Oversampling Delta-Sigma Data Converters*. IEEE Press ISBN 0-87942-258-8.

Cattermole, K. W. (1969). *Principles of Pulse Code Modulation*. New York NY: American Elsevier Publishing Company, Inc. ISBN 444-19747-8.

Demler, M. J. (1991). *High-Speed Analog-to-Digital Conversion*. Academic Press, Inc. ISBN 0-12-209048-9.

Dooley, D. J. (1980). *Data Conversion Integrated Circuits*. John Wiley-IEEE Press ISBN 0-471-08155-8.

Gordon, B. M. (1981). *The Analogic Data-Conversion Systems Digest* (Fourth Edition). Analogic Corporation.

Gustavsson, M., Wikner, J. J., & Tan, N. N. (2000). *CMOS Data Converters for Communications*. Kluwer Academic Publishers. ISBN 0-7923-7780-X.

Hnatek, E. R. (1976). *A User's Handbook of D/A and A/D Converters*. New York: John Wiley ISBN 0-471-40109-9.

Hoeschele, D. F., Jr. (1968). *Analog-to-Digital/Digital-to-Analog Conversion Techniques* John Wiley and Sons.

Hoeschele, D. F., Jr. (1994). *Analog-to-Digital and Digital-to-Analog Conversion Techniques* (Second Edition). John Wiley and Sons. ISBN-0-471-57147-4.

Jayant, N. S. (1976). *Waveform Quantizing and Coding*. John Wiley-IEEE Press ISBN 0-87942-074-X.

Johns, D. A., & Martin, K. (1997). *Analog Integrated Circuit Design*. John Wiley. ISBN 0-471-14448-7.

Kurth, C. F., editor. (July 1978). *IEEE Transactions on Circuits and Systems Special Issue on Analog/Digital Conversion*. CAS-25(7).

Mahoney, M. (1987). *DSP-Based Testing of Analog and Mixed-Signal Circuits*. IEEE Computer Society Press ISBN 0-8186-0785-8.

Owen, F. F. E. (1982). *PCM and Digital Transmission Systems*. McGraw-Hill. ISBN 0-07-047954-2.

Schmid, H. (1970). *Electronic Analog/Digital Conversions*. Van Nostrand Reinhold Co.

Sheingold, D. (1972). *Analog-Digital Conversion Handbook* (First Edition). Analog Devices.

Sheingold, D. (1977). *Analog-Digital Conversion Notes*. Analog Devices.

Sheingold, D. (1986). *Analog-Digital Conversion Handbook*. Analog Devices/Prentice-Hall ISBN 0-13-032848-0.

Susskind, A. K. (1957). *Notes on Analog-Digital Conversion Techniques*. John Wiley.

van de Plassche, R. (1994). *Integrated Analog-to-Digital and Digital-to-Analog Converters*. Kluwer Academic Publishers. ISBN 0-7923-9436-4.

van de Plassche, R. (2003). *CMOS Integrated Analog-to-Digital and Digital-to-Analog Converters* (Second Edition). Kluwer Academic Publishers. ISBN 1-4020-7500-6.

Zuch, E. L. (1982). *Data Acquisition and Conversion Handbook*. Datel-Intersil.

Analog Devices' Seminar Series

Kester, W. (1995). *Practical Analog Design Techniques*. Analog Devices. ISBN 0-916550-16-8, available for download at www.analog.com

Kester, W. (1996). *High Speed Design Techniques*. Analog Devices. ISBN 0-916550-17-6, available for download at www.analog.com

Kester, W. (1998). *Practical Design Techniques for Power and Thermal Management*. Analog Devices. ISBN 0-916550-19-2, available for download at www.analog.com

Kester, W. (1999). *Practical Design Techniques for Sensor Signal Conditioning*. Analog Devices. ISBN 0-916550-20-6, available for download at www.analog.com

Kester, W. (2000). *Mixed-Signal and DSP Design Techniques*. Analog Devices. ISBN 0-916550-22-2, available for download at www.analog.com

Kester, W. (2003). *Mixed-Signal and DSP Design Techniques*. Analog Devices and Newnes (An Imprint of Elsevier Science). ISBN 0-75067-611-6.

Jung, W. G. (2002). *Op Amp Applications*. Analog Devices.

CHAPTER 2

Digital Filters

Robert Meddins

Most introductory texts on DSP are very similar. Every author seems compelled to demonstrate that they can still derive the fundamental equations of DSP—never mind that you don't need to see these derivations! In this introduction to DSP, the author's stated goal is a genuine introduction without going into the usual derivations and examples. It's a refreshing approach. The text does a good job of quickly explaining the essential ideas without any fuss. If you want the derivations and more examples, the author provides plenty of references.

The chapter introduces the most basic principles of DSP, starting with the difference between analog and digital processing. It then moves on to sampling, anti-aliasing filters, and analog-to-digital converters (ADC). Finally, it wraps up with examples of basic digital systems such as finite-impulse-response (FIR) filters and infinite-impulse-response (IIR) filters.

The examples are very simple. For instance, ADCs are much more complex than the simple circuit given. But the example lets you know how ADCs operate, which is all most of us need to know. Wrap your brain around these examples and you will be well on your way to calling yourself a DSP engineer.

The text is showing its age in places (first published in 1995). It spends a few skippable paragraphs making the case for DSP over analog signal processing—a war long won by DSP. And its comments about DSP processors don't apply too well to the current generation of DSP processors.

All in all, however, this chapter should quickly give you the basic concepts needed to move on to the more advanced topics in this book.

—Kenton Williston

2.1 Chapter Preview

In this chapter you will be introduced to the basic principles of digital signal processing (DSP). We will look at how digital signal processing differs from the more conventional analog signal processing and also at its many advantages. Some simple digital processing systems will be described and analyzed. The main aim of this chapter is to set the scene and give a feel for what digital signal processing is all about.

2.2 Analog Signal Processing

You are probably very familiar with *analog* signal processing. Some obvious examples of this type of processing are amplification, rectification and filtering. With all analog processing, signals enter a system, are processed by passing them through circuits containing capacitors, resistors, inductors, op amps, transistors, etc. They are then outputted from the system with a different shape or size. Figure 2.1 shows a very elementary example of an analog signal processing system, consisting of just a resistor and a capacitor—you will probably recognize it as a simple type of lowpass filter. Analog signal processing circuits are commonplace and have been very important system building blocks since the early days of electrical engineering.

Unfortunately, as useful as they are, analog processing systems do have major defects. An obvious one is that they have to be physically modified if the processing needs to be changed. For example, if the gain of an amplifier has to be increased, then this usually means that at least a resistor has to be changed. What if a different cut-off frequency is required for a filter or, even worse, we want to replace a highpass filter with a lowpass filter? Once again, components must be changed. This can be very inconvenient to say the least—it's bad enough when a single system has to be adjusted but imagine the situation where a batch of several thousand is found to need modifying. How much better if changes could be achieved by altering a parameter or two in a computer program…

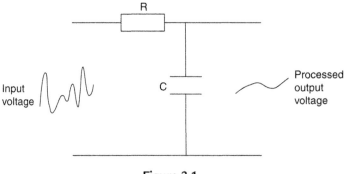

Figure 2.1

Another problem with analog systems is that of *"repeatability."* It is *very* unlikely that two analog systems will have identical performances, even though they have been made in exactly the same way, with supposedly the same value components. This is mainly because of component tolerances. Analog devices have two further disadvantages. The first is that their components age and so the device performance changes. The other is that components are also affected by temperature changes.

2.3 An Alternative Approach

So, having slightly dented the reputation of analog processors, what's the alternative? Luckily, signal processing systems do exist which work in a completely different way and do not have these problems. A major difference is that these systems first sample, at regular intervals, the signal to be processed (Figure 2.2). The sampled voltages are then converted to equivalent binary values, using an analog-to-digital converter (Figure 2.3). Next, these binary numbers are fed into a digital processor, containing a particular program, which will change the samples. The way in which the digital values are modified will obviously depend on the type of signal processing required—for example, do we want lowpass or highpass filtering and what cut-off frequency do we require? The transformed samples are then outputted, via a digital-to-analog converter, to produce the reconstituted but processed analog output signal.

Because computers can process data so quickly, the signal processing can be done almost in "real time," i.e., the processed output samples are fed out continuously, almost in step with the corresponding input samples. Alternatively, the processed data could be stored, perhaps on a chip or CD-ROM, and then read when required.

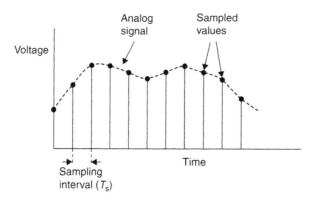

Figure 2.2

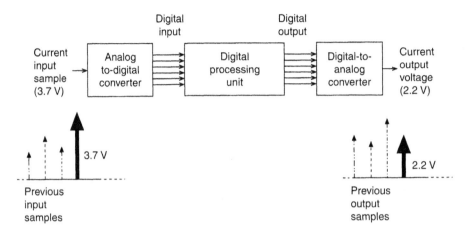

Figure 2.3

By now, you've probably guessed that this form of processing is called *digital signal processing*. Digital signal processing (DSP) does not have the drawbacks of analog signal processing, already mentioned. For example, the type of processing required can be modified very easily—if the specification of a filter needs to be changed then new parameters can simply be keyed into the DSP system, i.e., the processing is *programmable*. The performance of a digital filter is also constant, not changing with either time or temperature. DSP systems are also inherently repeatable—if several DSP systems have been programmed to process signals in a certain way then they will all behave identically. DSP systems can also process signals in ways impossible for analog systems.

To summarize:

- *Digital signal processing systems* are available that will do almost everything that analog signals can do, and much more—*versatile*.

- They can be easily changed—*programmable*.

- They can be made to process signals identically—*repeatable*.

- They are not affected by temperature or aging—*physically stable*.

2.4 The Complete DSP System

The heart of the digital signal processing system, the analog-to-digital converter (ADC), digital processor and the digital-to-analog converter (DAC), is shown in Figure 2.3. However,

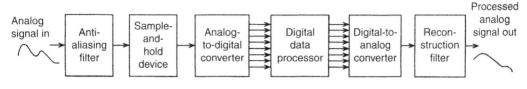

Figure 2.4

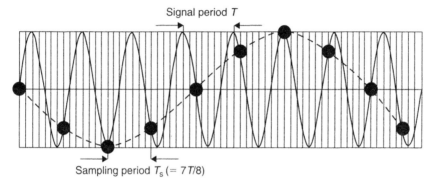

Figure 2.5

this sub-unit needs "topping and tailing" in order to create the complete system. An entire, general DSP system is shown in Figure 2.4.

Each block will now be described briefly.

2.4.1 The Anti-Aliasing Filter

If the analog input voltage is not sampled frequently enough then this results in something of a shambles. Basically, high frequency input signals will appear as low frequency signals at the output, which will be very confusing to say the least! This phenomenon is called *aliasing*. In other words, the high frequency input signals take on another identity, or "alias," on leaving the system.

To get a feel for the problem of aliasing, consider a sinusoidal signal, of fixed frequency, which is being sampled every 7/8 of a period, i.e., $7T/8$ (Figure 2.5). Having only the samples as a guide, it can be seen that the sampled signal appears to have a much lower frequency than it really has.

In practice, a signal will not usually have a single frequency but will consist of a very wide range of frequencies. For example, audio signals can contain frequency components in the range of about 20 Hz to 20 kHz.

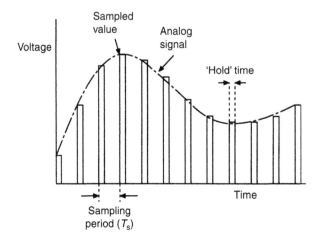

Figure 2.6

To prevent aliasing, it can be shown that the signal must be sampled at least twice as fast as the highest frequency component.

This very important rule is known as the Nyquist criterion, or Shannon's sampling theorem, after two distinguished pioneers from the world of signal processing.

If this sampling rate cannot be achieved, perhaps because the components used just cannot respond this quickly, then a lowpass filter must be used on the input end of the system. This has the job of removing signal frequencies greater than $f_s/2$, where f_s is the sampling frequency. This is the role of the *anti-aliasing filter*. An anti-aliasing filter is therefore a lowpass filter with a cut-off frequency of $f_s/2$.

The important frequency of $f_s/2$ is usually called the *Nyquist frequency.*

2.4.2 The Sample-and-Hold Device

An ADC should not be presented with a changing voltage to convert. The changing signal should be sampled and then this sampled voltage held while the conversion is carried out (Figure 2.6). (In practice, the sampled value is normally held until the next sample is taken.) If the voltage is *not* kept constant during conversion then, depending on the type of converter used, the digital output might not just be a little inaccurate but could be absolute rubbish, bearing no relationship to the true value.

At the heart of the *sample-and-hold* device is a capacitor (Figure 2.7). The electronic switch, S, is closed, causing the capacitor to charge to the current value of the input voltage. After a brief time interval the switch is reopened, so keeping the sampled voltage across the capacitor constant

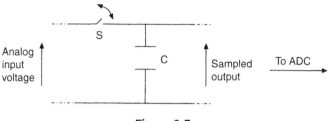

Figure 2.7

while the ADC carries out its conversion. The complete sample-and-hold device usually includes a voltage follower at both the input and the output of the basic system shown in Figure 2.7. The characteristically low output impedance and high input impedance of the voltage followers ensure that the capacitor is charged very quickly by the input voltage and discharges very slowly through the ADC connected to its output, so maintaining the stored voltage.

2.4.3 The Analog-to-Digital Converter

This converts the steady, sampled voltage, supplied by the sample-and-hold device, to an equivalent digital value in preparation for processing. The more output bits the converter has, the finer the resolution of the device, i.e., the smaller is the voltage change represented by the least significant output bit changing from 0 to 1 or from 1 to 0.

You are probably aware that there are many different types of ADC available. However, some of these are too slow for most DSP applications, e.g., single- and dual-slope and the basic counter-feedback versions. An ADC widely used in DSP systems is the sigma-delta converter. If you feel the need to do some extra reading in order to brush up on ADCs then some keywords to look out for are: single-slope, dual-slope, counter-feedback, successive approximation, flash, tracking and sigma-delta converters and also converter resolution. Millman and Grabel (1987) is just one of many books that give a good general treatment, while Marven and Ewers (1994) and also Proakis and Manolakis (1996) are two texts that give good coverage of the sigma-delta converter.

2.4.4 The Processor

This *could* be a general-purpose microprocessor chip, but this is unlikely. The data processing part of a purpose-built DSP chip is designed to be able to do a limited number of fairly simple operations, in particular addition and multiplication, *but they do these exceptionally quickly*. Most of the major chip-producing companies have developed their own DSP chips, e.g., Motorola, Texas Instruments and Analog Devices, and their user manuals are obvious reference sources for further reading.

2.4.5 The Digital-to-Analog Converter

This converts the processed digital value back to an equivalent analog voltage. Common types are the weighted resistor and the R-2R ladder converters, although the weighted resistor version is not a practical proposition, as it cannot be fabricated sufficiently accurately as an integrated circuit. Details of these two devices can be found in Millman and Grabel (1987), while Marven and Ewers (1994) describes the more sophisticated "bit-stream" DAC, often used in DSP systems.

2.4.6 The Reconstruction Filter

As the anti-aliasing filter ensures that there are no frequency components greater than $f_s/2$ entering the system, then it seems reasonable that the output signal will also have no frequency components greater than $f_s/2$. However, this is not so! The output from the DAC will be "steppy" because the DAC can only output certain voltage values. For example, an 8-bit DAC will have 256 different output voltage levels going from perhaps $-5\,V$ to $+5\,V$. When this quantized output is analyzed, frequency components of f_s, $2f_s$, $3f_s$, $4f_s$, etc. (harmonics of the sampling frequency) are found. The very action of sampling and converting introduces these harmonics of the sampling frequency into the output signal. It is these harmonics which give the output signal its steppy appearance. The *reconstruction filter* is a lowpass filter having a cut-off frequency of $f_s/2$, and is used to filter out these harmonics and so smooth the output signal.

2.5 Recap

- Analog signal processing systems have a variety of disadvantages, such as components needing to be changed in order to change the processor function, inaccuracies due to component aging and temperature changes, processors built in the same way not performing identically.

- Digital processing systems do not suffer from the problems above.

- Digital signal processing systems sample the input signal and convert the samples to equivalent digital values. These values are processed and the resulting digital outputs converted back to analog voltages. This series of discrete voltages is then smoothed to produce the processed analog output.

- The analog input signal must be sampled at a frequency which is at least twice as high as its highest frequency component, otherwise "aliasing" will take place.

2.6 Digital Data Processing

For the rest of this chapter we will concentrate on the processing of the digital values by the digital data processing unit—this is where the clever bit is done!

So, how does it all work? The digital data processor (Figure 2.4) is constantly being bombarded with digital values, one following the other at regular intervals. Its job is to output a suitable digital number in response to each digital input. This is something of an achievement as all that the processor has to work with is the current input value and the previous input and output samples. Somehow it has to use these to generate the output value corresponding to the current input value.

The mechanics of what happens is surprisingly simple. First, a number of the previous input and/or output values are stored in special data storage registers, the number stored depending on the nature of the signal processing to be done. Weighted versions of these stored values are then added to (or subtracted from) the current input sample to generate the corresponding output value—the actual algorithm obviously depending on the type of signal processing required. It is this processing algorithm which is at the heart of the whole system—arriving at this can be a *very* complicated business! This is something we will examine in detail in later chapters. Here we will look at some fairly simple examples of processing, just to get a feel for what is involved.

2.7 The Running Average Filter

A good example to start with is the *running (or moving) average filter*. This processing system merely outputs a value which is the average of the current input and a particular number of the *previous* input samples.

As an example, consider a simple running average filter that averages the current input and the *last three* input samples. Let's assume that the sampled input values are as shown in Table 2.1, where T represents the sampling period.

As we need to average the current sample and the previous *three* input samples, the processor will clearly need three registers to store the previous input samples, the contents of these registers being updated every time a new sample is taken. For simplicity, we will assume that these three registers have initially been reset, i.e., they contain the value zero.

Table 2.1

Time	Input sample
0	2
T	1
2T	4
3T	5
4T	7
5T	10
6T	8
7T	7
8T	4
9T	2

The following sequence shows how the first three samples of '2', '1', and '4' are processed:

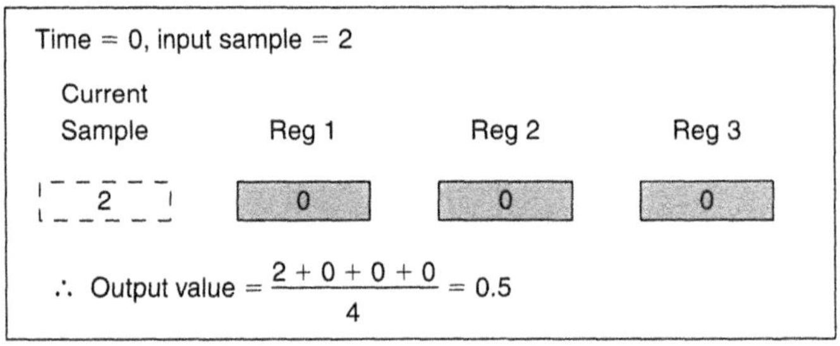

Time = 0, input sample = 2

Current Sample	Reg 1	Reg 2	Reg 3
2	0	0	0

$$\therefore \text{ Output value} = \frac{2 + 0 + 0 + 0}{4} = 0.5$$

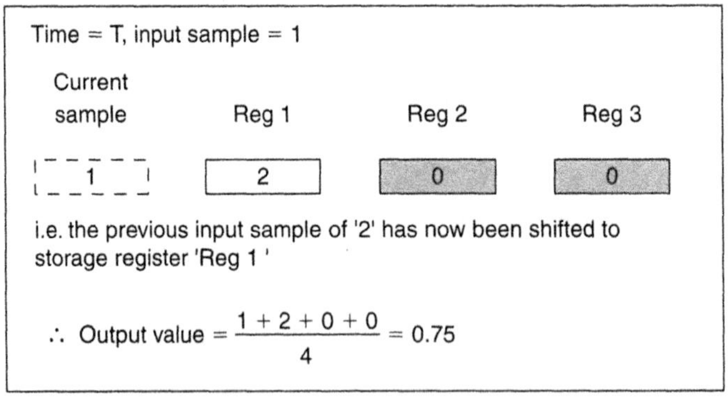

Time = T, input sample = 1

Current sample	Reg 1	Reg 2	Reg 3
1	2	0	0

i.e. the previous input sample of '2' has now been shifted to storage register 'Reg 1'

$$\therefore \text{ Output value} = \frac{1 + 2 + 0 + 0}{4} = 0.75$$

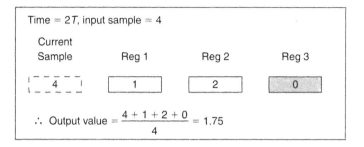

and so on.

Table 2.2

Time	Input sample	Output sample
0	2	0.5
T	1	0.75
2T	4	1.75
3T	5	3.00
4T	7	4.25
5T	10	6.50
6T	8	7.50
7T	7	8.00
8T	4	7.25
9T	2	5.25

Table 2.2 shows all of the output values—check that you agree with them before moving on.

N.B.1   The first three output values of 0.5, 0.75 and 1.75, represent the initial "transient," i.e., the part of the output signal where the initial three zeros are being shifted out of the three storage registers. The output values are only valid once these initial zeros have been cleared out of the storage registers.

N.B.2   A running average filter tends to smooth out any rapid changes in a signal and so is a form of lowpass filter.

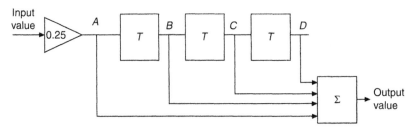

Figure 2.8

2.8 Representation of Processing Systems

The running average filter, just discussed, could be represented by the block diagram shown in Figure 2.8. Each of the three T blocks represents a time delay of one sample period, while the Σ box represents the summation of the four values. The 0.25 triangle is an attenuator which ensures that the average of the four values is outputted and not just the sum. So A is the current input divided by four, B the previous input, again divided by four, C the input before that, again divided by four, etc. If we catch the system at 6T say, then, from Table 2.2, $A=8/4$, $B=10/4$, $C=7/4$ and $D=5/4$, giving the output of 7.5, i.e., $A+B+C+D$.

N.B. The division by four could have been done *after* the summation rather than before, and this might seem the obvious thing to do. However, the option used is preferable as it means that, as we are processing smaller numbers, i.e., numbers already divided by four, we can get away with using smaller registers during processing. Here there were only four numbers to be added, but what if there had been a thousand? Dividing *before* addition, rather than after, would clearly makes a huge difference to the size of the registers needed.

2.9 Feedback (or Recursive) Filters

So far we have only met filters which make use of previous *inputs*. There is nothing to stop us from using the previous *outputs* instead—in fact, much more useful filters can be made in this way. A simple example is shown in Figure 2.9.

Because it is the previous *output* values which are fed back into the system and added to the current input, these filters are called *feedback* or *recursive* filters. Another name very commonly used is *infinite impulse response* filters—the reason for this particular name will become clear later.

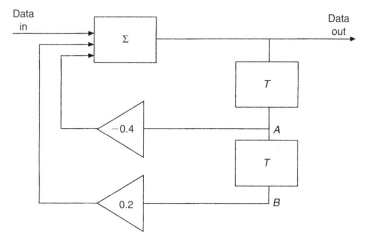

Figure 2.9

As we know, the *T* boxes represent time delays of one sampling period, and so *A* is the previous output and *B* the one before that. It is often useful to think of these boxes as the storage registers for the previous outputs, with *A* and *B* being their contents.

From Figure 2.9 you should see that:

$$\text{Data out} = \text{Data in} - 0.4A + 0.2B \qquad (2.1)$$

This is the simple processing that the digital processor needs to do for every input sample.

Imagine that this particular recursive filter is supplied with the data shown in Table 2.3, and that the two storage registers needed are initially reset.

From Eq. (2.1), as both *A* and *B* are initially zero, the first output must be the same as the input value, i.e., 10.

By the time the second input of 15 is received the previous output of 10 has been shifted into the storage register, appearing as *A* (Table 2.3). In the meantime, the previous *A* value (0) has been moved to *B*, while the previous value of *B* has been shifted right out of the system and lost, as it is no longer of any use.

So, when time=*T*, we have:

Input data=15, *A*=10, *B*=0

Table 2.3

Time	Input data	A	B	Output data
0	10	0.0	0.0	10.0
T	15	10.0	0.0	11.0
2T	20	11.0	10.0	17.6
3T	15	17.6	11.0	10.2
4T	8	10.2	17.6	7.5
5T	6	7.5	10.2	5.1
6T	9	5.1	7.5	8.5
7T	0	8.5	5.1	−2.4
8T	0	−2.4	8.5	2.7
9T	0	2.7	−2.4	−1.53

From Eq. (2.1), the new output value is given by:

$$15 - 0.4 \times 10 + 0.2 \times 0 = 11 \tag{2.1}$$

In preparation for generating the next output value, the current output of 11 is now shifted to *A*, the *A* value of 10 having already been moved to *B*. The third output value is therefore given by:

$$20 - 0.4 \times 11 + 0.2 \times 10 = 17.6$$

Before moving on it's best to check through the rest of Table 2.3 for yourself. (Spreadsheets lend themselves well to this application.)

You will notice from Table 2.3 that we are getting outputs *even when the input values are zero* (at times 7*T*, 8*T* and 9*T*). This makes sense as, at time 7*T*, we are pushing the previous output value of 8.5 back through the system to produce the next output of −2.4. This output value is, in turn, fed back into the system, and so on. Theoretically, the output could continue for ever, i.e. even if we put just a single pulse into the system we could get output values, every sampling period, *for an infinite time*. This explains the alternative name of *infinite impulse response* (*IIR*) filter for feedback filters. Note that this persisting output will not happen with processing systems which use only the input samples (nonrecursive)—with these, once the input stops, the output will continue for only a finite time. To be more specific, the output samples will continue for a time of $N \times T$, where N is the number of storage registers. This is why filters which make use of only the *previous* inputs are often called *finite impulse response* (FIR) filters (pronounced "F-I-R"), for short. A running average filter is therefore an example of an FIR filter. (Yet another name used for this type of filter is the *transversal* filter.)

IIR filters require fewer storage registers than equivalent FIR filters. For example, a particular highpass FIR filter might need 100 registers but an equivalent IIR filter might need as few as three or four. However, I must add a few words of warning here, as there are drawbacks to making use of the previous *outputs*. As with *any* system which uses feedback, we have to be *very* careful during the design as it is possible for the filter to become unstable. In other words, instead of acting as a well-behaved system, processing our signals in the required way, we might find that the output values very rapidly shoot up to the maximum possible and sit there. Another possibility is that the output oscillates between the maximum and minimum values. Not a pretty sight! We will look at this problem in more detail in later chapters.

So far we have looked at systems which make use of either previous inputs or previous outputs only. This restriction is rather artificial as, generally, the most effective DSP systems use both previous inputs *and* previous outputs.

2.10 Chapter Summary

Hopefully, you now have a reasonable understanding of the basics of digital signal processing. You should also realize that this type of signal processing is achieved in a very different way from "traditional" analog signal processing. In this chapter we have concentrated on the heart of the DSP system, i.e., the part that processes the digital samples of the original analog signal. Several processing systems have been analyzed. At this stage it will not be clear how these systems are designed to achieve a particular type of signal processing, or even the nature of the signal processing being carried out. This very important aspect will be dealt with in more detail in later chapters. We have met finite impulse response filters (those that make use of previous input samples only, such as running average filters) and also infinite impulse response filters (also called *feedback* or *recursive filters*)—these make use of the previous output samples. Although IIR systems generally need fewer storage registers than equivalent FIR systems, IIR systems can be unstable if not designed correctly, while FIR systems will *never* be unstable.

Bibliography

Marven, C., & Ewers, G. (1994). *A simple approach to digital signal processing*. Texas Instruments.

Millman, J., & Grabel, A. (1987). *Microelectronics*. London: McGraw-Hill.

Proakis, J. G., & Manolakis, D. G. (1996). *Digital Signal Processing – principles, algorithms and applications*. London: Prentice-Hall.

CHAPTER 3

Frequency Domain Processing

Nasser Kehtarnavaz

Frequency transforms are one of the most important concepts of DSP. Transforming a signal reveals a whole world of previously invisible information and enables the engineer to manipulate signal in ways that are not possible in the time domain. For example, frequency transforms are the basis for most modern audio and video compression algorithms. Taking audio and video into the frequency domain reveals data that can be "thrown away" without affecting the perceived quality of the signal.

Frequency transforms are also a powerful computational tool. Processing a signal in the frequency domain is often more computationally efficient than doing it in the time domain.

In this chapter, Nasser Kehtarnavaz provides the concepts you need to implement the most common transforms. He starts with the basic discrete Fourier transform (DFT) and its inverse, the IDFT. He then moves on to the fast Fourier transform (FFT), short time Fourier transform (STFT), and the discrete wavelet transform (DWT). Finally, he points us to a LabVIEW toolkit that uses these and other transforms for time-frequency analysis.

Of these transforms. the FFT is the most important—along with the FIR filter, it is an essential tool for every DSP engineer. There are many variants of the FFT, and the FFT presented in this chapter (taken from a Texas Instruments application note) may not be the best for your application. For a derivation of another common FFT variant, check out this informative article on the DFT and FFT:

http://www.dspdesignline.com/howto/206800602

The STFT is simply the FFT applied to a windowed time-domain signal. As noted in the chapter, choosing the right window and window size can be tricky. There are whole textbooks concerning this. For an introduction to windowing and time-frequency analysis, this page on power spectrum analysis is a good place to start:

http://www.dspdesignline.com/howto/206801391

The discrete wavelet transform is newer and less commonly used. However, it is often more efficient than the FFT and is being quickly adopted in signal processing applications. It is particularly popular for image processing.

For more on working in the frequency domain, see these online references:

Frequency domain tutorial

Transformation of signals to the frequency domain is widely used in signal processing. In many cases, such transformations provide a more effective representation and a more computationally efficient processing of signals as compared to time domain processing. For example, due to the equivalency of convolution operation in the time domain to multiplication in the frequency domain, one can find the output of a linear system by simply multiplying the Fourier transform of the input signal by the system transfer function.

This chapter presents an overview of three widely used frequency domain transformations, namely fast Fourier transform (FFT), short-time Fourier transform (STFT), and discrete wavelet transform (DWT). More theoretical details regarding these transformations can be found in many signal processing textbooks, e.g., [1].

3.1 Discrete Fourier Transform (DFT) and Fast Fourier Transform (FFT)

The discrete Fourier transform (DFT) $X[k]$ of an N-point signal $x[n]$ is given by:

$$\begin{cases} X[k] = \sum_{n=0}^{N-1} x[n] W_N^{nk}, & k = 0,1, \ldots, N-1 \\ x[n] = \dfrac{1}{N} \sum_{n=0}^{N-1} X[k] W_N^{-nk}, & n = 0,1, \ldots, N-1 \end{cases} \tag{3.1}$$

where $W_N = e^{-j2\pi/N}$. The above transform equations require N complex multiplications and $N-1$ complex additions for each term. For all N terms, N^2 complex multiplications and $N^2 - N$ complex additions are needed. As it is well known, the direct computation of Eq. (3.1) is not efficient.

To obtain a fast or real-time implementation of Eq. (3.1), one often uses a fast Fourier transform (FFT) algorithm, which makes use of the symmetry properties of DFT.

There are many approaches to finding a fast implementation of DFT; that is, there are many variations of FFT algorithms. Here, we mention the approach presented in the *TI Application Report SPRA291* for computing a 2N-point FFT [2]. This approach involves forming two new

N-point signals $x_1[n]$ and $x_2[n]$ from a $2N$-point signal $g[n]$ by splitting it into an even and an odd part as follows:

$$x_1[n] = g[2n] \quad 0 \leq n \leq N - 1$$
$$x_2[n] = g[2n + 1] \tag{3.2}$$

From the two sequences $x_1[n]$ and $x_2[n]$, a new complex sequence $x[n]$ is defined to be:

$$x[n] = x_1[n] + jx_2[n] \quad 0 \leq n \leq N - 1 \tag{3.3}$$

To get $G[k]$, the DFT of $g[n]$, the equation:

$$G[k] = X[k]A[k] + X*[N - k]B[k]$$
$$k = 0, 1, \ldots, N - 1, \text{ with } X[N] = X[0] \tag{3.4}$$

is used, where,

$$A[k] = \frac{1}{2}\left(1 - j\mathrm{W}_{2N}^k\right) \tag{3.5}$$

and,

$$B[k] = \frac{1}{2}\left(1 + j\mathrm{W}_{2N}^k\right) \tag{3.6}$$

Only N points of $G[k]$ are computed from Eq. (3.4). The remaining points are found by using the complex conjugate property of $G[k]$, that is, $G[2N - k] = G*[k]$. As a result, a $2N$-point transform is calculated based on an N-point transform, leading to a reduction in the number of operations.

3.2 Short-Time Fourier Transform (STFT)

Short-time Fourier transform (STFT) is a sequence of Fourier transforms of a windowed signal. STFT provides the time-localized frequency information for situations in which frequency components of a signal vary over time, whereas the standard Fourier transform provides the frequency information averaged over the entire signal time interval.

The STFT pair is given by:

$$\begin{cases} X_{STFT}[m,n] = \sum_{k=0}^{L-1} x[k]g[k - m]e^{-j2\pi nk/L} \\ x[k] = \sum_{m}\sum_{n} X_{STFT}[m,n]g[k - m]e^{j2\pi nk/L} \end{cases} \tag{3.7}$$

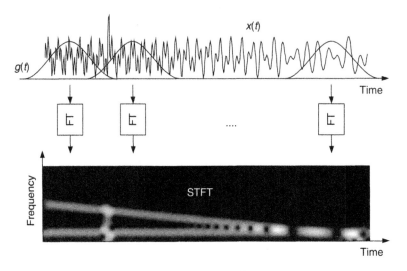

Figure 3.1: Short-time Fourier transform

where $x[k]$ denotes a signal and $g[k]$ denotes an L-point window function. From Eq. (3.7), the STFT of $x[k]$ can be interpreted as the Fourier transform of the product $x[k]g[k-m]$. Figure 3.1 illustrates computing STFT by taking Fourier transforms of a windowed signal.

There exists a trade-off between time and frequency resolution in STFT. In other words, although a narrow-width window results in a better resolution in the time domain, it generates a poor resolution in the frequency domain, and vice versa. Visualization of STFT is often realized via its spectrogram, which is an intensity plot of STFT magnitude over time. Three spectrograms illustrating different time-frequency resolutions are shown in Figure 3.2. The implementation details of STFT are described in Lab 3.

3.3 Discrete Wavelet Transform (DWT)

Wavelet transform offers a generalization of STFT. From a signal theory point of view, similar to DFT and STFT, wavelet transform can be viewed as the projection of a signal into a set of basis functions named wavelets. Such basis functions offer localization in the frequency domain. In contrast to STFT having equally spaced time-frequency localization, wavelet transform provides high frequency resolution at low frequencies and high time resolution at high frequencies. Figure 3.3 provides a tiling depiction of the time-frequency resolution of wavelet transform as compared to STFT and DFT.

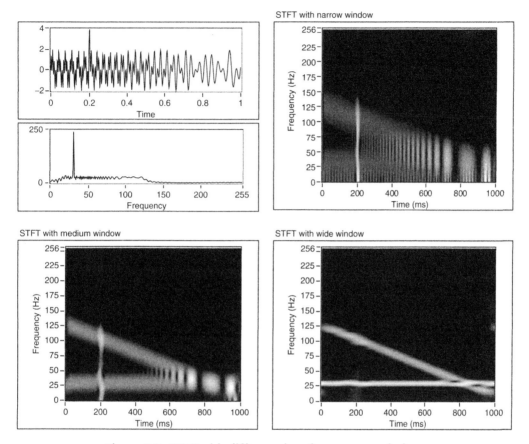

Figure 3.2: STFT with different time-frequency resolutions

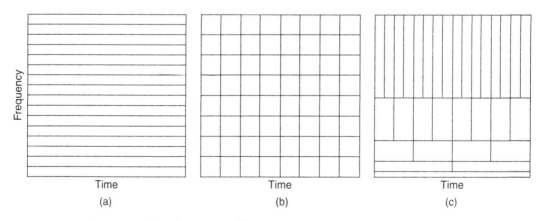

Figure 3.3: Time-frequency tiling for (a) DFT, (b) STFT, and (c) DWT

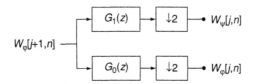

Figure 3.4: Discrete wavelet transform decomposition filter bank, G_0 lowpass and G_1 highpass decomposition filters

The discrete wavelet transform (DWT) of a signal $x[n]$ is defined based on approximation coefficients, $W_\phi[j_0,k]$ and detail coefficients, $W_\psi[j,k]$, as follows:

$$W_\phi[j_0,k] = \frac{1}{\sqrt{M}} \sum_n x[n]\phi_{j_0,k}[n]$$

$$W_\psi[j,k] = \frac{1}{\sqrt{M}} \sum_n x[n]\psi_{j,k}[n] \quad \text{for } j \geq j_0 \tag{3.8}$$

and the inverse DWT is given by:

$$x[n] = \frac{1}{\sqrt{M}} \sum_k W_\phi[j_0,k]\phi_{j_0,k}[n] + \frac{1}{\sqrt{M}} \sum_{j=j_0}^{J} \sum_k W_\psi[j,k]\psi_{j,k}[n] \tag{3.9}$$

where $n = 0, 1, 2, \dots, M-1, j = 0, 1, 2, \dots, J-1, k = 0, 1, 2, \dots, 2^j-1$, and M denotes the number of samples to be transformed. This number is selected to be $M = 2^J$, where J indicates the number of transform levels. The basis functions $\{\phi_{j,k}[n]\}$ and $\{\psi_{j,k}[n]\}$ are defined as:

$$\phi_{j,k}[n] = 2^{j/2}\phi[2^j n - k]$$

$$\psi_{j,k}[n] = 2^{j/2}\psi[2^j n - k] \tag{3.10}$$

where $\phi[n]$ is called the *scaling function* and $\psi[n]$ is called the *wavelet function*.

For an efficient implementation of DWT, the filter bank structure is often used. Figure 3.4 shows the decomposition or analysis filter bank for obtaining the forward DWT coefficients. The approximation coefficients at a higher level are passed through a highpass and a lowpass filter, followed by a downsampling by two to compute both the detail and approximation coefficients at a lower level. This tree structure is repeated for a multi-level decomposition.

Inverse DWT (IDWT) is obtained by using the reconstruction or synthesis filter bank shown in Figure 3.5. The coefficients at a lower level are upsampled by two and passed through

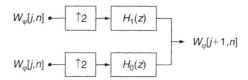

Figure 3.5: Discrete wavelet transform reconstruction filter bank, H_0 lowpass and H_1 highpass reconstruction filters

a highpass and a lowpass filter. The results are added together to obtain the approximation coefficients at a higher level.

3.4 Signal Processing Toolset

Signal Processing Toolset (SPT) is an add-on toolkit of LabVIEW that provides useful tools for performing time-frequency analysis [3]. SPT has three components: joint time-frequency analysis (JTFA), super-resolution spectral analysis (SRSA), and wavelet analysis.

The VIs associated with STFT are included as part of the JTFA component. The SRSA component is based on the model-based frequency analysis normally used for situations in which a limited number of samples is available. The VIs associated with the SRSA component include high-resolution spectral analysis and parameter estimation, such as amplitude, phase, damping factor, and damped sinusoidal estimation. The VIs associated with the wavelet analysis component include 1D and 2D wavelet transform as well as their filter bank implementations.

Lab 3: FFT, STFT, and DWT

This lab shows how to use the LabVIEW tools to perform FFT, STFT, and DWT as part of a frequency domain transformation system.

L3.1 FFT versus STFT

To illustrate the difference between FFT and STFT transformations, three signals are combined here to form a 512-point input signal: a 75 Hz sinusoidal signal sampled at 512 Hz, a chirp signal with linearly decreasing frequency from 200 to 120 Hz, and an impulse signal having an amplitude of 2 for 500 ms located at the 256th sample. This composite signal is shown in Figure L3.1. The FFT and STFT graphs are also shown in this figure. The FFT graph shows the time averaged spectrum reflecting the presence of a signal from 120 to 200 Hz, with one major peak at 75 Hz. As one can see from this graph, the impulse having the short time duration does not appear in the spectrum. The STFT graph shows the spectrogram for a time

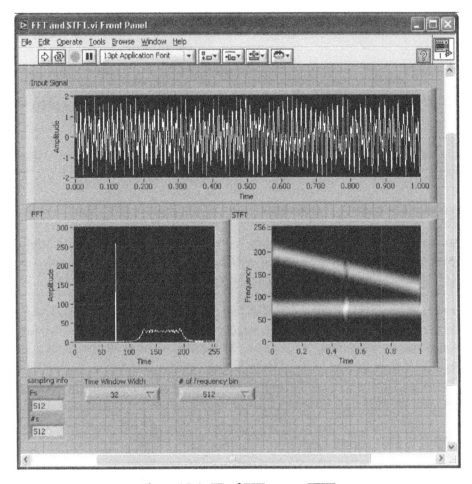

Figure L3.1: FP of FFT versus STFT

increment of 1 and a rectangular window of width 32 by which the presence of the impulse can be detected.

As far as the FP is concerned, two Menu Ring controls (**Controls** » **Modern** » **Ring & Enum** » **Menu Ring**) are used to input values via their labels. The labels and corresponding values of the ring controls can be modified by right-clicking and choosing **Edit Items**… from the shortcut menu. This brings up the dialog box shown in Figure L3.2.

An Enum (enumerate) control acts the same as a Menu Ring control, except that values of an Enum control cannot be modified and are assigned sequentially. A Menu Ring or Enum can be changed to a Ring Constant or Enum Constant when used on a BD.

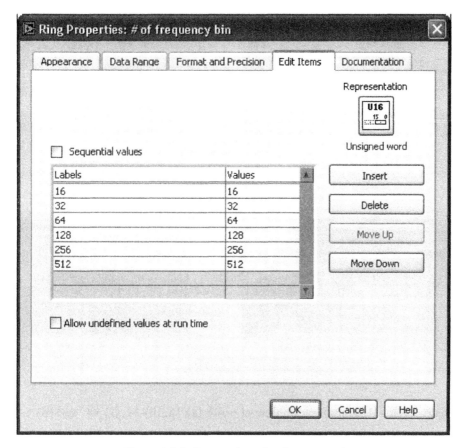

Figure L3.2: Properties of a ring control

Several spectrograms with different time window widths are shown in Figure L3.3. Figure L3.3(a) shows an impulse (vertical line) at time 500 ms because of the relatively time-localized characteristic of the window used. Even though a high resolution in the time domain is achieved with this window, the resolution in the frequency domain is so poor that the frequency contents of the sinusoidal and chirp signals cannot be easily distinguished. This is due to the Heisenberg's uncertainty principle [4], which states that if the time resolution is increased, the frequency resolution is decreased.

Now, let us increase the width of the time-frequency window. This causes the frequency resolution to become better while the time resolution becomes poorer. As a result, as shown in Figure L3.3(d), the frequency contents of the sinusoidal and chirp signals become better distinguished. One can also see that as the time resolution becomes poorer, the moment of occurrence of the impulse becomes more difficult to identify.

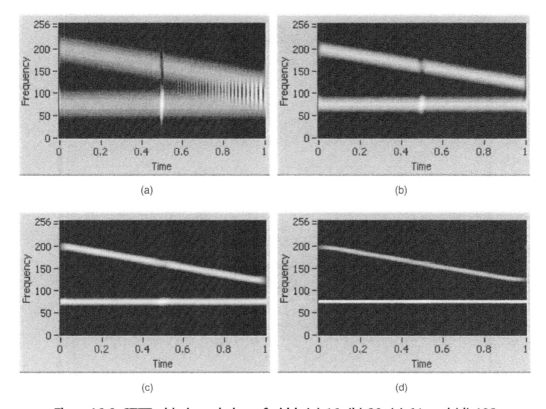

(a)

(b)

(c)

(d)

Figure L3.3: STFT with time window of width (a) 16, (b) 32, (c) 64, and (d) 128

The BD of this example is illustrated in Figure L3.4. To build this VI, let us first generate the input signal with the specifications stated previously. Figure L3.5(a) shows the generation of the input signal (512 samples generated with the sampling frequency of 512 Hz) using a `MathScript Node`. In order to use this VI as the signal source of the system, an output terminal in the connector pane is wired to a waveform indicator. Then, the VI is saved as *Composite Signal.vi*.

Alternatively, the three signals can be generated using the built-in LabVIEW VIs and added together to form a composite signal; see Figure L3.5(b). The sinusoidal waveform is generated by using the `Sine Waveform` VI (**Functions » Signal Processing » Waveform Generation » Sine Waveform**), and the chirp signal is generated by using the `Chirp Pattern` VI (**Functions » Signal Processing » Signal Generation » Chirp Pattern**). Also, the impulse is generated by using the `Impulse Pattern` VI (**Functions » Signal Processing » Signal Generation » Impulse Pattern**).

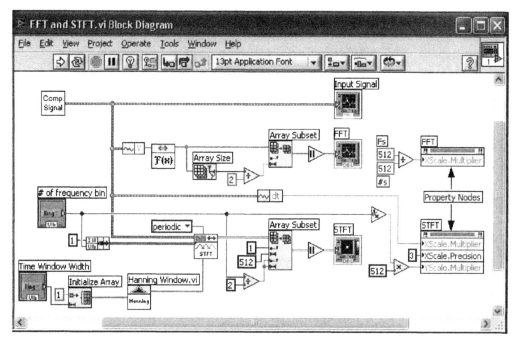

Figure L3.4: BD of FFT and STFT

Now, let us create the entire transformation system using the `Composite Signal` VI just made. Create a blank VI; then select **Functions** » **Select a VI…** This brings up a window for choosing and locating a VI. Click *Composite Signal.vi* to insert it into the BD. The composite signal output is connected to three blocks consisting of a waveform graph, an `FFT`, and an `STFT` VI. The waveform data (`Y` component) are connected to the input of the `FFT` VI (**Functions** » **Signal Processing** » **Transforms** » **FFT**). Only the first half of the output data from the `FFT` VI is taken, since the other half is a mirror image of the first half. This is done by placing an `Array Subset` function and wiring to it one half of the signal length. The magnitude of the FFT output is then displayed in the waveform graph. Properties of an FP object, such as scale multiplier of a graph, can be changed programmatically by using a property node. Property nodes are discussed in the next subsection.

Getting the STFT output is more involved than FFT. The `STFT` VI (**Functions** » **Addons** » **Time Frequency Analysis** » **Time Frequency Transform** » **STFT**), which is part of the Signal Processing Toolkit (SPT), is used here for this purpose. To utilize the `STFT` VI, one needs to connect several inputs as well as the input signal. These inputs are time-freq sampling info, extension, window info, and user-defined window. The time-freq sampling

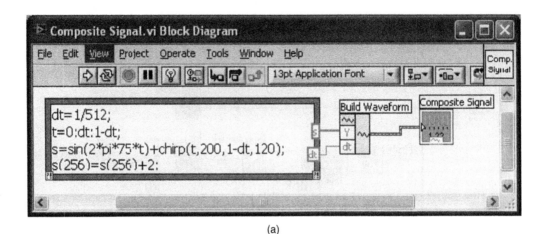

(a)

(b)

Figure L3.5: Composite signal (sine + chirp + impulse) generation using (a) MathScript Node and (b) graphical approach

info is a cluster of time steps and frequency bins where time steps specify the sampling period along the time axis and the frequency bins indicate the FFT block size of the STFT. A constant of 1 is used for time steps in the example shown in Figure L3.4. The extension input specifies the method to pad data at both ends of a signal to avoid abrupt changes in the transformed outcome. There exist three different extension options: zero padding, symmetric, and periodic. The periodic mode is used in the example shown in Figure L3.4. The window info input specifies which commonly used sliding window to apply and defines the resolution of the resulting time-frequency representation. On the other hand, the user-defined window input allows one to have a customized sliding window by specifying the coefficients. In our example, a Hanning window is considered by passing an array of all 1's whose width is adjustable by the user through the Hanning window VI (**Functions** » **Signal Processing** » **Window** » **Hanning Window**). Similar to FFT, only one-half of the frequency values are taken while the time values retain the original length. The start index of the array subset is set to one-half the number of frequency bins to access the positive frequency values, as shown in Figure L3.4. The reason is that the output of the STFT corresponding to the negative frequency values is followed by the output belonging to the positive frequency values. Additional details on using the STFT VI can be found in [5].

The output of the STFT is displayed in the Intensity Graph (**Controls** » **Modern** » **Graph** » **Intensity Graph**). Right-click on the Intensity Graph and then uncheck the **Loose Fit** option under both **X Scale** and **Y Scale** from the shortcut menu. When this is done, the STFT output graph gets fitted into the entire plotting area. Enable auto-scaling of intensity by right-clicking on the Intensity Graph and choosing **Z Scale** » **AutoScale Z**.

L3.1.1 Property Node

The number of FFT values varies based on the number of samples. Similarly, the number of frequency rows of STFT varies based on the number of frequency bins specified by the user. However, the scale of the frequency axis in FFT or STFT graphs should always remain between 0 and $f_s/2$, which is $256\,Hz$ in the example, regardless of the number of frequency bins, as illustrated in Figure L3.1 and Figure L3.3. For this reason, the multiplier for the spectrogram scale needs to be changed depending on the width of the time window during run time.

A property node can be used to modify the appearance of an FP object. A property node can be created by right-clicking either on a terminal icon in a BD or an object in an FP, and then by choosing the **visible** property element through **Create** » **Property Node.** This way, the default element of the chosen property gets created in a BD, which is linked to a corresponding FP object. Various property elements of the property node can be modified to reflect the read or the write mode. Note that, by default, a property node is set to read. To change to the write mode, right-click on a property element and choose **Change to Write**. The read/write mode of all elements can be changed together by choosing **Change all to Read/Write**.

To change the scale of the spectrogram graph, one needs to modify the value of the element **YScale.Multiplier**. Replace the element **visible** with **YScale.Multiplier** by clicking it and choosing **Y Scale » Offset and Multiplier » Multiplier**. The sampling frequency of the signal divided by the number of frequency bins, which defines the scale multiplier, is wired to the element **YScale.Multiplier** of the property node. Two more elements, **XScale.Multiplier** and **XScale.Precision,** are added to the property node for modifying the time axis multiplier and precision, respectively.

A property node of the FFT graph is also created and modified in a similar way considering that the resolution of FFT is altered depending on the sampling frequency and number of input signal samples. The property nodes of the STFT and FFT graphs are shown in Figure L3.4. More details on using property nodes can be found in *LabVIEW User Manual* [6].

L3.2 DWT

In this transformation, the time-frequency window has high frequency resolution for higher frequencies and high time resolution for lower frequencies. This is a great advantage over STFT where the window size is fixed for all frequencies.

The BD of a 1D decomposition and reconstruction wavelet transform is shown in Figure L3.6. Three VIs including WA Wavelet Filter VI (**Functions » Addons » Wavelet Analysis » Discrete Wavelet » Filter Banks**), WA Discrete Wavelet Transform VI, and WA Inverse Discrete Wavelet Transform VI (**Functions » Addons » Wavelet Analysis » Discrete Wavelet**) are used here from the wavelet analysis palette.

A chirp type signal, shown in Figure L3.7, is considered to be the input signal source. This signal is designed to consist of four sinusoidal signals, each consisting of 128 samples with increasing frequencies in this order: 250, 500, 1000, 2000 Hz. This makes the entire chirp signal 512 samples. The Fourier transform of this signal is also shown in Figure L3.7.

Figure L3.8(a) illustrates the BD of this signal generation process. Save this VI as *Chirp Signal.vi* to be used as a signal source subVI within the DWT VI. Note that the **Concatenate**

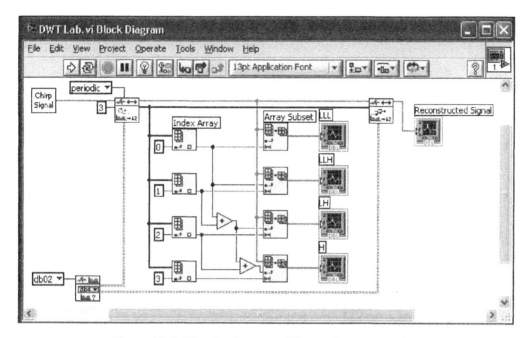

Figure L3.6: Wavelet decomposition and reconstruction

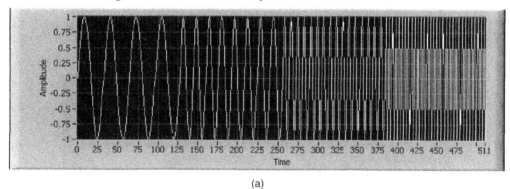

(a)

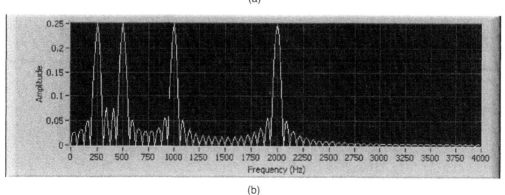

(b)

Figure L3.7: Waveforms of input signal: (a) time domain and (b) frequency domain

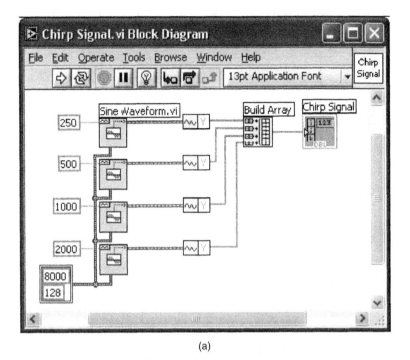

(a)

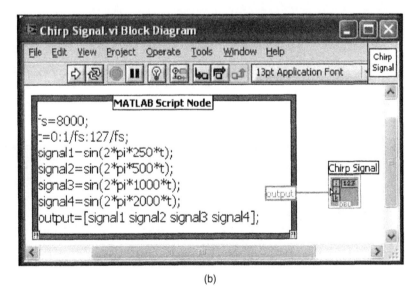

(b)

Figure L3.8: Generating input signal using (a) graphical approach and (b) textual approach

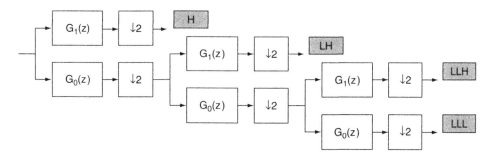

Figure L3.9: Waveform decomposition tree

Inputs option of the `Build Array` function should be chosen to build the 1D chirp signal. This VI has only one output terminal. As an alternative to the graphical approach, a `MATLAB Script Node` can be used to generate the chirp signal. This way, the four signals need to be concatenated using the operator [ ], as shown in Figure L3.8(b).

The `WA Discrete Wavelet Transform` VI requires four inputs, including input signal, extension, levels, and analysis filter. The input signal is provided by the `Chirp Signal` VI. For the extension input, the same options are available as mentioned earlier for STFT. The input levels specify the number of levels of decomposition. In the BD shown in Figure L3.6, a three-level decomposition is used via specifying a constant 3. The filter bank implementation for a three-level wavelet decomposition is illustrated in Figure L3.9. In this example, the Daubechies-2 wavelet is used. The coefficients of the filters are generated by the `Wavelet Filter` VI. This VI provides the coefficient sets for both the decomposition and reconstruction parts.

The result of the `WA Discrete Wavelet Transform` VI is structured into a 1D array corresponding to the components of the transformed signal in the order LLL, LLH, LH, H, where L stands for low and H for high. The length of each component is also available from this VI. The wavelet decomposed outcome for each stage of the filter bank is shown in Figure L3.10. From the outcome, it can be observed that lower frequencies occur earlier and higher frequencies occur later in time. This demonstrates the fact that wavelet transform provides both frequency and time resolution, a clear advantage over Fourier transform.

The decomposed signal can be reconstructed by the WA `Inverse Discrete Wavelet Transform` VI. From the reconstructed signal, shown in Figure L3.10, one can see that the wavelet decomposed signal is reconstructed perfectly by using the synthesis or reconstruction filter bank.

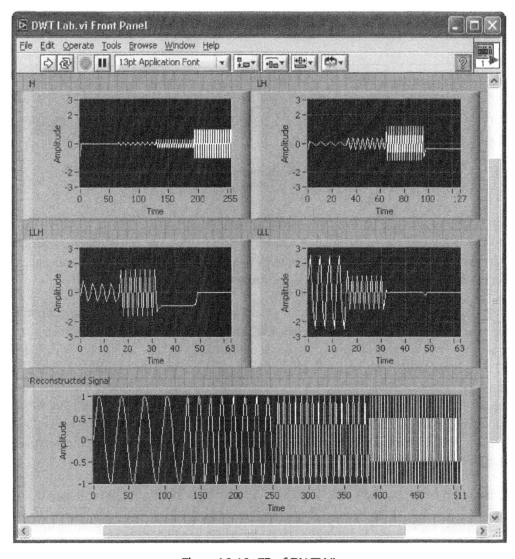

Figure L3.10: FP of DWT VI

References

1. Burrus C, Gopinath R, Gao H. *Wavelets and Wavelet Transforms—A Primer*: Prentice-Hall; 1998.

2. Texas Instruments, *TI Application Report SPRA291*.

3. National Instruments, *Signal Processing Toolset User Manual*, Part Number 322142C-01, 2002.

4. Burrus, *Wavelets and Wavelet Transforms—A Primer*.

5. National Instruments, *Signal Processing Toolset User Manual*, Part Number 322142C-01, 2002.

6. National Instruments, *LabVIEW User Manual*, Part Number 320999E-01, 2003.

CHAPTER 4

Audio Coding

Khalid Sayood

Audio processing has been an important part of DSP from the very beginning, and it is one of the few areas of DSP that is well-known among nonengineers. Recording studios have used DSP for over 20 years, and today every album on the shelves has gone through some form of digital signal processing. Audiophiles know about DSP thanks to the "DSP" button on receivers from Yamaha and other manufacturers. Even the average consumer knows a little bit about DSP thanks to the prevalence of audio compression. If you have an iPod, listen to satellite or Internet radio, or watch DVDs, you've experienced the wonders of audio compression.

In this chapter, Khalid Sayood gives an introduction to audio compression, focusing on popular compression standards. The author's stated goal is to help you understand these standards without getting bogged down in details. In this respect he succeeds. If you already understand basic DSP concepts such as frequency transforms (audio compression algorithms use the modified discrete cosine transform, or MDCT), you will walk away with a solid understanding of how these algorithms work, where they've been applied, and why.

Sayood starts with the basic principles of audio coding, and then dives into MPEG layer I and II—two of the first frequency-based codecs. This is primarily a setup for a discussion of the big kahuna of codecs: MPEG layer III, or MP3. Sayood describes MP3 in detail, giving us insight into why it is so successful. We also learn of its Achilles heel: it was designed for backwards compatibility with MPEG layer I/II, and is therefore inefficient. This leads to a discussion of more efficient codecs, including MPEG Advanced Audio Coding (AAC) and Dolby AC3 (Dolby Digital). Finally, we're given a brief rundown of other notable codecs such as DTS and the open-source Ogg Vorbis. (Yes, Ogg Vorbis is a real codec name—never doubt the creative power of the open-source community!)

—Kenton Williston

4.1 Overview

Lossy compression schemes can be based on a source model, as in the case of speech compression, or a user or sink model, as is somewhat the case in image compression. In this chapter we look at audio compression approaches that are explicitly based on the model of the user. We will look at audio compression approaches in the context of audio compression standards. Principally, we will examine the different MPEG standards for audio compression. These include MPEG Layer I, Layer II, Layer III (or *mp3*) and the Advanced Audio Coding Standard. As with other standards described in this book, the goal here is not to provide all the details required for implementation. Rather the goal is to provide the reader with enough familiarity so that they can then find it much easier to understand these standards.

4.2 Introduction

The various speech coding algorithms rely heavily on the speech production model to identify structures in the speech signal that can be used for compression. Audio compression systems have taken, in some sense, the opposite tack. Unlike speech signals, audio signals can be generated using a large number of different mechanisms. Lacking a unique model for audio production, the audio compression methods have focused on the unique model for audio perception, a psychoacoustic model for hearing. At the heart of the techniques described in this chapter is a psychoacoustic model of human perception. By identifying what can and, more importantly, what cannot be heard, the schemes described in this chapter obtain much of their compression by discarding information that cannot be perceived. The motivation for the development of many of these perceptual coders was their potential application in broadcast multimedia. However, their major impact has been in the distribution of audio over the Internet.

We live in an environment rich in auditory stimuli. Even an environment described as quiet is filled with all kinds of natural and artificial sounds. The sounds are always present and come to us from all directions. Living in this stimulus-rich environment, it is essential that we have mechanisms for ignoring some of the stimuli and focusing on others. Over the course of our evolutionary history we have developed limitations on what we can hear. Some of these limitations are physiological, based on the machinery of hearing. Others are psychological, based on how our brain processes auditory stimuli. The insight of researchers in audio coding has been the understanding that these limitations can be useful in selecting information that needs to be encoded and information that can be discarded. The limitations of human perception are incorporated into the compression process through the use of psychoacoustic models. We briefly describe the auditory model used by the most popular

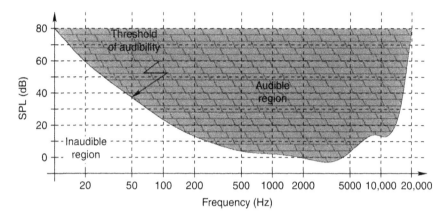

Figure 4.1: A typical plot of the audibility threshold

audio compression approaches. Our description is necessarily superficial and we refer readers interested in more detail to Moore, *An Introduction to the Psychology of Hearing* (Academic Press) and Bosi/Goldberg, *Introduction to Digital Audio Coding and Standards* (Kluwer Academic Press).

The machinery of hearing is frequency dependent. The variation of what is perceived as equally loud at different frequencies was first measured by Fletcher and Munson at Bell Labs in the mid-1930s. These measurements of perceptual equivalence were later refined by Robinson and Dadson. This dependence is usually displayed as a set of equal loudness curves, where the sound pressure level (SPL) is plotted as a function of frequency for tones perceived to be equally loud. Clearly, what two people think of as equally loud will be different. Therefore, these curves are actually averages and serve as a guide to human auditory perception. The particular curve that is of special interest to us is the threshold-of hearing curve. This is the SPL curve that delineates the boundary of audible and inaudible sounds at different frequencies. In Figure 4.1 we show a plot of this audibility threshold in quiet. Sounds that lie below the threshold are not perceived by humans. Thus, we can see that a low amplitude sound at a frequency of 3 kHz may be perceptible while the same level of sound at 100 Hz would not be perceived.

4.2.1 Spectral Masking

Lossy compression schemes require the use of quantization at some stage. Quantization can be modeled as as an additive noise process in which the output of the quantizer is the input plus the quantization noise. To hide quantization noise, we can make use of the fact that

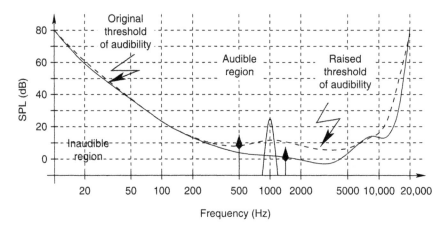

Figure 4.2: Change in the audibility threshold

signals below a particular amplitude at a particular frequency are not audible. If we select the quantizer step size such that the quantization noise lies below the audibility threshold, the noise will not be perceived. Furthermore, the threshold of audibility is not absolutely fixed and typically rises when multiple sounds impinge on the human ear. This phenomenon gives rise to *spectral masking*. A tone at a certain frequency will raise the threshold in a *critical band* around that frequency. These critical bands have a constant Q, which is the ratio of frequency to bandwidth. Thus, at low frequencies the critical band can have a bandwidth as low as 100 Hz, while at higher frequencies the bandwidth can be as large as 4 kHz. This increase of the threshold has major implications for compression. Consider the situation in Figure 4.2. Here a tone at 1 kHz has raised the threshold of audibility so that the adjacent tone above it in frequency is no longer audible. At the same time, while the tone at 500 Hz is audible, because of the increase in the threshold the tone can be quantized more crudely. This is because increase of the threshold will allow us to introduce more quantization noise at that frequency. The degree to which the threshold is increased depends on a variety of factors, including whether the signal is sinusoidal or atonal.

4.2.2 Temporal Masking

Along with spectral masking, the psychoacoustic coders also make use of the phenomenon of temporal masking. The temporal masking effect is the masking that occurs when a sound raises the audibility threshold for a brief interval preceding and following the sound.

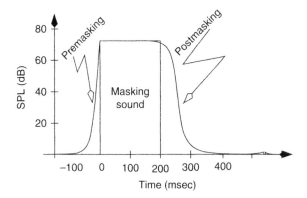

Figure 4.3: Change in the audibility threshold in time

In Figure 4.3 we show the threshold of audibility close to a masking sound. Sounds that occur in an interval around the masking sound (both after and before the masking tone) can be masked. If the masked sound occurs prior to the masking tone, this is called premasking or backward masking, and if the sound being masked occurs after the masking tone this effect is called postmasking or forward masking. The forward masking remains in effect for a much longer time interval than the backward masking.

4.2.3 Psychoacoustic Model

These attributes of the ear are used by all algorithms that use a psychoacoustic model. There are two models used in the MPEG audio coding algorithms. Although they differ in some details, the general approach used in both cases is the same. The first step in the psychoacoustic model is to obtain a spectral profile of the signal being encoded. The audio input is windowed and transformed into the frequency domain using a filter bank or a frequency domain transform. The Sound Pressure Level (SPL) is calculated for each spectral band. If the algorithm uses a subband approach, then the SPL for the band is computed from the SPL for each coefficient X_k. Because tonal and nontonal components have different effects on the masking level, the next step is to determine the presence and location of these components. The presence of any tonal components is determined by first looking for local maxima where a local maximum is declared at location k if $|X_k|^2 > |X_{k-1}|^2$ and $|X_k|^2 \geq |X_{k+1}|^2$. A local maximum is determined to be a tonal component if

$$20\log_{10}\frac{|X_k|}{|X_{k+j}|} \geq 7$$

where the values *j* depend on the frequency. The identified tonal maskers are removed from each critical band and the power of the remaining spectral lines in the band is summed to obtain the nontonal masking level. Once all the maskers are identified, those with SPL below the audibility threshold are removed. Furthermore, of those maskers that are very close to each other in frequency, the lower-amplitude masker is removed. The effects of the remaining maskers are obtained using a spreading function that models spectral masking. Finally, the masking due to the audibility level and the maskers is combined to give the final masking thresholds. These thresholds are then used in the coding process.

In the following sections we describe the various audio coding algorithms used in the MPEG standards. Although these algorithms provide audio that is perceptually noiseless, it is important to remember that even if we cannot perceive it, there is quantization noise distorting the original signal. This becomes especially important if the reconstructed audio signal goes through any postprocessing. Postprocessing may change some of the audio components, making the previously masked quantization noise audible. Therefore, if there is any kind of processing to be done, including mixing or equalization, the audio should be compressed only after the processing has taken place. This "hidden noise" problem also prevents multiple stages of encoding and decoding or tandem coding.

4.3 MPEG Audio Coding

We begin with the three separate, stand-alone audio compression strategies that are used in MPEG-1 and MPEG-2 and known as Layer I, Layer II, and Layer III. The Layer III audio compression algorithm is also referred to as *mp3*. Most standards have *normative* sections and *informative* sections. The *normative* actions are those that are required for compliance to the standard. Most current standards, including the MPEG standards, define the bitstream that should be presented to the decoder, leaving the design of the encoder to individual vendors. That is, the bitstream definition is normative, while most guidance about encoding is informative. Thus, two MPEG-compliant bitstreams that encode the same audio material at the same rate but on different encoders may sound very different. On the other hand, a given MPEG bitstream decoded on different decoders will result in essentially the same output.

A simplified block diagram representing the basic strategy used in all three layers is shown in Figure 4.4. The input, consisting of 16-bit PCM words, is first transformed to the frequency domain. The frequency coefficients are quantized, coded, and packed into an MPEG bitstream. Although the overall approach is the same for all layers, the details can vary

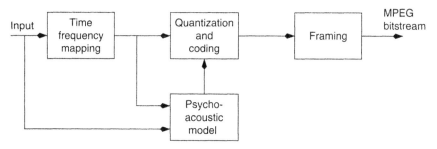

Figure 4.4: The MPEG audio coding algorithms

significantly. Each layer is progressively more complicated than the previous layer and also provides higher compression. The three layers are backward compatible. That is, a decoder for Layer III should be able to decode Layer I– and Layer II–encoded audio. A decoder for Layer II should be able to decode Layer I–encoded audio. Notice the existence of a block labeled *Psychoacoustic model* in Figure 4.4.

4.3.1 Layer I Coding

The Layer I coding scheme provides a 4:1 compression. In Layer I coding the time frequency mapping is accomplished using a bank of 32 subband filters. The output of the subband filters is critically sampled. That is, the output of each filter is down-sampled by 32. The samples are divided into groups of 12 samples each. Twelve samples from each of the 32 subband filters, or a total of 384 samples, make up one frame of the Layer I coder. Once the frequency components are obtained the algorithm examines each group of 12 samples to determine a *scalefactor*. The scalefactor is used to make sure that the coefficients make use of the entire range of the quantizer. The subband output is divided by the scalefactor before being linearly quantized. There are a total of 63 scalefactors specified in the MPEG standard. Specification of each scalefactor requires 6 bits.

To determine the number of bits to be used for quantization, the coder makes use of the psychoacoustic model. The inputs to the model include the *fast Fourier transform* (FFT) of the audio data as well as the signal itself. The model calculates the masking thresholds in each subband, which in turn determine the amount of quantization noise that can be tolerated and hence the quantization step size. As the quantizers all cover the same range, selection of the quantization stepsize is the same as selection of the number of bits to be used for quantizing the output of each subband. In Layer I the encoder has a choice of 14 different quantizers for each band (plus the option of assigning 0 bits). The quantizers are all midtread quantizers ranging from 3 levels to 65,535 levels. Each subband gets assigned a variable number of bits.

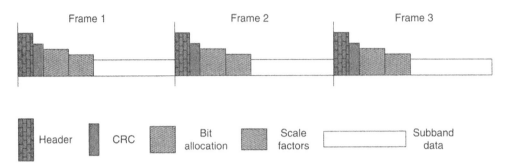

Figure 4.5: Frame structure for Layer 1

However, the total number of bits available to represent all the subband samples is fixed. Therefore, the bit allocation can be an iterative process. The objective is to keep the noise-to-mask ratio more or less constant across the subbands.

The output of the quantization and bit allocation steps is combined into a frame as shown in Figure 4.5. Because MPEG audio is a streaming format, each frame carries a header, rather than having a single header for the entire audio sequence. The header is made up of 32 bits. The first 12 bits comprise a sync pattern consisting of all 1 s. This is followed by a 1-bit version ID, a 2-bit layer indicator, a 1-bit CRC protection. The CRC protection bit is set to 0 if there is no CRC protection and is set to a 1 if there is CRC protection. If the layer and protection information is known, all 16 bits can be used for providing frame synchronization. The next 4 bits make up the bit rate index, which specifies the bit rate in kbits/sec. There are 14 specified bit rates to choose from. This is followed by 2 bits that indicate the sampling frequency. The sampling frequencies for MPEG-1 and MPEG-2 are different (one of the few differences between the audio coding standards for MPEG-1 and MPEG-2) and are shown in Table 4.1 These bits are followed by a single padding bit. If the bit is "1," the frame needs an additional bit to adjust the bit rate to the sampling frequency. The next two bits indicate the mode. The possible modes are "stereo," "joint stereo," "dual channel," and "single channel." The stereo mode consists of two channels that are encoded separately but intended to be played together. The joint stereo mode consists of two channels that are encoded together. The left and right channels are combined to form a *mid* and a *side* signal as follows:

$$M = \frac{L + R}{2}$$
$$S = \frac{L - R}{2}$$

Table 4.1: Allowable sampling frequencies in MPEG-1 and MPEG-2

Index	MPEG-1	MPEG-2
00	44.1 kHz	22.05 kHz
01	48 kHz	24 kHz
10	32 kHz	16 kHz
11	Reserved	

The dual-channel mode consists of two channels that are encoded separately and are not intended to be played together, such as a translation channel. These are followed by two mode extension bits that are used in the joint stereo mode. The next bit is a copyright bit ("1" if the material is copyrighted, "0" if it is not). The next bit is set to "1" for original media and "0" for copy. The final two bits indicate the type of de-emphasis to be used.

If the CRC bit is set, the header is followed by a 16-bit CRC. This is followed by the bit allocations used by each subband and is in turn followed by the set of 6-bit scalefactors. The scalefactor data is followed by the quantized 384 samples.

4.3.2 Layer II Coding

The Layer II coder provides a higher compression rate by making some relatively minor modifications to the Layer I coding scheme. These modifications include how the samples are grouped together, the representation of the scalefactors, and the quantization strategy. Where the Layer I coder puts 12 samples from each subband into a frame, the Layer II coder groups three sets of 12 samples from each subband into a frame. The total number of samples per frame increases from 384 samples to 1152 samples. This reduces the amount of overhead per sample. In Layer I coding a separate scalefactor is selected for each block of 12 samples. In Layer II coding the encoder tries to share a scale factor among two or all three groups of samples from each subband filter. The only time separate scalefactors are used for each group of 12 samples is when not doing so would result in a significant increase in distortion. The particular choice used in a frame is signaled through the *scalefactor selection information* field in the bitstream.

The major difference between the Layer I and Layer II coding schemes is in the quantization step. In the Layer I coding scheme the output of each subband is quantized using one of 14 possibilities; the same 14 possibilities for each of the subbands. In Layer II coding the quantizers used for each of the subbands can be selected from a different set of quantizers depending on the sampling rate and the bit rates. For some sampling rate and bit rate

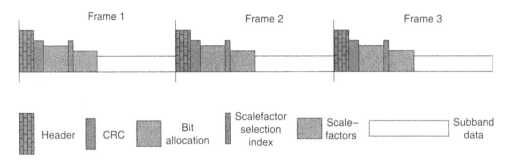

Figure 4.6: Frame structure for Layer 2

combinations, many of the higher subbands are assigned 0 bits. That is, the information from those subbands is simply discarded. Where the quantizer selected has 3, 5, or 9 levels, the Layer II coding scheme uses one more enhancement. Notice that in the case of 3 levels we have to use 2 bits per sample, which would have allowed us to represent 4 levels. The situation is even worse in the case of 5 levels, where we are forced to use 3 bits, wasting three codewords, and in the case of 9 levels where we have to use 4 bits, thus wasting 7 levels. To avoid this situation, the Layer II coder groups 3 samples into a *granule*. If each sample can take on 3 levels, a granule can take on 27 levels. This can be accommodated using 5 bits. If each sample had been encoded separately we would have needed 6 bits. Similarly, if each sample can take on 9 values, a granule can take on 729 values. We can represent 729 values using 10 bits. If each sample in the granule had been encoded separately, we would have needed 12 bits. Using all these savings, the compression ratio in Layer II coding can be increase from 4:1 to 8:1 or 6:1.

The frame structure for the Layer II coder can be seen in Figure 4.6. The only real difference between this frame structure and the frame structure of the Layer I coder is the scalefactor selection information field.

4.3.3 Layer III Coding–mp3

Layer III coding, which has become widely popular under the name *mp3*, is considerably more complex than the Layer I and Layer II coding schemes. One of the problems with the Layer I and coding schemes was that with the 32-band decomposition, the bandwidth of the subbands at lower frequencies is significantly larger than the critical bands. This makes it difficult to make an accurate judgement of the mask-to-signal ratio. If we get a high amplitude tone within a subband and if the subband was narrow enough, we could assume that it masked other tones in the band. However, if the bandwidth of the subband is

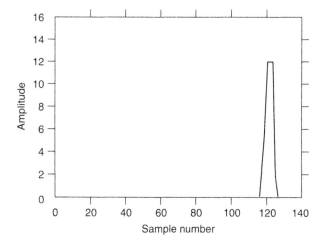

Figure 4.7: Source output sequence

significantly higher than the critical bandwidth at that frequency, it becomes more difficult to determine whether other tones in the subband will be be masked.

A simple way to increase the spectral resolution would be to decompose the signal directly into a higher number of bands. However, one of the requirements on the Layer III algorithm is that it be backward compatible with Layer I and Layer II coders. To satisfy this backward compatibility requirement, the spectral decomposition in the Layer III algorithm is performed in two stages. First the 32-band subband decomposition used in Layer I and Layer II is employed. The output of each subband is transformed using a modified discrete cosine transform (MDCT) with a 50% overlap. The Layer III algorithm specifies two sizes for the MDCT, 6 or 18. This means that the output of each subband can be decomposed into 18 frequency coefficients or 6 frequency coefficients.

The reason for having two sizes for the MDCT is that when we transform a sequence into the frequency domain, we lose time resolution even as we gain frequency resolution. The larger the block size the more we lose in terms of time resolution. The problem with this is that any quantization noise introduced into the frequency coefficients will get spread over the entire block size of the transform. Backward temporal masking occurs for only a short duration prior to the masking sound (approximately 20 msec). Therefore, quantization noise will appear as a *pre-echo*. Consider the signal shown in Figure 4.7. The sequence consists of 128 samples, the first 118 of which are 0, followed by a sharp increase in value. The 128-point DCT of this sequence is shown in Figure 4.8. Notice that many of these coefficients are quite large. If we were to send all these coefficients, we would have data expansion instead

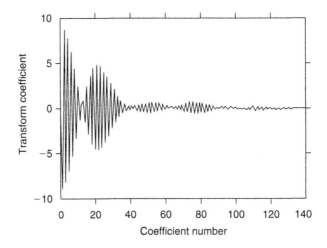

Figure 4.8: Transformed sequence

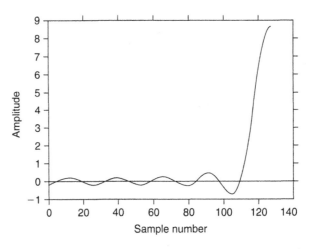

Figure 4.9: Reconstructed sequence from 10 DCT coefficients

of data compression. If we keep only the 10 largest coefficients, the reconstructed signal is shown in Figure 4.9. Notice that not only are the nonzero signal values not well represented, there is also error in the samples prior to the change in value of the signal. If this were an audio signal and the large values had occurred at the beginning of the sequence, the forward masking effect would have reduced the perceptibility of the quantization error. In the situation shown in Figure 4.9, backward masking will mask some of the quantization error. However, backward masking occurs for only a short duration prior to the masking sound. Therefore, if

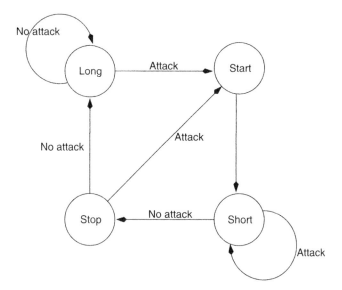

Figure 4.10: State diagram for the window switching process

the length of the block in question is longer than the masking interval, the distortion will be evident to the listener.

If we get a sharp sound that is very limited in time (such as the sound of castanets) we would like to keep the block size small enough that it can contain this sharp sound. Then, when we incur quantization noise it will not get spread out of the interval in which the actual sound occurred and will therefore get masked. The Layer III algorithm monitors the input and where necessary substitutes three short transforms for one long transform. What actually happens is that the subband output is multiplied by a window function of length 36 during the stationary periods (that is a blocksize of 18 plus 50% overlap from neighboring blocks). This window is called the *long window*. If a sharp attack is detected, the algorithm shifts to a sequence of three *short windows* of length 12 after a transition window of length 30. This initial transition window is called the *start* window. If the input returns to a more stationary mode, the short windows are followed by another transition window called the *stop* window of length 30 and then the standard sequence of long windows. The process of transitioning between windows is shown in Figure 4.10. A possible set of window transitions is shown in Figure 4.11. For the long windows we end up with 18 frequencies per subband, resulting in a total of 576 frequencies. For the short windows we get 6 coefficients per subband for a total of 192 frequencies. The standard allows for a mixed block mode in which the two lowest subbands use long windows while the remaining subbands use short windows. Notice that while

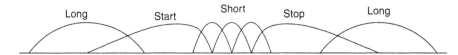

Figure 4.11: Sequence of windows

the number of frequencies may change depending on whether we are using long or short windows, the number of samples in a frame stays at 1152. That is 36 samples, or 3 groups of 12, from each of the 32 subband filters.

The coding and quantization of the output of the MDCT is conducted in an iterative fashion using two nested loops. There is an outer loop called the *distortion control loop* whose purpose is to ensure that the introduced quantization noise lies below the audibility threshold. The scalefactors are used to control the level of quantization noise. In Layer III scalefactors are assigned to groups or "bands" of coefficients in which the bands are approximately the size of critical bands. There are 21 scalefactor bands for long blocks and 12 scalefactor bands for short blocks.

The inner loop is called the *rate control loop*. The goal of this loop is to make sure that a target bit rate is not exceeded. This is done by iterating between different quantizers and Huffman codes. The quantizers used in *mp3* are companded nonuniform quantizers. The scaled MDCT coefficients are first quantized and organized into regions. Coefficients at the higher end of the frequency scale are likely to be quantized to zero. These consecutive zero outputs are treated as a single region and the run-length is Huffman encoded. Below this region of zero coefficients, the encoder identifies the set of coefficients that are quantized to 0 or ±1. These coefficients are grouped into groups of four. This set of quadruplets is the second region of coefficients. Each quadruplet is encoded using a single Huffman codeword. The remaining coefficients are divided into two or three subregions. Each subregion is assigned a Huffman code based on its statistical characteristics. If the result of using this variable length coding exceeds the bit budget, the quantizer is adjusted to increase the quantization stepsize. The process is repeated until the target rate is satisfied.

Once the target rate is satisfied, control passes back to the outer, distortion control loop. The psychoacoustic model is used to check whether the quantization noise in any band exceeds the allowed distortion. If it does, the scalefactor is adjusted to reduce the quantization noise. Once all scalefactors have been adjusted, control returns to the rate control loop. The iterations terminate either when the distortion and rate conditions are satisfied or the scalefactors cannot be adjusted any further.

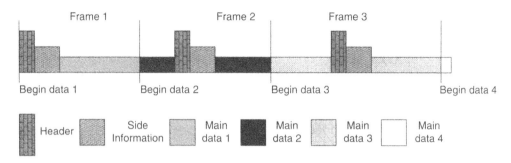

Figure 4.12: Sequence of windows

There will be frames in which the number of bits used by the Huffman coder is less than the amount allocated. These bits are saved in a conceptual *bit reservoir*. In practice what this means is that the start of a block of data does not necessarily coincide with the header of the frame. Consider the three frames shown in Figure 4.12. In this example, the main data for the first frame (which includes scalefactor information and the Huffman coded data) does not occupy the entire frame. Therefore, the main data for the second frame starts before the second frame actually begins. The same is true for the remaining data. The main data can begin in the *previous frame*. However, the main data for a particular frame cannot spill over into the *following* frame.

All this complexity allows for a very efficient encoding of audio inputs. The typical *mp3* audio file has a compression ratio of about 10:1. In spite of this high level of compression, most people cannot tell the difference between the original and the compressed representation.

We say most because trained professionals can at times tell the difference between the original and compressed versions. People who can identify very minute differences between coded and original signals have played an important role in the development of audio coders. By identifying where distortion may be audible they have helped focus effort onto improving the coding process. This development process has made *mp3* the format of choice for compressed music.

4.4 MPEG Advanced Audio Coding

The MPEG Layer III algorithm has been highly successful. However, it had some built-in drawbacks because of the constraints under which it had been designed. The principal constraint was the requirement that it be backward compatible. This requirement for

Table 4.2: AAC DecoderTools (ISO/IEC)

Tool Name	
Bitstream Formatter	Required
Huffman Decoding	Required
Inverse Quantization	Required
Rescaling	Required
M/S	Optional
Interblock Prediction	Optional
Intensity	Optional
Dependently Switched Coupling	Optional
TNS	Optional
Block switching/MDCT	Required
Gain Control	Optional
Independently Switched Coupling	Optional

backward compatibility forced the rather awkward decomposition structure involving a subband decomposition followed by an MDCT decomposition. The period immediately following the release of the MPEG specifications also saw major developments in hardware capability. The Advanced Audio Coding (AAC) standard was approved as a higher quality multichannel alternative to the backward compatible MPEG Layer III in 1997.

The AAC approach is a modular approach based on a set of self-contained tools or modules. Some of these tools are taken from the earlier MPEG audio standard while others are new. As with previous standards, the AAC standard actually specifies the decoder. The decoder tools specified in the AAC standard are listed in Table 4.2. As shown in the table, some of these tools are required for all profiles while others are only required for some profiles. By using some or all of these tools, the standard describes three profiles. These are the *main* profile, the *low complexity* profile, and the *sampling-rate-scalable* profile. The AAC approach used in MPEG-2 was later enhanced and modified to provide an audio coding option in MPEG-4. In the following section we first describe the MPEG-2 AAC algorithm, followed by the MPEG-4 AAC algorithm.

4.4.1 MPEG-2 AAC

A block diagram of an MPEG-2 AAC encoder is shown in Figure 4.13. Each block represents a tool. The psychoacoustic model used in the AAC encoder is the same as the model used in

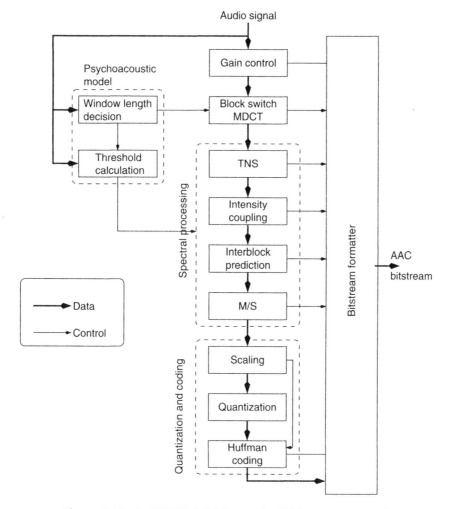

Figure 4.13: An MPEG-2 AAC encoder (ISO/IEC IS 14496)

the MPEG Layer III encoder. As in the Layer III algorithm, the psychoacoustic model is used to trigger switching in the blocklength of the MDCT transform and to produce the threshold values used to determine scalefactors and quantization thresholds. The audio data is fed in parallel to both the acoustic model and to the modified Discrete Cosine Transform.

4.4.1.1 Block Switching and MDCT

Because the AAC algorithm is not backward compatible it does away with the requirement of the 32-band filterbank. Instead, the frequency decomposition is accomplished by a

Modified Discrete Cosine Transform (MDCT). The AAC algorithm allows switching between a window length of 2048 samples and 256 samples. These window lengths include a 50% overlap with neighboring blocks. So 2048 time samples are used to generate 1024 spectral coefficients, and 256 time samples are used to generate 128 frequency coefficients. The kth spectral coefficient of block i, $X_{i,k}$ is given by:

$$X_{i,k} = 2\sum_{n=0}^{N-1} z_{i,n} \cos\left(\frac{2\pi(n + n_0)}{N}\left(k + \frac{1}{2}\right)\right)$$

where $z_{i,n}$ is the nth time sample of the ith block, N is the window length and

$$n_0 = \frac{N/2 + 1}{2}.$$

The longer block length allows the algorithm to take advantage of stationary portions of the input to get significant improvements in compression. The short block length allows the algorithm to handle sharp attacks without incurring substantial distortion and rate penalties. Short blocks occur in groups of eight in order to avoid framing issues. As in the case of MPEG Layer III, there are four kinds of windows: long, short, start, and stop. The decision about whether to use a group of short blocks is made by the psychoacoustic model. The coefficients are divided into scalefactor bands in which the number of coefficients in the bands reflects the critical bandwidth. Each scalefactor band is assigned a single scalefactor. The exact division of the coefficients into scalefactor bands for the different windows and different sampling rates is specified in the standard (ISO/IEC IS 14496).

4.4.1.2 Spectral Processing

In MPEG Layer III coding the compression gain is mainly achieved through the unequal distribution of energy in the different frequency bands, the use of the psychoacoustic model, and Huffman coding. The unequal distribution of energy allows use of fewer bits for spectral bands with less energy. The psychoacoustic model is used to adjust the quantization step size in a way that masks the quantization noise. The Huffman coding allows further reductions in the bit rate. All these approaches are also used in the AAC algorithm. In addition, the algorithm makes use of prediction to reduce the dynamic range of the coefficients and thus allow further reduction in the bit rate.

Recall that prediction is generally useful only in stationary conditions. By their very nature, transients are almost impossible to predict. Therefore, generally speaking, predictive coding would not be considered for signals containing significant amounts of transients. However, music signals have exactly this characteristic. Although they may contain long periods of stationary signals, they also generally contain a significant amount of transient signals. The AAC algorithm makes clever use of the time frequency duality to handle this situation. The standard contains two kinds of predictors, an intrablock predictor, referred to as temporal noise shaping (TNS), and an interblock predictor. The interblock predictor is used during stationary periods. During these periods it is reasonable to assume that the coefficients at a certain frequency do not change their value significantly from block to block. Making use of this characteristic, the AAC standard implements a set of parallel DPCM systems. There is one predictor for each coefficient up to a maximum number of coefficients. The maximum is different for different sampling frequencies. Each predictor is a backward adaptive two-tap predictor. This predictor is really useful only in stationary periods. Therefore, the psychoacoustic model monitors the input and determines when the output of the predictor is to be used. The decision is made on a scalefactor band by scalefactor band basis. Because notification of the decision that the predictors are being used has to be sent to the decoder, this would increase the rate by one bit for each scalefactor band. Therefore, once the preliminary decision to use the predicted value has been made, further calculations are made to check if the savings will be sufficient to offset this increase in rate. If the savings are determined to be sufficient, a *predictor_data_present* bit is set to 1 and one bit for each scalefactor band (called the *prediction_used* bit) is set to 1 or 0 depending on whether prediction was deemed effective for that scalefactor band. If not, the *predictor_data_present* bit is set to 0 and the *prediction_used* bits are not sent. Even when a predictor is disabled, the adaptive algorithm is continued so that the predictor coefficients can track the changing coefficients. However, because this is a streaming audio format it is necessary from time to time to reset the coefficients. Resetting is done periodically in a staged manner and also when a short frame is used.

When the audio input contains transients, the AAC algorithm uses the intraband predictor. Recall that narrow pulses in time correspond to wide bandwidths. The narrower a signal in time, the broader its Fourier transform will be. This means that when transients occur in the audio signal, the resulting MDCT output will contain a large number of correlated coefficients. Thus, unpredictability in time translates to a high level of predictability in terms of the frequency components. The AAC uses neighboring coefficients to perform prediction. A target set of coefficients is selected in the block. The standard suggests a range of 1.5 kHz to the uppermost scalefactor band as specified for different profiles and sampling rates. A set of linear predictive coefficients is obtained using any of the standard approaches, such as the Levinson-Durbin algorithm. The maximum order of the filter ranges from 12 to 20 depending on the profile. The process of obtaining the filter coefficients also provides the expected prediction gain g_p. This expected prediction gain is compared against a threshold to determine if intrablock prediction is going to be used. The standard suggests a value of

1.4 for the threshold. The order of the filter is determined by the first PARCOR coefficient with a magnitude smaller than a threshold (suggested to be 0.1). The PARCOR coefficients corresponding to the predictor are quantized and coded for transfer to the decoder. The reconstructed LPC coefficients are then used for prediction. In the time domain predictive coders, one effect of linear prediction is the spectral shaping of the quantization noise. The effect of prediction in the frequency domain is the *temporal* shaping of the quantization noise, hence the name Temporal Noise Shaping. The shaping of the noise means that the noise will be higher during time periods when the signal amplitude is high and lower when the signal amplitude is low. This is especially useful in audio signals because of the masking properties of human hearing.

4.4.1.3 Quantization and Coding

The quantization and coding strategy used in AAC is similar to what is used in MPEG Layer III. Scalefactors are used to control the quantization noise as a part of an outer *distortion control loop*. The quantization step size is adjusted to accommodate a target bit rate in an inner *rate control loop*. The quantized coefficients are grouped into *sections*. The section boundaries have to coincide with scalefactor band boundaries. The quantized coefficients in each section are coded using the same Huffman codebook. The partitioning of the coefficients into sections is a dynamic process based on a greedy merge procedure. The procedure starts with the maximum number of sections. Sections are merged if the overall bit rate can be reduced by merging. Merging those sections will result in the maximum reduction in bit rate. This iterative procedure is continued until there is no further reduction in the bit rate.

4.4.1.4 Stereo Coding

The AAC scheme uses multiple approaches to stereo coding. Apart from independently coding the audio channels, the standard allows Mid/Side (M/S) coding and intensity stereo coding. Both stereo coding techniques can be used at the same time for different frequency ranges. Intensity coding makes use of the fact that at higher frequencies two channels can be represented by a single channel plus some directional information. The AAC standard suggests using this technique for scalefactor bands above 6 kHz. The M/S approach is used to reduce noise imaging. As described previously in the joint stereo approach, the two channels (L and R) are combined to generate sum and difference channels.

4.4.1.5 Profiles

The main profile of MPEG-2 AAC uses all the tools except for the gain control tool of Figure 4.13. The low complexity profile in addition to the gain control tool the interblock prediction tool is also dropped. In addition the maximum prediction order for intra-band prediction (TNS) for long windows is 12 for the low complexity profile as opposed to 20 for the main profile.

The Scalable Sampling Rate profile does not use the coupling and interband prediction tools. However this profile does use the gain control tool. In the scalable-sampling profile the MDCT block is preceded by a bank of four equal width 96 tap filters. The filter coefficients are provided in the standard. The use of this filterbank allows for a reduction in rate and decoder complexity. By ignoring one or more of the filterbank outputs the output bandwidth can be reduced. This reduction in bandwidth and sample rate also leads to a reduction in the decoder complexity. The gain control allows for the attenuation and amplification of different bands in order to reduce perceptual distortion.

4.4.2 MPEG-4 AAC

The MPEG-4 AAC adds a perceptual noise substitution (PNS) tool and substitutes a long term prediction (LTP) tool for the interband prediction tool in the spectral coding block. In the quantization and coding section the MPEG-4 AAC adds the options of Transform-Domain Weighted Interleave Vector Quantization (TwinVQ) and Bit Sliced Arithmetic Coding (BSAC).

4.4.2.1 Perceptual Noise Substitution (PNS)

There are portions of music that sound like noise. Although this may sound like a harsh (or realistic) subjective evaluation, that is not what is meant here. What is meant by noise here is a portion of audio where the MDCT coefficients are stationary without containing tonal components (Watkinson, *The MPEG Handbook*). This kind of noise-like signal is the hardest to compress. However, at the same time it is very difficult to distinguish one noise-like signal from another. The MPEG-4 AAC makes use of this fact by not transmitting such noise-like scalefactor bands. Instead the decoder is alerted to this fact and the power of the noise-like coefficients in this band is sent. The decoder generates a noise-like sequence with the appropriate power and inserts it in place of the unsent coefficients.

4.4.2.2 Long Term Prediction

The interband prediction in MPEG-2 AAC is one of the more computationally expensive parts of the algorithm. MPEG-4 AAC replaces that with a cheaper long term prediction (LTP) module.

4.4.2.3 TwinVQ

The Transform-Domain Weighted Interleave Vector Quantization (TwinVQ) (Iwakami et al.) option is suggested in the MPEG-4 AAC scheme for low bit rates. Developed at NTT in the early 1990s, the algorithm uses a two-stage process for flattening the MDCT coefficients. In the first stage, a linear predictive coding algorithm is used to obtain the LPC coefficients for the audio data corresponding to the MDCT coefficients. These coefficients are used to obtain the spectral envelope for the audio data. Dividing the MDCT coefficients with this spectral envelope results in some degree of "flattening" of the coefficients. The spectral envelope computed from the LPC coefficients reflects the gross features of the envelope of the MDCT coefficients. However, it does not reflect any of the fine structure. This fine structure is predicted from the previous frame and provides further flattening of the MDCT coefficients. The flattened coefficients are interleaved and grouped into subvectors and quantized. The flattening process reduces the dynamic range of the coefficients, allowing them to be quantized using a smaller VQ codebook than would otherwise have been possible. The flattening process is reversed in the decoder as the LPC coefficients are transmitted to the decoder.

4.4.2.4 Bit Sliced Arithmetic Coding (BSAC)

In addition to the Huffman coding scheme of the MPEG-2 AAC scheme, the MPEG-4 AAC scheme also provides the option of using binary arithmetic coding. The binary arithmetic coding is performed on the bitplanes of the magnitudes of the quantized MDCT coefficients. By bitplane we mean the corresponding bit of each coefficient. Consider the sequence of 4-bit coefficients x_n: 5, 11, 8, 10, 3, 1. The most significant bitplane would consist of the MSBs of these numbers, 011100. The next bitplane would be 100000. The next bitplane is 010110. The least significant bitplane is 110011.

The coefficients are divided into *coding bands* of 32 coefficients each. One probability table is used to encode each coding band. Because we are dealing with binary data, the probability table is simply the number of zeros. If a coding band contains only zeros, this is indicated to the decoder by selecting the probability table 0. The sign bits associated with the nonzero coefficients are sent after the arithmetic code when the coefficient has a 1 for the the first time.

The scalefactor information is also arithmetic coded. The maximum scalefactor is coded as an 8-bit integer. The differences between scalefactors are encoded using an arithmetic code. The first scalefactor is encoded using the difference between it and the maximum scalefactor.

4.5 Dolby AC3 (Dolby Digital)

Unlike the MPEG algorithms described in the previous section, the Dolby AC-3 method became a de facto standard. It was developed in response to the standardization activities

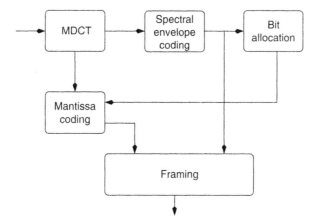

Figure 4.14: The Dolby AC3 algorithm

of the *Grand Alliance*, which was developing a standard for HDTV in the United States. However, even before it was accepted as the recommendation for HDTV audio, Dolby-AC3 had already made its debut in the movie industry. It was first released in a few theaters during the showing of *Star Trek IV* in 1991 and was formally released with the movie *Batman Returns* in 1992. It was accepted by the *Grand Alliance* in October of 1993 and became an Advanced Television Systems Committee (ATSC) standard in 1995. Dolby AC-3 had the multichannel capability required by the movie industry along with the ability to downmix the channels to accommodate the varying capabilities of different applications. The 5.1 channels include right, center, left, left rear, and right rear, and a narrowband low-frequency effects channel (the 0.1 channel). The scheme supports downmixing the 5.1 channels to 4, 3, 2, or 1 channel. It is now the standard used for DVDs as well as for Direct Broadcast Satellites (DBS) and other applications.

A block diagram of the Dolby-AC3 algorithm is shown in Figure 4.14. Much of the Dolby-AC3 scheme is similar to what we have already described for the MPEG algorithms. As in the MPEG schemes, the Dolby-AC3 algorithm uses the modified DCT (MDCT) with 50% overlap for frequency decomposition. As in the case of MPEG, there are two different sizes of windows used. For the stationary portions of the audio a window of size 512 is used to get a 256 coefficient. A surge in the power of the high frequency coefficients is used to indicate the presence of a transient and the 512 window is replaced by two windows of size 256. The one place where the Dolby-AC3 algorithm differs significantly from the algorithm described is in the bit allocation.

4.5.1 Bit Allocation

The Dolby-AC3 scheme has a very interesting method for bit allocation. Like the MPEG schemes, it uses a psychoacoustic model that incorporates the hearing thresholds and the presence of noise and tone maskers. However, the input to the model is different. In the MPEG schemes the audio sequence being encoded is provided to the bit allocation procedure and the bit allocation is sent to the decoder as side information. In the Dolby-AC3 scheme the signal itself is not provided to the bit allocation procedure. Instead a crude representation of the spectral envelope is provided to both the decoder and the bit allocation procedure. As the decoder then possesses the information used by the encoder to generate the bit allocation, the allocation itself is not included in the transmitted bitstream.

The representation of the spectral envelope is obtained by representing the MDCT coefficients in binary exponential notation. The binary exponential notation of a number 110.101 is 0.110101×2^3, where 110101 is called the *mantissa* and 3 is the exponent. Given a sequence of numbers, the exponents of the binary exponential representation provide an estimate of the relative magnitude of the numbers. The Dolby-AC3 algorithm uses the exponents of the binary exponential representation of the MDCT coefficients as the representation of the spectral envelope. This encoding is sent to the bit allocation algorithm, which uses this information in conjunction with a psychoacoustic model to generate the number of bits to be used to quantize the mantissa of the binary exponential representation of the MDCT coefficients. To reduce the amount of information that needs to be sent to the decoder, the spectral envelope coding is not performed for every audio block. Depending on how stationary the audio is, the algorithm uses one of three strategies (Bosi/Goldberg).

4.5.1.1 The D15 Method

When the audio is relatively stationary, the spectral envelope is coded once for every six audio blocks. Because a frame in Dolby-AC3 consists of six blocks, during each block we get a new spectral envelope and hence a new bit allocation. The spectral envelope is coded differentially. The first exponent is sent as is. The difference between exponents is encoded using one of five values $\{0, \pm 1, \pm 2\}$. Three differences are encoded using a 7-bit word. Note that three differences can take on 125 different combinations. Therefore, using 7 bits, which can represent 128 different values, is highly efficient.

4.5.1.2 The D25 and D45 Methods

If the audio is not stationary, the spectral envelope is sent more often. To keep the bit rate down, the Dolby-AC3 algorithm uses one of two strategies. In the D25 strategy, which is used for moderate spectral activity, every other coefficient is encoded. In the D45 strategy, used during transients, every fourth coefficient is encoded. These strategies make use of the fact that during a transient the fine structure of the spectral envelope is not that important, allowing for a more crude representation.

4.6 Other Standards

We have described a number of audio compression approaches that make use of the limitations of human audio perception. These are by no means the only ones. Competitors to Dolby Digital include Digital Theater Systems (DTS) and Sony Dynamic Digital Sound (SDDS). Both of these proprietary schemes use psychoacoustic modeling. The Adaptive TRansform Acoustic Coding (ATRAC) algorithm (Tsutsui et al.) was developed for the minidisc by Sony in the early 1990s, followed by enhancements in ATRAC3 and ATRAC3plus. As with the other schemes described in this chapter, the ATRAC approach uses MDCT for frequency decomposition, though the audio signal is first decomposed into three bands using a two-stage decomposition. As in the case of the other schemes, the ATRAC algorithm recommends the use of the limitations of human audio perception in order to discard information that is not perceptible.

Another algorithm that also uses MDCT and a psychoacoustic model is the open source encoder Vorbis. The Vorbis algorithm also uses vector quantization and Huffman coding to reduce the bit rate.

4.7 Summary

The audio coding algorithms described in this chapter take, in some sense, the opposite tack from speech coding algorithms. Instead of focusing on the source of information, as is the case with the speech coding algorithm, the focus in the audio coding algorithm is on the sink, or user, of the information. By identifying the components of the source signal that are not perceptible, the algorithms reduce the amount of data that needs to be transmitted.

Bibliography

Bosi, M., & Goldberg, R. E. (2003). *Introduction to Digital Audio Coding and Standards.* Kluwer Academic Press.

Fletcher, H., & Munson, W. A. (1933). Loudness, its measurement, definition, and calculation. *Journal of the Acoustical Society of America, 5,* 82–108.

ISO/IEC IS 14496, Coding of Moving Pictures and Audio.

Iwakami, N., Moriya, T., & Miki, S. (1985). High Quality Audio-Coding at Less Than 64 kbit/s by Using Transform Domain Weighted Interleave Vector Quantization TwinVQ. In *Proceedings ICASSP. '95, Vol. 5* (pp. 3095–3098). IEEE.

Moore, B. C. J. (1989). *An Introduction to the Psychology of Hearing* (3rd ed.). Academic Press.

Painter, T., & Spanias, A. (2000). Perceptual Coding of Digital Audio. *Proceedings of the IEEE, 88*, 451–513.

Pan, D. (1995). A Tutorial on MPEG/Audio Compression. *IEEE Multimedia, 2*, 60–74.

Tsutsui, K., Suzuki, H., Shimoyoshi, O., Sonohara, M., Agagiri, K., & Heddle, R. M. (October 1992). ATRAC: Adaptive Transform Acoustic Coding for MiniDisc. In *Conference Records Audio Engineering Society Convention.* AES.

Watkinson, J. (2001). *The MPEG Handbook.* Focal Press.

The website *http://www.tnt.uni-hannover.de/project/mpeg/audio/faq/* contains information about all the audio coding schemes described here as well as an overview of MPEG-7 audio.

CHAPTER 5

Video Processing

Keith Jack

Video processing has become one of the main applications of signal processing technology. One reason for the increasing importance of digital video is that it is everywhere, from YouTube to unpiloted military drones. The other reason is that digital video is hard to do—the algorithms are complicated, and they demand huge computational power. Thus, video applications will keep many of us employed for years to come!

In this chapter, Keith Jack gives an in-depth treatment of digital video processing. The main topics include display enhancement, video mixing and graphics overlay, frame rate conversion, interlacing/deinterlacing, video scaling, and user controls (brightness, contrast, etc).

You may have noticed that this list does not include compression and decompression algorithms (i.e., codecs) such as H.264. The end of the chapter does give some insight into discrete cosine transform (DCT) based compression, but otherwise I have decided to leave the details out of this book. Few DSP engineers design or implement video codecs, so most of us just need to know the basics. For a good general overview of video codecs, I recommend reading "How Video Compression Works":

http://www.dspdesignline.com/howto/201202637

Jack brings a wealth of experience to the topic. He's currently director of product marketing at Sigma Designs, which develops system-on-chips (SoC) for IPTV, set-top-boxes, HD-DVD players, portable media players, and more. Before that he was involved in designing or marketing over forty multimedia chips for the consumer market. In this text, his experience shows. Each topic is chock-full of practical technical detail one would need to implement the video processing covered.

—Kenton Williston

In addition to encoding and decoding MPEG, NTSC/PAL, and many other types of video, a typical system usually requires considerable additional video processing.

Since many consumer displays, and most computer displays, are progressive (noninterlaced), interlaced video must be converted to progressive ("deinterlaced"). Progressive video must be converted to interlaced to drive a conventional analog VCR or interlaced TV, requiring noninterlaced-to-interlaced conversion.

Many computer displays support refresh rates up to at least 75 frames per second. CRT-based televisions have a refresh rate of 50 or 59.94 (60/1.001) fields per second. Refresh rates of up to 120 frames per second are becoming common for flat-panel televisions. For film-based compressed content, the source may only be 24 frames per second. Thus, some form of frame rate conversion must be done.

Another not-so-subtle problem includes video scaling. SDTV and HDTV support multiple resolutions, yet the display may be a single, fixed resolution.

Alpha mixing and chroma keying are used to mix multiple video signals or video with computer-generated text and graphics. Alpha mixing ensures a smooth crossover between sources, allows subpixel positioning of text, and limits source transition bandwidths to simplify eventual encoding to composite video signals.

Since no source is perfect, even digital sources, user controls for adjustable brightness, contrast, saturation, and hue are always desirable.

5.1 Rounding Considerations

When two 8-bit values are multiplied together, a 16-bit result is generated. At some point, a result must be rounded to some lower precision (for example, 16 bits to 8 bits or 32 bits to 16 bits) in order to realize a cost-effective hardware implementation. There are several rounding techniques: truncation, conventional rounding, error feedback rounding, and dynamic rounding.

5.1.1 Truncation

Truncation drops any fractional data during each rounding operation. As a result, after only a few operations, a significant error may be introduced. This may result in contours being visible in areas of solid colors.

5.1.2 Conventional Rounding

Conventional rounding uses the fractional data bits to determine whether to round up or round down. If the fractional data is 0.5 or greater, rounding up should be performed—positive numbers should be made more positive and negative numbers should be made more negative.

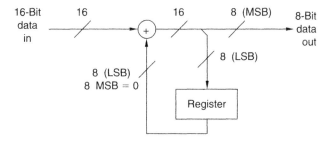

Figure 5.1: Error feedback rounding

If the fractional data is less than 0.5, rounding down should be performed—positive numbers should be made less positive and negative numbers should be made less negative.

5.1.3 Error Feedback Rounding

Error feedback rounding follows the principle of "never throw anything away." This is accomplished by storing the residue of a truncation and adding it to the next video sample. This approach substitutes less visible noise-like quantizing errors in place of contouring effects caused by simple truncation. An example of an error feedback rounding implementation is shown in Figure 5.1. In this example, 16 bits are reduced to 8 bits using error feedback.

5.1.4 Dynamic Rounding

This technique (a licensable Quantel patent) dithers the LSB according to the weighting of the discarded fractional bits. The original data word is divided into two parts, one representing the resolution of the final output word and one dealing with the remaining fractional data. The fractional data is compared to the output of a random number generator equal in resolution to the fractional data. The output of the comparator is a 1-bit random pattern weighted by the value of the fractional data, and serves as a carry-in to the adder. In all instances, only one LSB of the output word is changed, in a random fashion. An example of a dynamic rounding implementation is shown in Figure 5.2.

5.2 SDTV-HDTV YCbCr Transforms

SDTV and HDTV applications have different colorimetric characteristics. Thus, when SDTV (HDTV) data is displayed on an HDTV (SDTV) display, the YCbCr data should be processed to compensate for the different colorimetric characteristics.

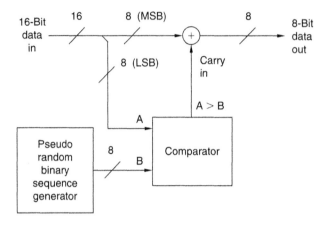

Figure 5.2: Dynamic rounding

5.2.1 SDTV to HDTV

A 3×3 matrix can be used to convert from $Y_{601}CbCr$ (SDTV) to $Y_{709}CbCr$ (HDTV):

1   −0.11554975   −0.20793764

0   1.01863972   0.11461795

0   0.07504945   1.02532707

Note that before processing, the 8-bit DC offset (16 for Y and 128 for CbCr) must be removed, then added back in after processing.

5.2.2 HDTV to SDTV

A 3×3 matrix can be used to convert from $Y_{709}CbCr$ (HDTV) to $Y_{601}CbCr$ (SDTV):

1   0.09931166   0.19169955

0   0.98985381   −0.11065251

0   −0.07245296   0.98339782

Note that before processing, the 8-bit DC offset (16 for Y and 128 for CbCr) must be removed, then added back in after processing.

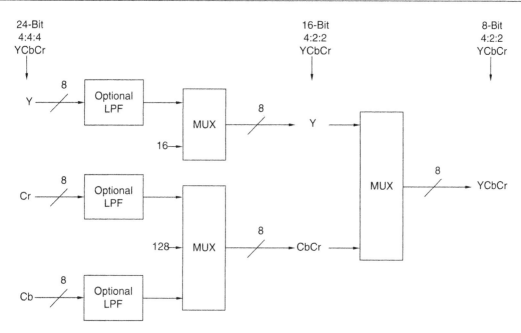

Figure 5.3: 4:4:4 to 4:2:2 YCbCr conversion

5.3 4:4:4 to 4:2:2 YCbCr Conversion

Converting 4:4:4 YCbCr to 4:2:2 YCbCr (Figure 5.3) is a common function in digital video. 4:2:2 YCbCr is the basis for many digital video interfaces, and requires fewer connections to implement than 4:4:4.

Saturation logic should be included in the Y, Cb, and Cr data paths to limit the 8-bit range to 1–254. The 16 and 128 values shown in Figure 5.3 are used to generate the proper levels during blanking intervals.

5.3.1 Y Filtering

A template for the Y lowpass filter is shown in Figure 5.4 and Table 5.1.

Because there may be many cascaded conversions (up to 10 were envisioned), the filters were designed to adhere to very tight tolerances to avoid a buildup of visual artifacts. Departure from flat amplitude and group delay response due to filtering is amplified through successive stages. For example, if filters exhibiting −1 dB at 1 MHz and −3 dB at 1.3 MHz were employed, the overall response would be −8 dB (at 1 MHz) and −24 dB (at 1.3 MHz) after four conversion stages (assuming two filters per stage).

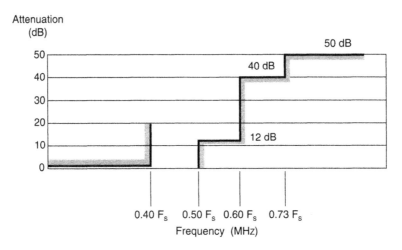

Figure 5.4: Y filter template. F$_s$ = Y1 $\times$ sample rate

Table 5.1: Y filter ripple and group delay tolerances. F$_s$ = Y1 $\times$ sample rate. T = 1/F$_s$

Frequency Range	Typical SDTV Tolerances	Typical HDTV Tolerances
Passband Ripple Tolerance		
0 to 0.40F$_s$	$\pm$0.01 dB increasing to $\pm$0.05 dB	$\pm$0.05 dB
Passband Group Delay Tolerance		
0 to 0.27F$_s$	0 increasing to $\pm$1.35 ns	$\pm$0.075T
0.27F$_s$ to 0.40F$_s$	$\pm$1.35 ns increasing to $\pm$2 ns	$\pm$0.110T

Although the sharp cut-off results in ringing on Y edges, the visual effect should be minimal provided that group-delay performance is adequate. When cascading multiple filtering operations, the passband flatness and group-delay characteristics are very important. The passband tolerances, coupled with the sharp cut-off, make the template very difficult (some say impossible) to match. As a result, there is usually a temptation to relax passband accuracy, but the best approach is to reduce the rate of cut-off and keep the passband as flat as possible.

5.3.2 CbCr Filtering

Cb and Cr are lowpass filtered and decimated. In a standard design, the lowpass and decimation filters may be combined into a single filter, and a single filter may be used for both Cb and Cr by multiplexing.

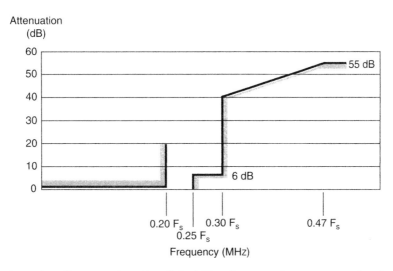

Figure 5.5: Cb and Cr filter template for digital filter for sample rate conversion from 4:4:4 to 4:2:2. F$_s$ = Y1 × sample rate

Table 5.2: CbCr filter ripple and group delay tolerances. F$_s$ = Y1 × sample rate. T = 1/F$_s$

Frequency Range	Typical SDTV Tolerances	Typical HDTV Tolerances
Passband Ripple Tolerance		
0 to 0.20F$_s$	0 dB increasing to ±0.05 dB	±0.05 dB
Passband Group Delay Tolerance		
0 to 0.20F$_s$	delay distortion is zero by design	

As with Y filtering, the Cb and Cr lowpass filtering requires a sharp cut-off to prevent repeated conversions from producing a cumulative resolution loss. However, due to the low cut-off frequency, the sharp cut-off produces ringing that is more noticeable than for Y.

A template for the Cb and Cr filters is shown in Figure 5.5 and Table 5.2.

Since aliasing is less noticeable in color difference signals, the attenuation at half the sampling frequency is only 6 dB. There is an advantage in using a skew-symmetric response passing through the −6 dB point at half the sampling frequency—this makes alternate coefficients in the digital filter zero, almost halving the number of taps, and also allows using a single digital filter for both the Cb and Cr signals. Use of a transversal digital filter has the advantage of providing perfect linear phase response, eliminating the need for group-delay correction.

As with the Y filter, the passband flatness and group-delay characteristics are very important, and the best approach again is to reduce the rate of cut-off and keep the passband as flat as possible.

5.4 Display Enhancement

5.4.1 Brightness, Contrast, Saturation (Color), and Hue (Tint)

Working in the YCbCr color space simplifies the implementation of brightness, contrast, saturation, and hue controls, as shown in Figure 5.6. Also illustrated are multiplexers to allow the output of black screen, blue screen, and color bars.

The design should ensure that no overflow or underflow wraparound errors occur, effectively saturating results to the 0 and 255 values.

5.4.1.1 Y Processing

16 is subtracted from the Y data to position the black level at zero. This removes the DC offset so adjusting the contrast does not vary the black level. Since the Y input data may have values below 16, negative Y values should be supported at this point.

The contrast (or *picture* or *white level*) control is implemented by multiplying the YCbCr data by a constant. If Cb and Cr are not adjusted, a color shift will result whenever the contrast is changed. A typical 8-bit contrast adjustment range is 0–1.992×.

The brightness (or *black level*) control is implemented by adding or subtracting from the Y data. Brightness is done after the contrast to avoid introducing a varying DC offset due to adjusting the contrast. A typical 8-bit brightness adjustment range is −128 to +127.

Finally, 16 is added to position the black level at 16.

5.4.1.2 CbCr Processing

128 is subtracted from Cb and Cr to position the range about zero.

The hue (or *tint*) control is implemented by mixing the Cb and Cr data:

$$Cb' = Cb\cos\theta + Cr\sin\theta$$
$$Cr' = Cr\cos\theta - Cb\sin\theta$$

where θ is the desired hue angle. A typical 8-bit hue adjustment range is −30° to +30°.

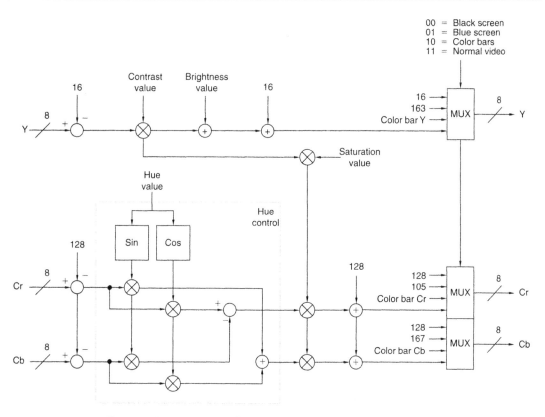

Figure 5.6: Hue, saturation, contrast, and brightness controls

The saturation (or *color*) control is implemented by multiplying both Cb and Cr by a constant. A typical 8-bit saturation adjustment range is 0–1.992×. In the example shown in Figure 5.6, the contrast and saturation values are multiplied together to reduce the number of multipliers in the CbCr datapath.

Finally, 128 is added to both Cb and Cr.

Many displays also use separate hue and saturation controls for each of the red, green, blue, cyan, yellow, and magenta colors. This enables tuning the image at production time to better match the display's characteristics.

5.4.2 Color Transient Improvement

YCbCr transitions should be aligned. However, the Cb and Cr transitions are usually slower and time-offset due to the narrower bandwidth of color difference information.

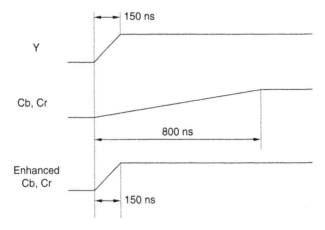

Figure 5.7: Color transient improvement

By monitoring coincident Y transitions, faster horizontal and vertical transitions may be synthesized for Cb and Cr. Small pre- and after-shoots may also be added to the Cb and Cr signals.

The new Cb and Cr edges are then aligned with the Y edge, as shown in Figure 5.7.

Displays commonly use this technique to provide a sharper-looking picture.

5.4.3 Luma Transient Improvement

In this case, the Y horizontal and vertical transitions are shortened, and small pre- and after-shoots may also be added, to artificially sharpen the image.

Displays commonly use this technique to provide a sharper-looking picture.

5.4.4 Sharpness

The apparent sharpness of a picture may be increased by increasing the amplitude of high-frequency luminance information.

As shown in Figure 5.8, a simple bandpass filter with selectable gain (also called a *peaking filter*) may be used. The frequency where maximum gain occurs is usually selectable to be either at the color subcarrier frequency or at about 2.6 MHz. A coring circuit is typically used after the filter to reduce low-level noise.

Figure 5.9 illustrates a more complex sharpness control circuit. The high-frequency luminance is increased using a variable bandpass filter, with adjustable gain. The coring

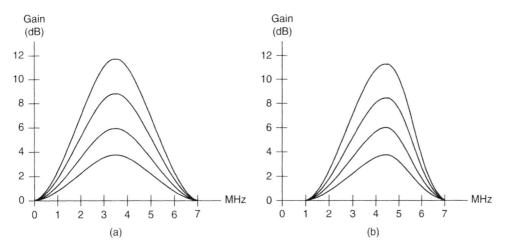

Figure 5.8: Simple adjustable sharpness control: (A) NTSC, (B) PAL

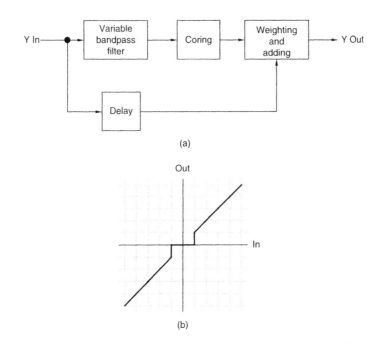

Figure 5.9: More complex sharpness control: (A) Typical implementation, (B) Coring function

function (typically $\pm 1\,\mathrm{LSB}$) removes low-level noise. The modified luminance is then added to the original luminance signal.

In addition to selectable gain, selectable attenuation of high frequencies should also be supported. Many televisions boost high-frequency gain to improve the apparent sharpness of the picture. Although the sharpness control on the television may be turned down, this affects the picture quality of analog broadcasts.

5.4.5 Blue Stretch

Blue stretch increases the blue value of white and near-white colors in order to make whites appear brighter. When applying blue stretch, only colors only within a specified color range should be processed.

Colors with a Y value of ~80% or more of the maximum, have a low saturation value, and fall within a white detection area in the CbCr-plane, have their blue components increased by ~4% (the blue gain factor) and their red components decreased the same amount. For more complex designs, the white detection area and blue gain factor can be dependent on the color's Y value and saturation level.

A transition boundary can be used around the white detection area for gradually decreasing the blue gain factor as colors move away from the white detection area boundary. This can prevent hard transitions between areas that are blue stretched and areas that are not. If a color falls inside the transition boundary area, it is blue stretched using a fraction of the blue gain factor, with the fraction decreasing as the distance from the edge of the detection area boundary increases.

5.4.6 Green Enhancement

Green enhancement creates a richer, more saturated green color when the level of green is low. Displays commonly use this technique to provide greener looking grass, plants, etc. When applying green enhancement, only colors only within a specified color range should be processed.

Colors with a low green saturation value, and fall within a green detection area in the CbCr-plane, have their saturation increased. Rather then centering the green detection area about the green axis (241° in Figure 9.28) some designs use ~213° for the green detection axis so the same design can also easily be used to implement skin tone correction.

Simple implementations have the maximum saturation gain ($\sim1.2\times$) occurring on the green detection axis, with the saturation gain decreasing to $1\times$ as the distance from the green detection axis increases. For more complex designs, the green detection area and maximum saturation gain can be dependent on the color's Y value and saturation level

Some displays also use this technique to implement blue enhancement, used to make the sky appear more blue.

5.4.7 Dynamic Contrast

Using dynamic contrast (also called *adaptive contrast enhancement*), the differences between dark and light portions of the image are artificially enhanced based on the content in the image. Displays commonly use this technique to improve their contrast ratio.

Bright colors in mostly dark images are enhanced by making them brighter (white stretch). This is typically done by using histogram information to modify the upper portion of the gamma curve.

Dark colors in mostly light images are enhanced by making them darker (black stretch). This is typically done by using histogram information to modify the lower portion of the gamma curve.

For a medium-bright image, both techniques may be applied.

A minor gamma correction adjustment may also be applied to colors that are between dark and light, resulting in a more detailed and contrasting picture.

5.4.8 Color Correction

The RGB chromaticities are usually slightly different between the source video and what the display uses. This results in red, green and blue colors that are not completely accurate.

Color correction can be done on the source video to compensate for the display characteristics, enabling more accurate red, green and blue colors to be displayed.

An alternate type of color correction is to perform color expansion, taking advantage of the greater color reproduction capabilities of modern displays. This can result in greener greens, bluer blues, etc. One common technique of implementing color expansion is to use independent hue and saturation controls for each primary and complementary color, plus the skin color.

5.4.9 Color Temperature Correction

In an uncalibrated television, the color temperature (white color) varies based on the brightness level.

The color temperature of D_{65}, the white point specified by most video standards, is 6500°K. Color temperatures above 6500°K. are more bluish (cool); color temperatures below 6500°K. are more reddish (warm).

Many televisions ship from the factory with a very high average color temperature (7000–8000°K.) to emphasize the brightness of the set. Viewers can select from two or three factory presets (warm, cool, etc.) or viewing modes (movies, sports, etc.) which are a reference to the color temperature. A "cool" setting is brighter (like what you see in midday light) and is better for daylight viewing, such as sporting events, because of the enhanced brightness. A "warm" setting is softer (like what you see in a softly lit indoor environment) and is better for viewing movies, or in darkened environments.

The color temperature may be finely adjusted by using a 3×3 matrix multiplier to process the YCbCr or R′G′B′ data. 10 registers (one for every 10 IRE step from 10–100 IRE) provide the nine coefficients for the 3×3 matrix multiplier. The values of the registers are determined by a calibrating process. YCbCr or R′G′B′ values for intermediate IRE levels may be determined using interpolation.

5.5 Video Mixing and Graphics Overlay

Mixing video signals may be as simple as switching between two video sources. This is adequate if the resulting video is to be displayed on a computer monitor.

For most other applications, a technique known as *alpha mixing* should be used. Alpha mixing may also be used to fade to or from a specific color (such as black) or to overlay computer-generated text and graphics onto a video signal.

Alpha mixing must be used if the video is to be encoded to composite video. Otherwise, ringing and blurring may appear at the source switching points, such as around the edges of computer-generated text and graphics. This is due to the color information being lowpass filtered within the NTSC/PAL encoder. If the filters have a sharp cut-off, a fast color transition will produce ringing. In addition, the intensity information may be bandwidth-limited to about 4–5 MHz somewhere along the video path, slowing down intensity transitions.

Mathematically, with alpha normalized to have values of 0–1, alpha mixing is implemented as:

$$out = (alpha\_0)(in\_0) + (alpha\_1)(in\_1) + \ldots$$

In this instance, each video source has its own alpha information. The alpha information may not total to one (unity gain).

Figure 5.10 shows mixing of two YCbCr video signals, each with its own alpha information. As YCbCr uses an offset binary notation, the offset (16 for Y and 128 for Cb and Cr) is removed prior to mixing the video signals. After mixing, the offset is added back in. Note that two 4:2:2 YCbCr streams may also be processed directly; there is no need to convert them to 4:4:4 YCbCr, mix, then convert the result back to 4:2:2 YCbCr.

When only two video sources are mixed and alpha_0 + alpha_1 = 1 (implementing a crossfader), a single alpha value may be used mathematically shown as:

$$out = (alpha)(in\_0) + (1 - alpha)(in\_1)$$

When alpha = 0, the output is equal to the in_1 video signal; when alpha=1, the output is equal to the in_0 video signal. When alpha is between 0 and 1, the two video signals are proportionally multiplied, and added together.

Expanding and rearranging the previous equation shows how a two-channel mixer may be implemented using a single multiplier:

$$out = (alpha)(in\_0 - in\_1) + in\_1$$

Fading to and from a specific color is done by setting one of the input sources to a constant color.

Figure 5.11 illustrates mixing two YCbCr sources using a single alpha channel. Figures 5.12 and 5.13 illustrate mixing two R′G′B′ video sources (R′G′B′ has a range of 0–255). Figures 5.14 and 5.15 show mixing two digital composite video signals.

A common problem in computer graphics systems that use alpha is that the frame buffer may contain preprocessed R′G′B′ or YCbCr data; that is, the R′G′B′ or YCbCr data in the frame buffer has already been multiplied by alpha. Assuming an alpha (A) value of 0.5, nonprocessed R′G′B′A values for white are (255, 255, 255, 128); preprocessed R′G′B′A values for white are (128, 128, 128, 128). Therefore, any mixing circuit that accepts R′G′B′ or YCbCr data from a frame buffer should be able to handle either format.

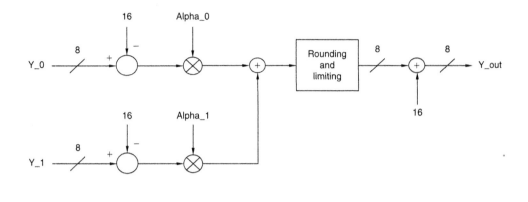

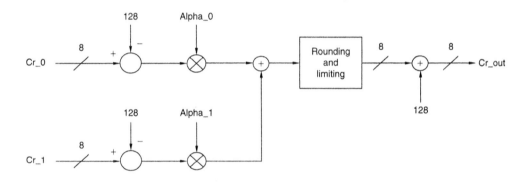

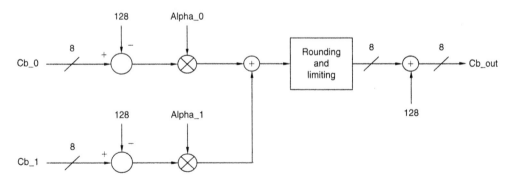

Figure 5.10: Mixing two YCbCr video signals, each with its own alpha channel

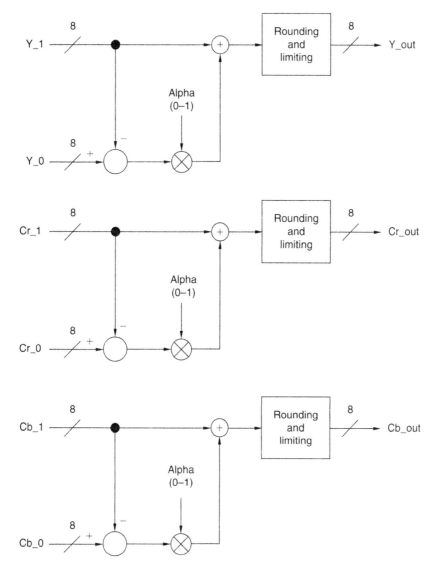

**Figure 5.11: Simplified mixing (crossfading) of two YCbCr video signals using a
single alpha channel**

By adjusting the alpha values, slow to fast crossfades are possible, as shown in Figure 5.16.
Large differences in alpha between samples result in a fast crossfade; smaller differences
result in a slow crossfade. If using alpha mixing for special effects, such as wipes, the
switching point (where 50% of each video source is used) must be able to be adjusted to

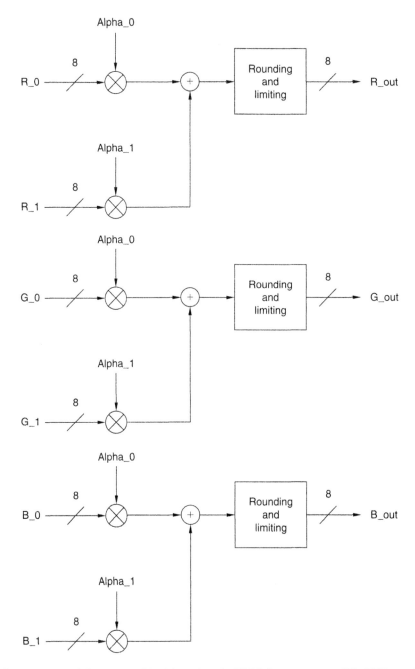

Figure 5.12: Mixing two RGB video signals (RGB has a range of 0–255), each with its own alpha channel

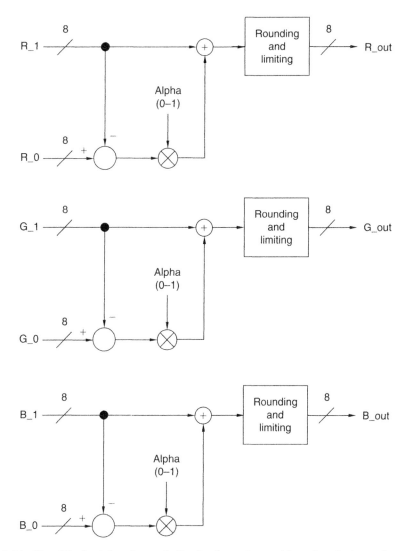

Figure 5.13: Simplified mixing (crossfading) of two RGB video signals (RGB has a range of 0–255) using a single alpha channel

an accuracy of less than one sample to ensure smooth movement. By controlling the alpha values, the switching point can be effectively positioned anywhere, as shown in Figure 5.16a.

Text can be overlaid onto video by having a character generator control the alpha inputs. By setting one of the input sources to a constant color, the text will assume that color.

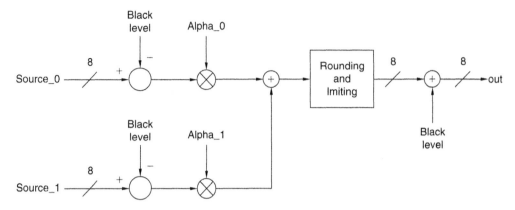

Figure 5.14: Mixing two digital composite video signals, each with its own alpha channel

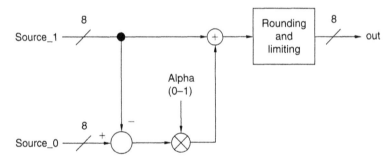

Figure 5.15: Simplified mixing (crossfading) of two digital composite video signals using a single alpha channel

Note that for those designs that subtract 16 (the black level) from the Y channel before processing, negative Y values should be supported after the subtraction. This allows the design to pass through real-world and test video signals with minimum artifacts.

5.6 Luma and Chroma Keying

Keying involves specifying a desired foreground color; areas containing this color are replaced with a background image. Alternately, an area of any size or shape may be specified; foreground areas inside (or outside) this area are replaced with a background image.

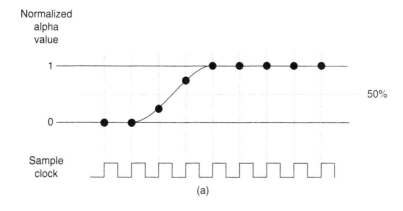

(a)

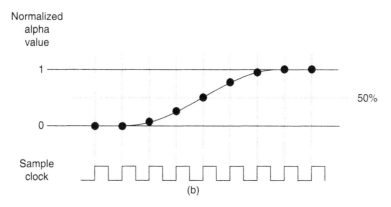

(b)

Figure 5.16: Controlling alpha values to implement (A) Fast or (B) Slow keying. In (A), the effective switching point lies between two samples. In (B), the transition is wider and is aligned at a sample instant.

5.6.1 Luminance Keying

Luminance keying involves specifying a desired foreground luminance level; foreground areas containing luminance levels above (or below) the keying level are replaced with the background image.

Alternately, this hard keying implementation may be replaced with soft keying by specifying two luminance values of the foreground image: Y_H and Y_L ($Y_L < Y_H$). For keying the background into white foreground areas, foreground luminance values (Y_{FG}) above Y_H are replaced with the background image; Y_{FG} values below Y_L contain the foreground image. For Y_{FG} values between Y_L and Y_H, linear mixing is done between the foreground and background images. This operation may be expressed as:

$$\text{if } Y_{FG} > Y_H$$
$$K = 1 = \text{background only}$$

$$\text{if } Y_{FG} < Y_L$$
$$K = 0 = \text{foreground only}$$

$$\text{if } Y_H \geq Y_{FG} \geq Y_L$$
$$K = (Y_{FG} - Y_L)/(Y_H - Y_L) = \text{mix}$$

By subtracting K from 1, the new luminance keying signal for keying into black foreground areas can be generated.

Figure 5.17 illustrates luminance keying for two YCbCr sources. Although chroma keying typically uses a suppression technique to remove information from the foreground image, this is not done when luminance keying as the magnitudes of Cb and Cr are usually not related to the luminance level.

Figure 5.18 illustrates luminance keying for R′G′B′ sources, which is more applicable for computer graphics. Y_{FG} may be obtained by the equation:

$$Y_{FG} = 0.299R' + 0.587G' + 0.114B'$$

In some applications, the red and blue data is ignored, resulting in Y_{FG} being equal to only the green data.

Figure 5.19 illustrates one technique of luminance keying between two digital composite video sources.

5.6.2 Chroma Keying

Chroma keying involves specifying a desired foreground key color; foreground areas containing the key color are replaced with the background image. Cb and Cr are used to specify the key color; luminance information may be used to increase the realism of the chroma keying function. The actual mixing of the two video sources may be done in the component or composite domain, although component mixing reduces artifacts.

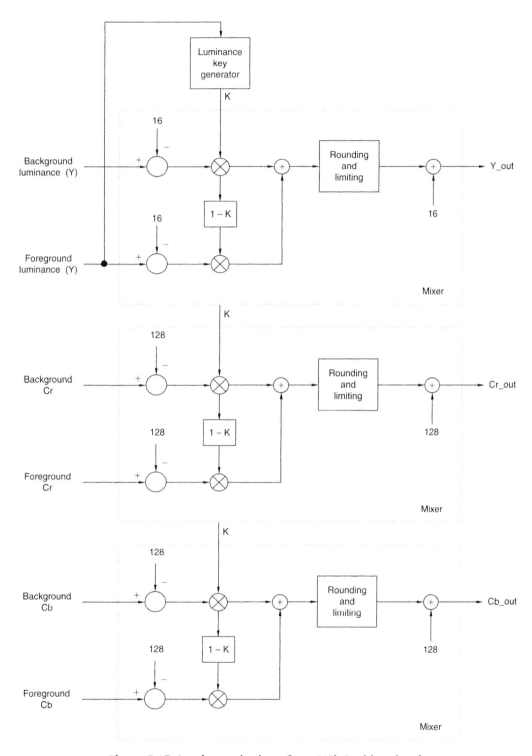

Figure 5.17: Luminance keying of two YCbCr video signals

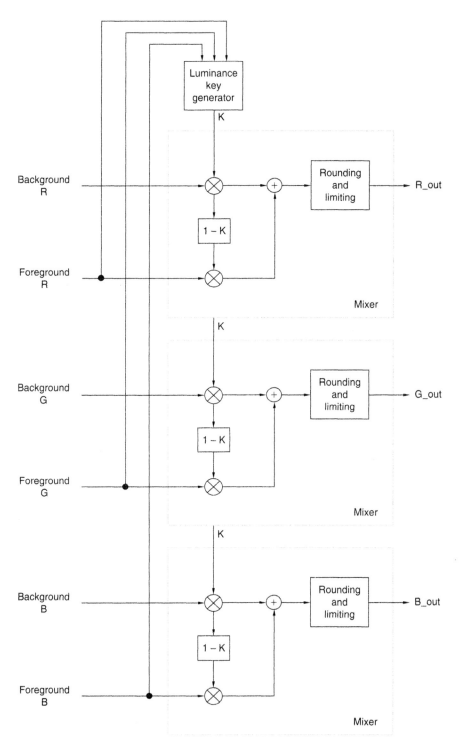

Figure 5.18: Luminance keying of two RGB video signals. RGB range is 0–255

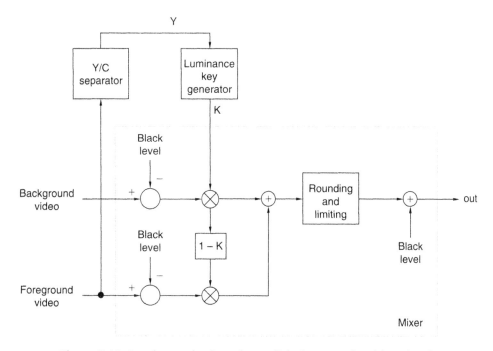

Figure 5.19: Luminance keying of two digital composite video signals

Early chroma keying circuits simply performed a hard or soft switch between the foreground and background sources. In addition to limiting the amount of fine detail maintained in the foreground image, the background was not visible through transparent or translucent foreground objects, and shadows from the foreground were not present in areas containing the background image.

Linear keyers were developed that combine the foreground and background images in a proportion determined by the key level, resulting in the foreground image being attenuated in areas containing the background image. Although allowing foreground objects to appear transparent, there is a limit on the fineness of detail maintained in the foreground. Shadows from the foreground are not present in areas containing the background image unless additional processing is done—the luminance levels of specific areas of the background image must be reduced to create the effect of shadows cast by foreground objects.

If the blue or green backing used with the foreground scene is evenly lit except for shadows cast by the foreground objects, the effect on the background will be that of shadows cast by the foreground objects. This process, referred to as shadow chroma keying, or luminance modulation, enables the background luminance levels to be adjusted in proportion to the brightness of the blue or green backing in the foreground scene. This results in more realistic keying of transparent or translucent foreground objects by preserving the spectral highlights.

Note that green backgrounds are now more commonly used due to lower chroma noise.

Chroma keyers are also limited in their ability to handle foreground colors that are close to the key color without switching to the background image. Another problem may be a bluish tint to the foreground objects as a result of blue light reflecting off the blue backing or being diffused in the camera lens. Chroma spill is difficult to remove since the spill color is not the original key color; some mixing occurs, changing the original key color slightly.

One solution to many of the chroma keying problems is to process the foreground and background images individually before combining them, as shown in Figure 5.20. Rather than choosing between the foreground and background, each is processed individually and then combined. Figure 5.21 illustrates the major processing steps for both the foreground and background images during the chroma key process. Not shown in Figure 5.20 is the circuitry to initially subtract 16 (Y) or 128 (Cb and Cr) from the foreground and background video signals and the addition of 16 (Y) or 128 (Cb and Cr) after the final output adder. Any DC

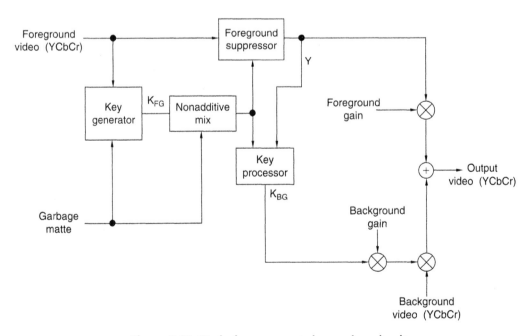

Figure 5.20: Typical component chroma key circuit

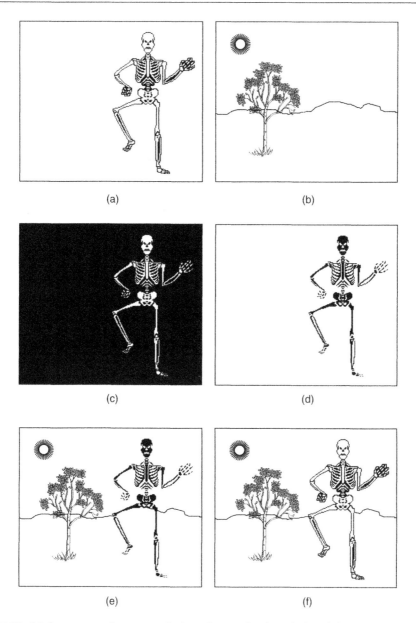

Figure 5.21: Major processing steps during chroma keying. (A) Original foreground scene
(B) Original background scene (C) Suppressed foreground scene (D) Background keying signal
(E) Background scene after multiplication by background key (F) Composite scene generated
by adding (C) and (E)

offset not removed will be amplified or attenuated by the foreground and background gain factors, shifting the black level.

The foreground key (K_{FG}) and background key (K_{BG}) signals have a range of 0 to 1. The garbage matte key signal (the term *matte* comes from the film industry) forces the mixer to output the foreground source in one of two ways.

The first method is to reduce K_{BG} in proportion to increasing K_{FG}. This provides the advantage of minimizing black edges around the inserted foreground.

The second method is to force the background to black for all nonzero values of the matte key, and insert the foreground into the background hole. This requires a cleanup function to remove noise around the black level, as this noise affects the background picture due to the straight addition process.

The garbage matte is added to the foreground key signal (K_{FG}) using a non-additive mixer (NAM). A non-additive mixer takes the brighter of the two pictures, on a sample-by-sample basis, to generate the key signal. Matting is ideal for any source that generates its own keying signal, such as character generators, and so on.

The key generator monitors the foreground Cb and Cr data, generating the foreground keying signal, K_{FG}. A desired key color is selected, as shown in Figure 5.22. The foreground Cb and Cr data are normalized (generating Cb′ and Cr′) and rotated θ degrees to generate the X and Z data, such that the positive X axis passes as close as possible to the desired key color. Typically, θ may be varied in 1° increments, and optimum chroma keying occurs when the X axis passes through the key color.

X and Z are derived from Cb and Cr using the equations:

$$X = Cb'\cos\theta + Cr'\sin\theta$$
$$Z = Cr'\cos\theta - Cb'\sin\theta$$

Since Cb′ and Cr′ are normalized to have a range of ± 1, X and Z have a range of ± 1.

The foreground keying signal (K_{FG}) is generated from X and Z and has a range of 0–1:

$$K_{FG} = X - (|Z|/(\tan(\alpha/2)))$$
$$K_{FG} = 0 \text{ if } X < (|Z|/(\tan(\alpha/2)))$$

where α is the acceptance angle, symmetrically centered about the positive X axis, as shown in Figure 5.23. Outside the acceptance angle, K_{FG} is always set to zero. Inside the acceptance

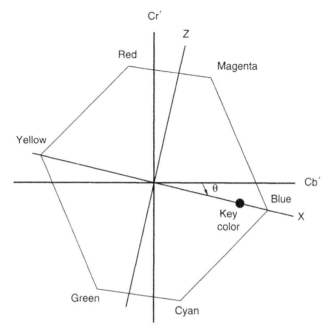

Figure 5.22: Rotating the normalized Cb and Cr (Cb′ and Cr′) axes by θ to obtain the X and Z axes, such that the X axis passes through the desired key color (blue in this example)

angle, the magnitude of K_{FG} linearly increases the closer the foreground color approaches the key color and as its saturation increases. Colors inside the acceptance angle are further processed by the foreground suppressor.

The foreground suppressor reduces foreground color information by implementing $X = X - K_{FG}$, with the key color being clamped to the black level. To avoid processing Cb and Cr when $K_{FG} = 0$, the foreground suppressor performs the operations:

$$Cb_{FG} = Cb - K_{FG} \cos \theta$$
$$Cr_{FG} = Cr - K_{FG} \sin \theta$$

where Cb_{FG} and Cr_{FG} are the foreground Cb and Cr values after key color suppression. Early implementations suppressed foreground information by multiplying Cb and Cr by a clipped version of the K_{FG} signal. This, however, generated in-band alias components due to the multiplication and clipping process and produced a hard edge at key color boundaries.

Unless additional processing is done, the Cb_{FG} and Cr_{FG} components are set to zero only if they are exactly on the X axis. Hue variations due to noise or lighting will result in areas

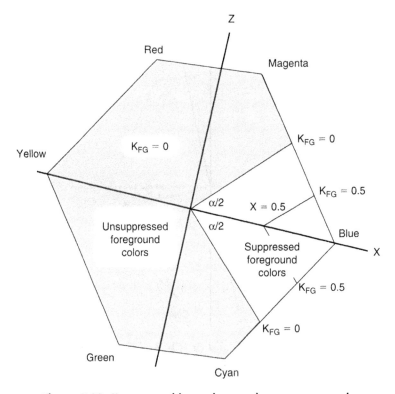

Figure 5.23: Foreground key values and acceptance angle

of the foreground not being entirely suppressed. Therefore, a suppression angle is set, symmetrically centered about the positive X axis. The suppression angle (β) is typically configurable from a minimum of zero degrees, to a maximum of about one-third the acceptance angle (α). Any CbCr components that fall within this suppression angle are set to zero. Figure 5.24 illustrates the use of the suppression angle.

Foreground luminance, after being normalized to have a range of 0–1, is suppressed by:

$$Y_{FG} = Y' - y_S K_{FG}$$
$$Y_{FG} = 0 \quad \text{if } y_S K_{FG} > Y'$$

Here, y_S is a programmable value and used to adjust Y_{FG} so that it is clipped at the black level in the key color areas.

The foreground suppressor also removes key-color fringes on wanted foreground areas caused by chroma spill, the overspill of the key color, by removing discolorations of the wanted foreground objects.

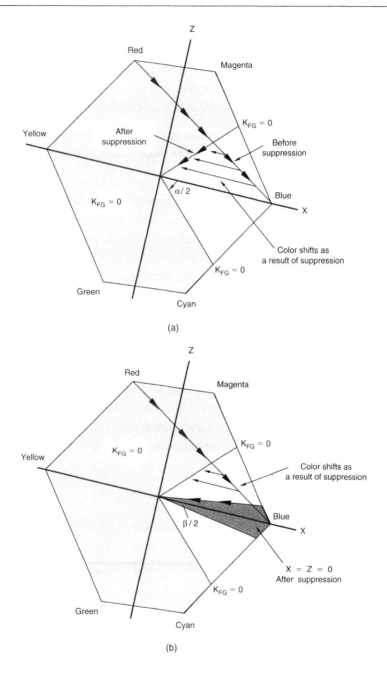

Figure 5.24: Suppression angle operation for a gradual change from a red foreground object to the blue key color. (A) Simple suppression (B) Improved suppression using a suppression angle

Ultimatte® improves on this process by measuring the difference between the blue and green colors, as the blue backing is never pure blue and there may be high levels of blue in the foreground objects. Pure blue is rarely found in nature, and most natural blues have a higher content of green than red. For this reason, the red, green, and blue levels are monitored to differentiate between the blue backing and blue in wanted foreground objects.

If the difference between blue and green is great enough, all three colors are set to zero to produce black; this is what happens in areas of the foreground containing the blue backing.

If the difference between blue and green is not large, the blue is set to the green level unless the green exceeds red. This technique allows the removal of the bluish tint caused by the blue backing while being able to reproduce natural blues in the foreground. As an example, a white foreground area normally would consist of equal levels of red, green, and blue. If the white area is affected by the key color (blue in this instance), it will have a bluish tint—the blue levels will be greater than the red or green levels. Since the green does not exceed the red, the blue level is made equal to the green, removing the bluish tint.

There is a price to pay, however. Magenta in the foreground is changed to red. A green backing can be used, but in this case, yellow in the foreground is modified. Usually, the clamping is released gradually to increase the blue content of magenta areas.

The key processor generates the initial background key signal (K'_{BG}) used to remove areas of the background image where the foreground is to be visible. K'_{BG} is adjusted to be zero in desired foreground areas and unity in background areas with no attenuation. It is generated from the foreground key signal (K_{FG}) by applying lift (k_L) and gain (k_G) adjustments followed by clipping at zero and unity values:

$$K'_{BG} = (K_{FG} - k_L)k_G$$

Figure 5.25 illustrates the operation of the background key signal generation. The transition between $K'_{BG} = 0$ and $K'_{BG} = 1$ should be made as wide as possible to minimize discontinuities in the transitions between foreground and background areas.

For foreground areas containing the same CbCr values, but different luminance (Y) values, as the key color, the key processor may also reduce the background key value as the foreground luminance level increases, allowing turning off the background in foreground areas containing a lighter key color, such as light blue. This is done by:

$$K_{BG} = K'_{BG} - y_C Y_{FG}$$
$$K_{BG} = 0 \quad \text{if} \quad y_C Y_{FG} > K_{FG}$$

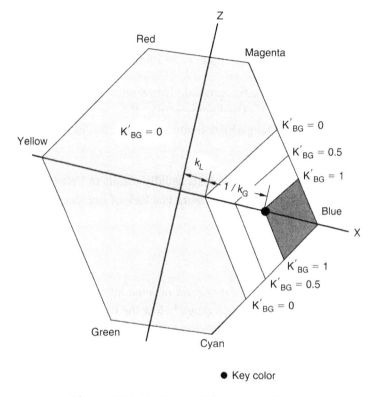

Figure 5.25: Background key generation

To handle shadows cast by foreground objects, and opaque or translucent foreground objects, the luminance level of the blue backing of the foreground image is monitored. Where the luminance of the blue backing is reduced, the luminance of the background image also is reduced. The amount of background luminance reduction must be controlled so that defects in the blue backing (such as seams or footprints) are not interpreted as foreground shadows.

Additional controls may be implemented to enable the foreground and background signals to be controlled independently. Examples are adjusting the contrast of the foreground so it matches the background or fading the foreground in various ways (such as fading to the background to make a foreground object vanish or fading to black to generate a silhouette).

In the computer environment, there may be relatively slow, smooth edges—especially edges involving smooth shading. As smooth edges are easily distorted during the chroma keying process, a wide keying process is usually used in these circumstances. During wide keying, the keying signal starts before the edge of the graphic object.

5.6.2.1 Composite Chroma Keying

In some instances, the component signals (such as YCbCr) are not directly available. For these situations, composite chroma keying may be implemented, as shown in Figure 5.26.

To detect the chroma key color, the foreground video source must be decoded to produce the Cb and Cr color difference signals. The keying signal, K_{FG}, is then used to mix between the two composite video sources. The garbage matte key signal forces the mixer to output the background source by reducing K_{FG}.

Chroma keying using composite video signals usually results in unrealistic keying, since there is inadequate color bandwidth. As a result, there is a lack of fine detail, and halos may be present on edges.

5.6.3 Superblack and Luma Keying

Video systems also may make use of *superblack* or *luma* keying. Areas of the foreground video that have a value within a specified range below the blanking level (analog video) or black level (digital video) are replaced with the background video information.

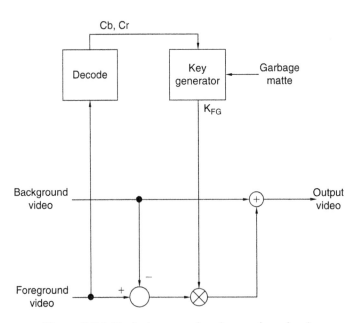

Figure 5.26: Typical composite chroma key circuit

5.7 Video Scaling

With all the various video resolutions (Table 5.3), scaling is usually needed in almost every solution.

When generating objects that will be displayed on SDTV, computer users must be concerned with such things as text size, line thickness, and so forth. For example, text readable on a 1280 × 1024 computer display may not be readable on an SDTV display due to the large amount of downscaling involved. Thin horizontal lines may either disappear completely or flicker at a 25 or 29.97 Hz rate when converted to interlaced SDTV.

Note that scaling must be performed on component video signals (such as R′G′B′ or YCbCr). Composite color video signals cannot be scaled directly due to the color subcarrier phase information present, which would be meaningless after scaling.

In general, the spacing between output samples can be defined by a Target Increment (tarinc) value:

$$\text{tarinc} = I/O$$

where I and O are the number of input (I) and output (O) samples, either horizontally or vertically.

The first and last output samples may be aligned with the first and last input samples by adjusting the equation to be:

$$\text{tarinc} = (I - 1)/(O - 1)$$

Table 5.3: Common active resolutions for consumer displays and broadcast sources

Displays		SDTV Sources		HDTV Sources
704 × 480	640 × 480	704 × 360[1]	704 × 432[1]	1280 × 720
854 × 480	800 × 600	480 × 480	480 × 576	1440 × 816[2]
704 × 576	1024 × 768	528 × 480		1440 × 1040[3]
854 × 576	1280 × 768	544 × 480	544 × 576	1280 × 1080
1280 × 720	1366 × 768	640 × 480		1440 × 1080
1280 × 768	1024 × 1024	704 × 480	704 × 576	1920 × 1080
1920 × 1080	1280 × 1024		768 × 576	

[1]16:9 letterbox on a 4:3 display. [2]2.35:1 anamorphic for a 16:9 1920 × 1080 display. [3]1.85:1 anamorphic for a 16:9 1920 × 1080 display

5.7.1 Pixel Dropping and Duplication

This is also called *nearest neighbor* scaling since only the input sample closest to the output sample is used.

The simplest form of scaling down is pixel dropping, where (m) out of every (n) samples are thrown away both horizontally and vertically. A modified version of the Bresenham line-drawing algorithm (described in most computer graphics books) is typically used to determine which samples not to discard.

Simple upscaling can be accomplished by pixel duplication, where (m) out of every (n) samples are duplicated both horizontally and vertically. Again, a modified version of the Bresenham line-drawing algorithm can be used to determine which samples to duplicate.

Scaling using pixel dropping or duplication is not recommended due to the visual artifacts and the introduction of aliasing components.

5.7.2 Linear Interpolation

An improvement in video quality of scaled images is possible using linear interpolation. When an output sample falls between two input samples (horizontally or vertically), the output sample is computed by linearly interpolating between the two input samples. However, scaling to images smaller than one-half of the original still results in deleted samples.

Figure 5.27 illustrates the vertical scaling of a 16:9 image to fit on a 4:3 display. A simple bi-linear vertical filter is commonly used, as shown in Figure 5.28a. Two source samples, L_n and L_{n+1}, are weighted and added together to form a destination sample, D_m.

$$D_0 = 0.75L_0 + 0.25L_1$$
$$D_1 = 0.5L_1 + 0.5L_2$$
$$D_2 = 0.25L_2 + 0.75L_3$$

However, as seen in Figure 5.28a, this results in uneven line spacing, which may result in visual artifacts. Figure 5.28b illustrates vertical filtering that results in the output lines being more evenly spaced:

$$D_0 = L_0$$
$$D_1 = (2/3)L_1 + (1/3)L_2$$
$$D_2 = (1/3)L_2 + (2/3)L_3$$

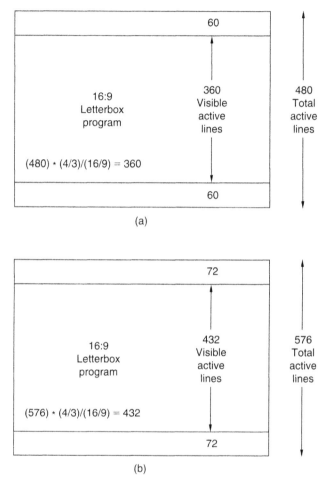

Figure 5.27: Vertical scaling of 16:9 images to fit on a 4:3 display. (A) 480-line systems (B) 576-line systems

The linear interpolator is a poor bandwidth-limiting filter. Excess high-frequency detail is removed unnecessarily and too much energy above the Nyquist limit is still present, resulting in aliasing.

5.7.3 Anti-Aliased Resampling

The most desirable approach is to ensure the frequency content scales proportionally with the image size, both horizontally and vertically.

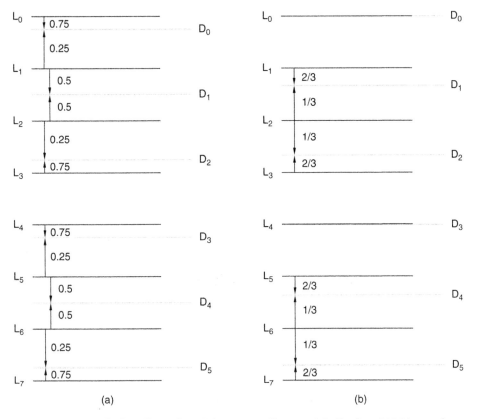

Figure 5.28: 75% vertical scaling of 16:9 images to fit on a 4:3 display. (A) Unevenly spaced results (B) Evenly spaced results

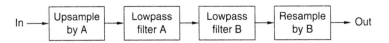

Figure 5.29: General anti-aliased resampling structure

Figure 5.29 illustrates the fundamentals of an anti-aliased resampling process. The input data is upsampled by A and lowpass filtered to remove image frequencies created by the interpolation process. Filter B bandwidth-limits the signal to remove frequencies that will alias in the resampling process B. The ratio of B/A determines the scaling factor.

Filters A and B are usually combined into a single filter. The response of the filter largely determines the quality of the interpolation. The ideal lowpass filter would have a very flat passband, a sharp cutoff at half of the lowest sampling frequency (either input or output), and very high attenuation in the stopband. However, since such a filter generates ringing on sharp edges, it is usually desirable to roll off the top of the passband. This makes for slightly softer pictures, but with less pronounced ringing.

Passband ripple and stopband attenuation of the filter provide some measure of scaling quality, but the subjective effect of ringing means a flat passband might not be as good as one might think. Lots of stopband attenuation is almost always a good thing.

There are essentially three variations of the general resampling structure. Each combines the elements of Figure 5.29 in various ways.

One approach is a variable-bandwidth anti-aliasing filter followed by a combined interpolator/resampler. In this case, the filter needs new coefficients for each scale factor—as the scale factor is changed, the quality of the image may vary. In addition, the overall response is poor if linear interpolation is used. However, the filter coefficients are time-invariant and there are no gain problems.

A second approach is a combined filter/interpolator followed by a resampler. Generally, the higher the order of interpolation, n, the better the overall response. The center of the filter transfer function is always aligned over the new output sample. With each scaling factor, the filter transfer function is stretched or compressed to remain aligned over n output samples. Thus, the filter coefficients, and the number of input samples used, change with each new output sample and scaling factor. Dynamic gain normalization is required to ensure the sum of the filter coefficients is always equal to one.

A third approach is an interpolator followed by a combined filter/resampler. The input data is interpolated up to a common multiple of the input and output rates by the insertion of zero samples. This is filtered with a low-pass finite-impulse-response (FIR) filter to interpolate samples in the zero-filled gaps, then resampled at the required locations. This type of design is usually achieved with a "polyphase" filter which switches its coefficients as the relative position of input and output samples change.

5.7.4 Display Scaling Examples

Figures 5.30 through 5.38 illustrate various scaling examples for displaying 16:9 and 4:3 pictures on 4:3 and 16:9 displays, respectively.

1920 Samples

1080
Scan
lines

Figure 5.30: 16:9 source example

720 Samples

480
Scan
lines

Figure 5.31: Scaling 16:9 content for a 4:3 display: "Normal" or pan-and-scan mode. Results in some of the 16:9 content being ignored (indicated by gray regions)

How content is displayed is a combination of user preferences and content aspect ratio. For example, when displaying 16:9 content on a 4:3 display, many users prefer to have the entire display filled with the cropped picture (Figure 5.31) rather than seeing black or gray bars with the letterbox solution (Figure 5.32). In addition, some displays incorrectly assume any progressive video signal on their YPbPr inputs is from an "anamorphic" source. As a

720 Samples

360
Scan
lines

480
Scan
lines

Figure 5.32: Scaling 16:9 content for a 4:3 display: "Letterbox" mode. Entire 16:9 program visible, with black bars at top and bottom of display

720 Samples

480
Scan
lines

Figure 5.33: Scaling 16:9 content for a 4:3 display: "Squeezed" mode. Entire 16:9 program horizontally squeezed to fit 4:3 display, resulting in a distorted picture

result, they horizontally upscale progressive 16:9 programs by 25% when no scaling should be applied. Therefore, for set-top boxes it is useful to include a "16:9 (Compressed)" mode, which horizontally downscales the progressive 16:9 program by 25% to precompensate for the horizontal upscaling being done by the 16:9 display.

720 Samples

480
Scan
lines

Figure 5.34: 4:3 Source example

1920 Samples

1080
Scan
lines

Figure 5.35: Scaling 4:3 content for a 16:9 display: "Normal" mode. Left and right portions of 16:9 display not used, so made black or gray

5.8 Scan Rate Conversion

In many cases, some form of scan rate conversion (also called *temporal rate conversion, frame rate conversion,* or *field rate conversion*) is needed. Multi-standard analog VCRs and scan converters use scan rate conversion to convert between various video standards.

1920 Samples

1080 Scan lines

Figure 5.36: Scaling 4:3 content for a 16:9 display: "Wide" mode. Entire picture linearly scaled horizontally to fill 16:9 display, resulting in distorted picture unless used with anamorphic content

1920 Samples

1080 Scan lines

Figure 5.37: Scaling 4:3 content for a 16:9 display: "Zoom" mode. Top and bottom portion of 4:3 picture deleted, then scaled to fill 16:9 display

1920 Samples

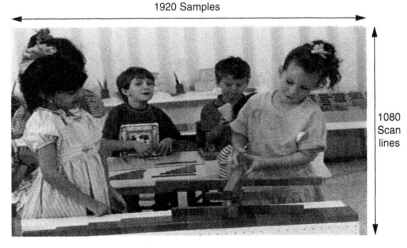

1080 Scan lines

Figure 5.38: Scaling 4:3 content for a 16:9 display: "Panorama" mode. Left and right 25% edges of picture are nonlinearly scaled horizontally to fill 16:9 display, distorted picture on left and right sides

Computers usually operate the display at about 75 Hz noninterlaced, yet need to display 50 and 60 Hz interlaced video. With digital television, multiple frame rates can be supported.

Note that processing must be performed on component video signals (such as R′G′B′ or YCbCr). Composite color video signals cannot be processed directly due to the color subcarrier phase information present, which would be meaningless after processing.

5.8.1 Frame or Field Dropping and Duplicating

Simple scan-rate conversion may be done by dropping or duplicating one out of every N fields. For example, the conversion of 60 Hz to 50 Hz interlaced operation may drop one out of every six fields, as shown in Figure 5.39, using a single field store.

The disadvantage of this technique is that the viewer may see jerky motion, or *motion judder*. In addition, some video decompression products use top-field only to convert from 60 Hz to 50 Hz, degrading the vertical resolution.

The worst artifacts are present when a noninteger scan rate conversion is done—for example, when some frames are displayed three times, while others are displayed twice. In this instance, the viewer will observe double or blurred objects. As the human brain tracks an object in successive frames, it expects to see a regular sequence of positions, and has trouble

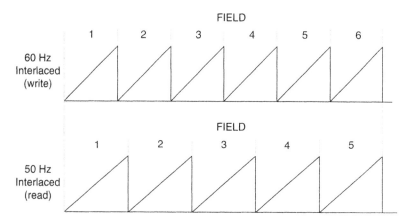

Figure 5.39: 60 Hz to 50 Hz conversion using a single field store by dropping one out of every six fields

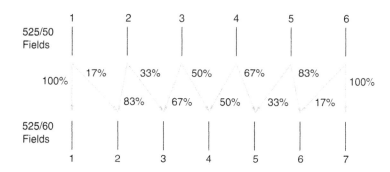

Figure 5.40: 50 Hz to 60 Hz conversion using temporal interpolation with no motion compensation

reconciling the apparent stop-start motion of objects. As a result, it incorrectly concludes that there are two objects moving in parallel.

5.8.2 Temporal Interpolation

This technique generates new frames from the original frames as needed to generate the desired frame rate. Information from both past and future input frames should be used to optimally handle objects appearing and disappearing.

Conversion of 50 Hz to 60 Hz operation using temporal interpolation is illustrated in Figure 5.40. For every five fields of 50 Hz video, there are six fields of 60 Hz video.

After both sources are aligned, two adjacent 50 Hz fields are mixed together to generate a new 60 Hz field. This technique is used in some inexpensive standards converters to convert between 50 Hz and 60 Hz standards. Note that no motion analysis is done. Therefore, if the camera operating at 50 Hz pans horizontally past a narrow vertical object, you see one object once every six 60 Hz fields, and for the five fields in between, you see two objects, one fading in while the other fades out.

5.8.2.1 50 Hz to 60 Hz Examples

Figure 5.41 illustrates a scan rate converter that implements vertical, followed by temporal, interpolation. Figure 5.42 illustrates the spectral representation of the design in Figure 5.41.

Many designs now combine the vertical and temporal interpolation into a single design, as shown in Figure 5.43, with the corresponding spectral representation shown in Figure 5.44. This example uses vertical, followed by temporal, interpolation. If temporal, followed by vertical, interpolation were implemented, the field stores would be half the size. However, the number of line stores would increase from four to eight.

In either case, the first interpolation process must produce an intermediate, higher-resolution progressive format to avoid interlace components that would interfere with the second interpolation process. It is insufficient to interpolate, either vertically or temporally, using a mixture of lines from both fields, due to the interpolation process not being able to compensate for the temporal offset of interlaced lines.

5.8.2.2 Motion Compensation

Higher-quality scan rate converters using temporal interpolation incorporate motion compensation to minimize motion artifacts. This results in extremely smooth and natural motion, and images appear sharper and do not suffer from *motion judder*.

Motion estimation for scan rate conversion differs from that used by MPEG. In MPEG, the goal is to minimize the displaced frame difference (error) by searching for a high correlation between areas in subsequent frames. The resulting motion vectors do not necessarily correspond to true motion vectors.

For scan rate conversion, it is important to determine true motion information to perform correct temporal interpolation. The interpolation should be tolerant of incorrect motion vectors to avoid introducing artifacts as unpleasant as those the technique is attempting to remove. Motion vectors could be incorrect for several reasons, such as insufficient time to track the motion, out-of-range motion vectors, and estimation difficulties due to aliasing.

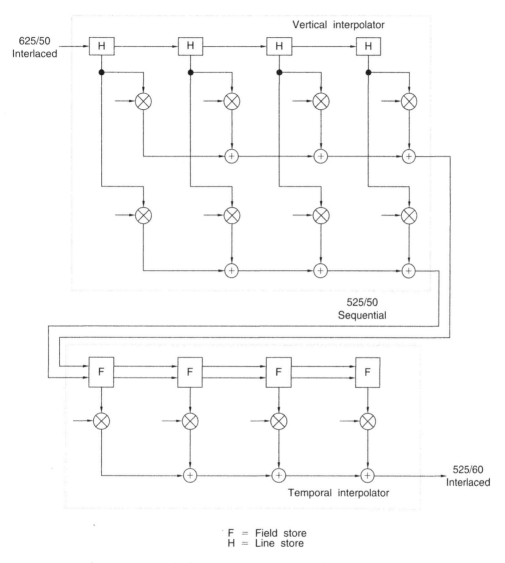

F = Field store
H = Line store

**Figure 5.41: Typical 50 Hz to 60 Hz conversion using vertical,
followed by temporal, interpolation**

5.8.2.3 100 Hz Interlaced Television Example

A standard 50 Hz interlaced television shows 50 fields per second. The images flicker, especially when you look at large areas of highly saturated color. A much improved picture can be achieved using a 100 Hz interlaced frame rate (also called *double scan*).

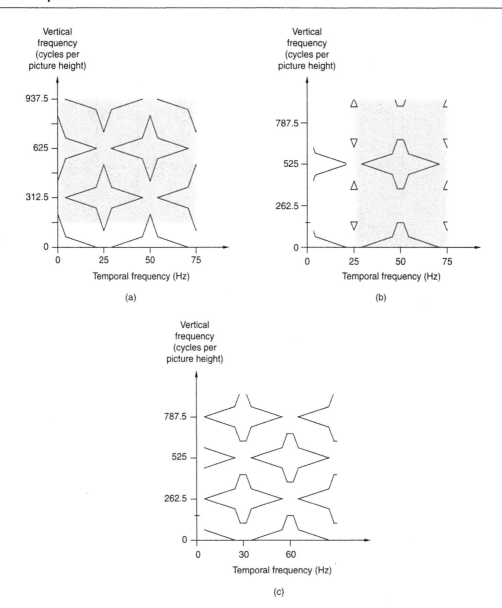

Figure 5.42: Spectral representation of vertical, followed by temporal, interpolation (A) Vertical lowpass filtering (B) Resampling to intermediate sequential format and temporal lowpass filtering (C) Resampling to final standard

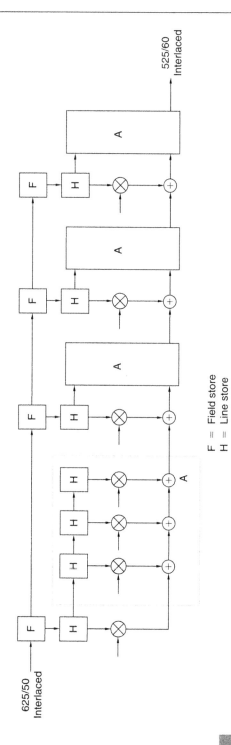

Figure 5.43: Typical 50 Hz to 60 Hz conversion using combined vertical and temporal interpolation

F = Field store
H = Line store

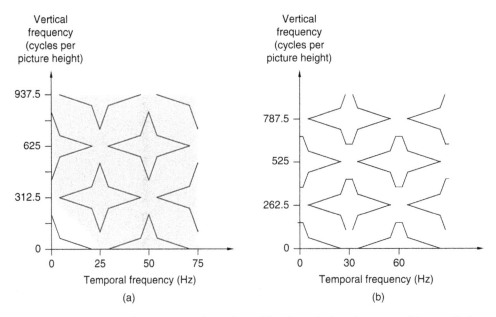

Figure 5.44: Spectral representation of combined vertical and temporal interpolation (A) Two-dimensional lowpass filtering (B) Resampling to final standard

Early 100 Hz televisions simply repeated fields ($F_1F_1F_2F_2F_3F_3F_4F_4...$), as shown in Figure 5.45(A). However, they still had line flicker, where horizontal lines constantly jumped between the odd and even lines. This disturbance occurred once every twenty-fifth of a second.

The field sequence $F_1F_2F_1F_2F_3F_4F_3F_4...$ can be used, which solves the line flicker problem. Unfortunately, this gives rise to the problem of judder in moving images. This can be compensated for by using the $F_1F_2F_1F_2F_3F_4F_3F_4...$ sequence for static images, and the $F_1F_1F_2F_2F_3F_3F_4F_4...$ sequence for moving images.

An ideal picture is still not obtained when viewing programs created for film. They are subject to judder, owing to the fact that each film frame is transmitted twice. Instead of the field sequence $F_1F_1F_2F_2F_3F_3F_4F_4...$, the situation calls for the sequence $F_1F_1F_2F_2F_3F_3F_4F_4...$ (Figure 5.45(B)), where $F_{n'}$ is a motion-compensated generated image between F_n and F_{n+1}.

5.8.3 2:2 Pulldown

This technique is used with some film-based compressed content for 50 Hz regions. Film is usually recorded at 24 frames per second.

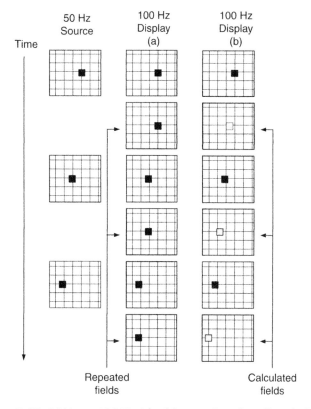

Figure 5.45: 50 Hz to 100 Hz (double scan interlaced) techniques

During compression, the telecine machine is sped up from 24 to 25 frames per second, making the content 25 frames per second progressive. During decompression, each film frame is simply mapped into two video fields (resulting in 576i25 or 1080i25 video) or two video frames (resulting in 576p50, 720p50, or 1080p50 video).

This technique provides higher video quality and avoids motion judder artifacts. However, it shortens the duration of the program by about 4%, cutting the duration of a 2-hour movie by ~5 minutes. Some audio decoders cannot handle the 4% faster audio data via S/PDIF (IEC 60958).

To compensate the audio changing pitch due to the telecine speedup, it may be resampled during decoding to restore the original pitch (costly to do in a low-cost consumer product) or resampling may be done during the program authoring.

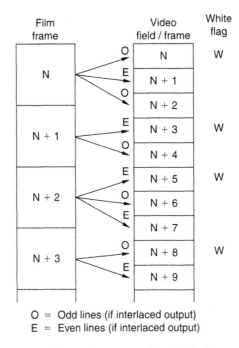

O = Odd lines (if interlaced output)
E = Even lines (if interlaced output)

Figure 5.46: 3:2 pulldown for converting 24 Hz film to 60 Hz video

5.8.4 3:2 Pulldown

When converting 24 frames per second content to 60 Hz, 3:2 pulldown is commonly used, as shown in Figure 5.46. During compression, the film speed is slowed down by 0.1% to 23.976 (24/1.001) frames per second since 59.94 Hz is used for NTSC timing compatibility. During decompression, 2 film frames generate 5 video fields (resulting in 480i30 or 1080i30 video) or 5 video frames (resulting in 480p60, 720p60, or 1080p60 video).

In scenes of high-speed motion of objects, the specific film frame used for a particular video field or frame may be manually adjusted to minimize motion artifacts.

3:2 pulldown may also be used during video decompression to simply to increase the frame rate from 23.976 (24/1.001) to 59.94 (60/1.001) frames per second, avoiding the deinterlacing issue.

Varispeed may be used to cover up problems such as defects, splicing, censorship cuts, or to change the running time of a program. Rather than repeating film frames and causing a *stutter*, the 3:2 relationship between the film and video is disrupted long enough to ensure a smooth temporal rate.

Analog laserdiscs used a white flag signal to indicate the start of another sequence of related fields for optimum still-frame performance. During still-frame mode, the white flag signal tells the system to back up two fields (to use two fields that have no motion between them) to re-display the current frame.

5.8.5 3:3 Pulldown

This technique is used in some displays that support 72 Hz frame rate. The 24 frames per second film-based content is converted to 72 Hz progressive by simply duplicating each film frame three times.

5.8.6 24:1 Pulldown

This technique, also called *12:1 pulldown*, can be used to convert 24 frames/second content to 50 fields per second.

Two video fields are generated from every film frame, except every 12th film frame generates 3 video fields. Although the audio pitch is correct, motion judder is present every one-half second when smooth motion is present.

5.9 Noninterlaced-to-Interlaced Conversion

In some applications, it is necessary to display a noninterlaced video signal on an interlaced display. Thus, some form of noninterlaced-to-interlaced conversion may be required.

Noninterlaced-to-interlaced conversion must be performed on component video signals (such as R′G′B′ or YCbCr). Composite color video signals (such as NTSC or PAL) cannot be processed directly due to the presence of color subcarrier phase information, which would be meaningless after processing. These signals must be decoded into component color signals, such as R′G′B′ or YCbCr, prior to conversion.

There are essentially two techniques: scan line decimation and vertical filtering.

5.9.1 Scan Line Decimation

The easiest approach is to throw away every other active scan line in each noninterlaced frame, as shown in Figure 5.47. Although the cost is minimal, there are problems with this approach, especially with the top and bottom of objects.

If there is a sharp vertical transition of color or intensity, it will flicker at one-half the frame rate. The reason is that it is only displayed every other field as a result of the decimation.

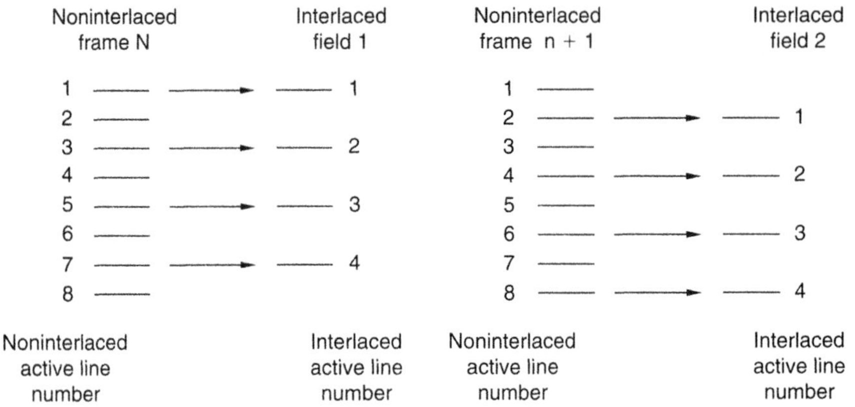

Noninterlaced frame N	Interlaced field 1	Noninterlaced frame n + 1	Interlaced field 2

```
Noninterlaced      Interlaced      Noninterlaced      Interlaced
   frame N           field 1          frame n + 1        field 2

   1 ─── ────────── ─── 1       1 ───
   2 ───                        2 ─── ────────── ─── 1
   3 ─── ────────── ─── 2       3 ───
   4 ───                        4 ─── ────────── ─── 2
   5 ─── ────────── ─── 3       5 ───
   6 ───                        6 ─── ────────── ─── 3
   7 ─── ────────── ─── 4       7 ───
   8 ───                        8 ─── ────────── ─── 4

Noninterlaced      Interlaced      Noninterlaced      Interlaced
 active line       active line      active line       active line
   number            number           number            number
```

Figure 5.47: Noninterlaced-to-interlaced conversion using scan line decimation

For example, a horizontal line that is one noninterlaced scan line wide will flicker on and off. Horizontal lines that are two noninterlaced scan lines wide will oscillate up and down.

Simple decimation may also add aliasing artifacts. While not necessarily visible, they will affect any future processing of the picture.

5.9.2 Vertical Filtering

A better solution is to use two or more lines of noninterlaced data to generate one line of interlaced data. Fast vertical transitions are smoothed out over several interlaced lines.

For a 3-line filter, such as shown in Figure 5.48, typical coefficients are [0.25, 0.5, 0.25]. Using more than three lines usually results in excessive blurring, making small text difficult to read.

An alternate implementation uses IIR rather than FIR filtering. In addition to averaging, this technique produces a reduction in brightness around objects, further reducing flicker.

Note that care must be taken at the beginning and end of each frame in the event that fewer scan lines are available for filtering.

5.10 Interlaced-to-Noninterlaced Conversion

In some applications, it is necessary to display an interlaced video signal on a noninterlaced display. Thus, some form of *deinterlacing* or *progressive scan conversion* may be required.

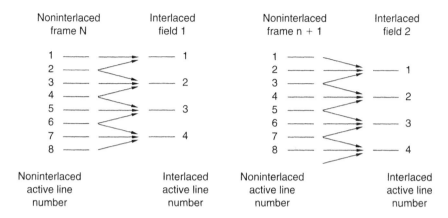

Figure 5.48: Noninterlaced-to-interlaced conversion using 3-line vertical filtering

Note that deinterlacing must be performed on component video signals (such as R′G′B′ or YCbCr). Composite color video signals (such as NTSC or PAL) cannot be deinterlaced directly due to the presence of color subcarrier phase information, which would be meaningless after processing. These signals must be decoded into component color signals, such as R′G′B′ or YCbCr, prior to deinterlacing.

There are two fundamental deinterlacing algorithms: video mode and film mode. Video mode deinterlacing can be further broken down into inter-field and intra-field processing. The goal of a good deinterlacer is to correctly choose the best algorithm needed at a particular moment.

In systems where the vertical resolution of the source and display do not match (due to, for example, displaying SDTV content on an HDTV), the deinterlacing and vertical scaling can be merged into a single process.

5.10.1 Video Mode: Intra-Field Processing

This is the simplest method for generating additional scan lines using only information in the original field. The computer industry has coined this technique as *bob*.

Although there are two common techniques for implementing intra-field processing, scan line duplication and scan line interpolation, the resulting vertical resolution is always limited by the content of the original field.

5.10.1.1 Scan Line Duplication

Scan line duplication (Figure 5.49) simply duplicates the previous active scan line. Although the number of active scan lines is doubled, there is no increase in the vertical resolution.

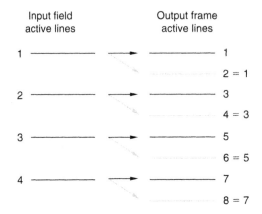

Figure 5.49: Deinterlacing using scan line duplication. New scan lines are generated by duplicating the active scan line above it

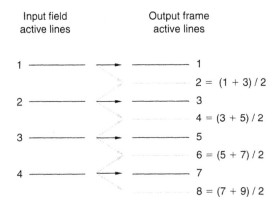

Figure 5.50: Deinterlacing using scan line interpolation. New scan lines are generated by averaging the previous and next active scan lines.

5.10.1.2 Scan Line Interpolation

Scan line interpolation generates interpolated scan lines between the original active scan lines. Although the number of active scan lines is doubled, the vertical resolution is not.

The simplest implementation, shown in Figure 5.50, uses linear interpolation to generate a new scan line between two input scan lines:

$$\text{out}_n = (\text{in}_{n-1} + \text{in}_{n+1})/2$$

Better results, at additional cost, may be achieved by using a FIR filter:

$$\begin{aligned}
\text{out}_n = {} & (160{*}(\text{in}_{n-1} + \text{in}_{n+1}) \\
& - 48{*}(\text{in}_{n-3} + \text{in}_{n+3}) \\
& + 24{*}(\text{in}_{n-5} + \text{in}_{n+5}) \\
& - 12{*}(\text{in}_{n-7} + \text{in}_{n+7}) \\
& + 6{*}(\text{in}_{n-9} + \text{in}_{n+9}) \\
& - 2{*}(\text{in}_{n-11} + \text{in}_{n+11})
\end{aligned}$$

5.10.1.3 Fractional Ratio Interpolation

In many cases, there is a periodic, but nonintegral, relationship between the number of input scan lines and the number of output scan lines. In this case, fractional ratio interpolation may be necessary, similar to the polyphase filtering used for scaling only performed in the vertical direction. This technique combines deinterlacing and vertical scaling into a single process.

5.10.1.4 Variable Interpolation

In a few cases, there is no periodicity in the relationship between the number of input and output scan lines. Therefore, in theory, an infinite number of filter phases and coefficients are required. Since this is not feasible, the solution is to use a large, but finite, number of filter phases. The number of filter phases determines the interpolation accuracy. This technique also combines deinterlacing and vertical scaling into a single process.

5.10.2 Video Mode: Inter-Field Processing

In this method, video information from more than one field is used to generate a single progressive frame. This method can provide higher vertical resolution since it uses content from more than a single field.

5.10.2.1 Field Merging

This technique merges two consecutive fields together to produce a frame of video (Figure 5.51). At each field time, the active scan lines of that field are merged with the active scan lines of the previous field. The result is that for each input field time, a pair of fields combine to generate a frame (see Figure 5.52). Although simple to implement, the vertical resolution is doubled only in regions of no movement.

Field 1   Field 2   Deinterlaced
active   active   frame
line   line   active line

```
1 ─────────────────    ───▶   ─────────── 1
        1 ──────────    ───▶   ..........  2
2 ─────────────────    ───▶   ─────────── 3
        2 ──────────    ───▶   ..........  4
3 ─────────────────    ───▶   ─────────── 5
        3 ──────────    ───▶   ..........  6
4 ─────────────────    ───▶   ─────────── 7
        4 ──────────    ───▶   ..........  8
```

Figure 5.51: Deinterlacing using field merging. Shaded scan lines are generated by using the input scan line from the next or previous field.

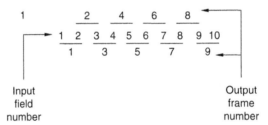

Input
field
number

Output
frame
number

Figure 5.52: Producing deinterlaced frames at field rates

Moving objects will have artifacts, also called *combing*, due to the time difference between two fields—a moving object is located in a different position from one field to the next. When the two fields are merged, moving objects will have a double image (see Figure 5.53).

It is common to soften the image slightly in the vertical direction to attempt to reduce the visibility of combing. When implemented, it causes a loss of vertical resolution and jitter on movement and pans.

The computer industry refers to this technique as *weave*, but weave also includes the inverse telecine process to remove any 3:2 pull-down present in the source. Theoretically, this eliminates the double image artifacts since two identical fields are now being merged.

5.10.2.2 Motion Adaptive Deinterlacing

A good deinterlacing solution is to use field merging for still areas of the picture and scan line interpolation for areas of movement. To accomplish this, motion, on a sample-by-sample

Object position
in field one

Object position
in field two

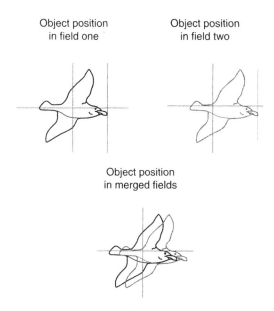

Object position
in merged fields

Figure 5.53: Movement artifacts when field merging is used

basis, must be detected over the entire picture in real time, requiring processing several fields of video.

As two fields are combined, full vertical resolution is maintained in still areas of the picture, where the eye is most sensitive to detail. The sample differences may have any value, from 0 (no movement and noise-free) to maximum (for example, a change from full intensity to black). A choice must be made when to use a sample from the previous field (which is in the wrong location due to motion) or to interpolate a new sample from adjacent scan lines in the current field. Sudden switching between methods is visible, so crossfading (also called *soft switching*) is used. At some magnitude of sample difference, the loss of resolution due to a double image is equal to the loss of resolution due to interpolation. That amount of motion should result in the crossfader being at the 50% point. Less motion will result in a fade towards field merging and more motion in a fade towards the interpolated values.

Rather than "per pixel" motion adaptive deinterlacing, which makes decisions for every sample, some low-cost solutions use "per field" motion adaptive deinterlacing. In this case, the algorithm is selected each field, based on the amount of motion between the fields. "Per pixel" motion adaptive deinterlacing, although difficult to implement, looks quite good when properly done. "Per field" motion adaptive deinterlacing rarely looks much better than vertical interpolation.

5.10.3 Motion-Compensated Deinterlacing

Motion-compensated (or motion vector steered) deinterlacing is several orders of magnitude more complex than motion adaptive deinterlacing, and is commonly found in pro-video format converters.

Motion-compensated processing requires calculating motion vectors between fields for each sample, and interpolating along each sample's motion trajectory. Motion vectors must also be found that pass through each of any missing samples. Areas of the picture may be covered or uncovered as you move between frames. The motion vectors must also have sub-pixel accuracy, and be determined in two temporal directions between frames.

The motion vector errors used by MPEG are self-correcting since the residual difference between the predicted macroblocks is encoded. As motion-compensated deinterlacing is a single-ended system, motion vector errors will produce artifacts, so different search and verification algorithms must be used.

5.10.4 Film Mode (Using Inverse Telecine)

For sources that have 3:2 pulldown (i.e., 60 fields/second video converted from 24 frames/second film), higher deinterlacing performance may be obtained by removing duplicate fields prior to processing.

The inverse telecine process detects the 3:2 field sequence and the redundant third fields are removed. The remaining field pairs are merged (since there is no motion between them) to form progressive frames at 24 frames/second. These are then repeated in a 3:2 sequence to get to 60 frames/second.

Although this may seem to be the ideal solution, some content uses both 60 fields/second video and 24 frames/second video (film-based) within a program. In addition, some content may occasionally have both video types present simultaneously. In other cases, the 3:2 pulldown timing (cadence) doesn't stay regular, or the source was never originally from film. Thus, the deinterlacer has to detect each video type and process it differently (video mode vs. film mode). Display artifacts are common due to the delay between the video type changing and the deinterlacer detecting the change.

5.10.5 Frequency Response Considerations

Various two-times vertical upsampling techniques for deinterlacing may be implemented by stuffing zero values between two valid lines and filtering, as shown in Figure 5.54.

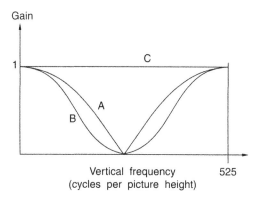

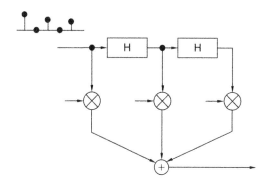

Figure 5.54: Frequency response of various deinterlacing filters (A) Line duplication (B) Line interpolation (C) Field merging

Line A shows the frequency response for line duplication, in which the lowpass filter coefficients for the filter shown are 1, 1, and 0.

Line interpolation, using lowpass filter coefficients of 0.5, 1.0, and 0.5, results in the frequency response curve of Line B. Note that line duplication results in a better high-frequency response. Vertical filters with a better frequency response than the one for line duplication are possible, at the cost of more line stores and processing.

The best vertical frequency response is obtained when field merging is implemented. The spatial position of the lines is already correct and no vertical processing is required, resulting in a flat curve (Line C). Again, this applies only for stationary areas of the image.

5.11 DCT-Based Compression

The transform process of many video compression standards is based on the Discrete Cosine Transform, or DCT. The easiest way to envision it is as a filter bank with all the filters computed in parallel.

During encoding, the DCT is usually followed by several other operations, such as quantization, zig-zag scanning, run-length encoding, and variable-length encoding. During decoding, this process flow is reversed.

Many times, the terms *macroblocks* and *blocks* are used when discussing video compression. Figure 5.55 illustrates the relationship between these two terms, and shows why transform processing is usually done on 8 × 8 samples. MPEG-4.10 (H.264) also supports 8 × 4, 4 × 8, and 4 × 4 blocks.

5.11.1 DCT

The 8 × 8 DCT processes an 8 × 8 block of samples to generate an 8 × 8 block of DCT coefficients. The input may be samples from an actual frame of video or motion-compensated difference (error) values, depending on the encoder mode of operation. Each DCT coefficient indicates the amount of a particular horizontal or vertical frequency within the block.

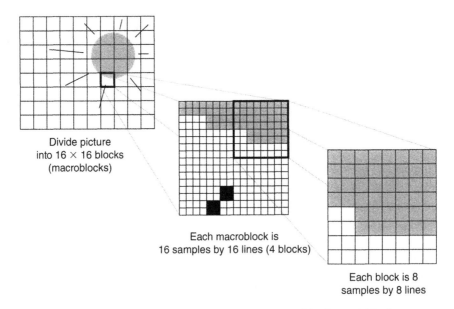

Divide picture
into 16 × 16 blocks
(macroblocks)

Each macroblock is
16 samples by 16 lines (4 blocks)

Each block is 8
samples by 8 lines

Figure 5.55: The relationship between macroblocks and blocks

DCT coefficient (0,0) is the DC coefficient, or average sample value. Since natural images tend to vary only slightly from sample to sample, low frequency coefficients are typically larger values and high frequency coefficients are typically smaller values.

A reconstructed 8 × 8 block of samples is generated using an 8 × 8 inverse DCT (IDCT). Although exact reconstruction is theoretically achievable, it is not practical due to finite-precision arithmetic, quantization and differing IDCT implementations. As a result, there are mismatches between different IDCT implementations.

Mismatch control attempts to reduce the drift between encoder and decoder IDCT results by eliminating bit patterns having the greatest contribution towards mismatches.

MPEG-1 mismatch control is known as *oddification* since it forces all quantized DCT coefficients to negative values. MPEG-2 and MPEG-4.2 use an improved method called *LSB toggling*, which affects only the LSB of the 63rd DCT coefficient after inverse quantization.

H.264 (also known as *MPEG-4.10*) neatly sidesteps the issue by using an "exact-match inverse transform." Every decoder will produce exactly the same pictures, all else being equal.

5.11.2 Quantization

The 8 × 8 block of DCT coefficients is quantized, which reduces the overall precision of the integer coefficients and tends to eliminate high frequency coefficients, while maintaining

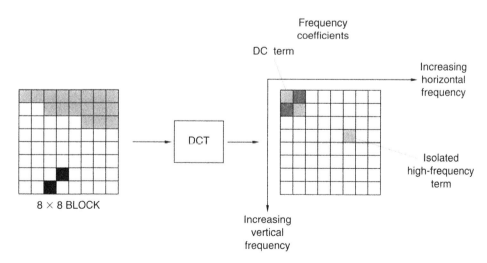

Figure 5.56: The DCT processes the 8 × 8 block of samples or error terms to generate an 8 × 8 block of DCT coefficients

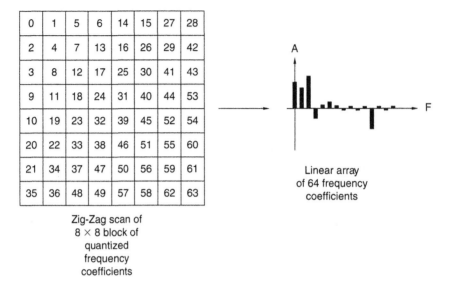

Zig-Zag scan of
8 × 8 block of
quantized
frequency
coefficients

Linear array
of 64 frequency
coefficients

Figure 5.57: The 8 × 8 block of quantized DCT coefficients are zig-zag scanned to arrange in order of increasing frequency. This scanning order is used for H.261, H.263, MPEG-1, MPEG-2, MPEG-4.2, ITU-R BT.1618, ITU-R BT.1620, SMPTE 314M, and SMPTE 370M

perceptual quality. Higher frequencies are usually quantized more coarsely (fewer values allowed) than lower frequencies, due to visual perception of quantization error. The quantizer is also used for constant bit-rate applications where it is varied to control the output bit-rate.

5.11.3 Zig-Zag Scanning

The quantized DCT coefficients are re-arranged into a linear stream by scanning them in a zig-zag order. This rearrangement places the DC coefficient first, followed by frequency coefficients arranged in order of increasing frequency, as shown in Figures 5.57, 5.58, and 5.59. This produces long runs of zero coefficients.

5.11.4 Run Length Coding

The linear stream of quantized frequency coefficients is converted into a series of [run, amplitude] pairs. [run] indicates the number of zero coefficients, and [amplitude] the nonzero coefficient that ended the run.

0	4	6	20	22	36	38	52
1	5	7	21	23	37	39	53
2	8	19	24	34	40	50	54
3	9	18	25	35	41	51	55
10	17	26	30	42	46	56	60
11	16	27	31	43	47	57	61
12	15	28	32	44	48	58	62
13	14	29	33	45	49	59	63

Zig-Zag scan of
8 × 8 block of
quantized
frequency
coefficients

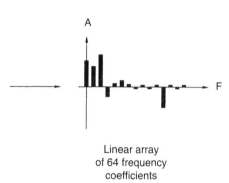

Linear array
of 64 frequency
coefficients

Figure 5.58: H.263, MPEG-2, and MPEG-4.2 "alternate-vertical" scanning order

0	1	2	3	10	11	12	13
4	5	8	9	17	16	15	14
6	7	19	18	26	27	28	29
20	21	24	25	30	31	32	33
22	23	34	35	42	43	44	45
36	37	40	41	46	47	48	49
38	39	50	51	56	57	58	59
52	53	54	55	60	61	62	63

Zig-Zag scan of
8 × 8 block of
quantized
frequency
coefficients

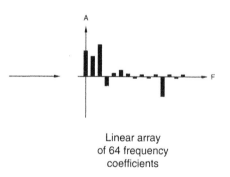

Linear array
of 64 frequency
coefficients

Figure 5.59: H.263 and MPEG-4.2 "alternate-horizontal" scanning order

5.11.5 Variable-Length Coding

The [run, amplitude] pairs are coded using a variable-length code, resulting in additional lossless compression. This produces shorter codes for common pairs and longer codes for less common pairs.

This coding method produces a more compact representation of the DCT coefficients, as a large number of DCT coefficients are usually quantized to zero and the re-ordering results (ideally) in the grouping of long runs of consecutive zero values.

5.12 Fixed Pixel Display Considerations

The unique designs and color reproduction gamuts of fixed-pixel displays have resulted in new video processing technologies being developed. The result is brighter, sharper, more colorful images regardless of the video source.

5.12.1 Expanded Color Reproduction

Broadcast stations are usually tuned to meet the limited color reproduction characteristics of CRT-based televisions. To fit the color reproduction capabilities of PDP and LCD, manufacturers have introduced various color expansion technologies. These include using independent hue and saturation controls for each primary and complementary color, plus the flesh color.

5.12.2 Detail Correction

In CRT-based televisions, enhancing the image is commonly done by altering the electron beam diameter. With fixed-pixel displays, adding overshoot and undershoot to the video signals causes distortion. An acceptable implementation is to gradually change the brightness of the images before and after regions needing contour enhancement.

5.12.3 Nonuniform Quantization

Rather than simply increasing the number of quantization levels, the quantization steps can be changed in accordance with the intensity of the image. This is possible since people better detect small changes in brightness for dark images than for bright images. In addition, the brighter the image, the less sensitive people are to changes in brightness. This means that more quantization steps can be used for dark images than for bright ones. This technique can also be used to increase the quantization steps for shades that appear frequently.

5.12.4 Scaling and Deinterlacing

Fixed-pixel displays, such as LCD and plasma, usually upscale then downscale during deinterlacing to minimize moiré noise due to folded distortion. For example, a 1080i source is deinterlaced to 2160p, scaled to 1536p, then finally scaled to 768p (to drive a 1024 × 768 display). Alternately, some solutions deinterlace and upscale to 1500p, then scale to the display's native resolution.

5.13 Application Example

Figures 5.60 and 5.61 illustrate the typical video processing done after video decompression and deinterlacing.

In addition to the primary video source, additional video sources typically include an on-screen-display (OSD), content navigation graphics, closed captioning or subtitles, and a second video for picture-in-picture (PIP).

The *OSD plane* displays configuration menus for the box, such as video output format and resolution, audio output format, etc. OSD design is unique to each product, so the OSD plane usually supports a wide variety of RGB/YCbCr formats and resolutions. Lookup tables gamma-correct linear RGB data, convert 2-, 4-, or 8-bit indexed color to 32-bit YCbCrA data, or translate 0–255 graphics levels to the 16–235 video levels.

The *content navigation plane* displays graphics generated by Blu-ray BD-J, HD DVD HDi, electronic program guides, etc. It should support the same formats and capabilities as the OSD plane.

The *subtitle plane* is a useful region for rendering closed captioning, DVB subtitles, DVD subpictures., etc. Lookup tables convert 2-, 4-, or 8-indexed color to 32-bit YCbCrA data.

The *secondary video plane* is usually used to support a second video source for picture-in-picture (PIP) or graphics (such as JPEG images). For graphics data, lookup tables can gamma-correct linear RGB data, convert 2-, 4-, or 8-indexed color to 32-bit YCbCrA data, or translate 0–255 graphics levels to the 16–235 video levels.

Being able to scale each source independently offers maximum flexibility. In addition to being able to output any resolution regard-less of the source resolutions, special effects can also be accommodated.

Chromaticity correction ensures colors are accurate independent of the sources and display (SDTV vs. HDTV).

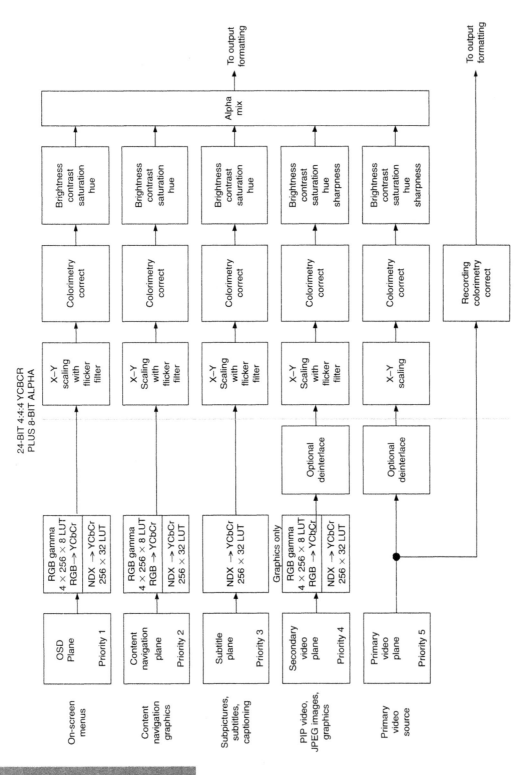

Figure 5.60: Video composition simplified block diagram

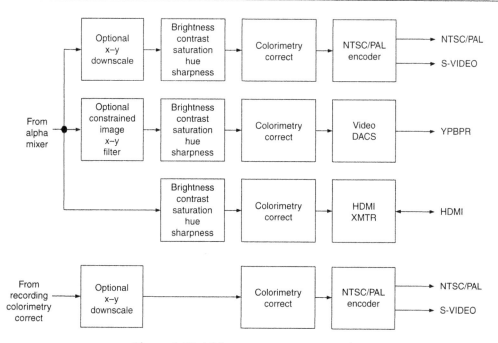

Figure 5.61: Video output port processing

Independent brightness, contrast, saturation, hue, and sharpness controls for each source and video output interface offers the most flexibility. For example, PIP can be adjusted without affecting the main picture, video can be adjusted without affecting still picture video quality, etc.

The optional downscaling and progressive-to-interlaced conversion block for the top NTSC/PAL encoder in Figure 5.61 enables simultaneous HD and SD outputs, or simultaneous progressive and interlaced outputs, without affecting the HD or progressive video quality.

The second NTSC/PAL encoder shown at the bottom of Figure 5.61 is useful for recording a program without any OSD or subtitle information being accidently recorded.

Bibliography

Clarke, C. K. P. (1989). *Digital Video: Studio Signal Processing*. BBC Research Department Report BBC RD1989/14.

Devereux, V. G. (1984). *Filtering of the Colour-Difference Signals in 4:2:2 YUV Digital Video Coding Systems*. BBC Research Department Report BBC RD1984/4.

ITU-R BT.601–5. (1995). *Studio Encoding Parameters of Digital Television for Standard 4:3 and Widescreen 16:9 Aspect Ratios.*

ITU-R BT.709–5. (2002). *Parameter Values for the HDTV Standards for Production and International Programme Exchange.*

ITU-R BT.1358. (1998). *Studio Parameters of 625 and 525 Line Progressive Scan Television Systems.*

Janssen, J. G. W. M., Stessen, J. H., & de With, P. H. N. *An Advanced Sampling Rate Conversion Technique for Video and Graphics Signals.* Philips Research Labs.

Sandbank, C. P. (1990). *Digital Television.* New York: John Wiley & Sons, Ltd..

SMPTE 274M–2005. *Television—1920 × 1080 Image Sample Structure, Digital Representation and Digital Timing Reference Sequences for Multiple Picture Rates.*

SMPTE 293M–2003. *Television—720 × 483 Active Line at 59.94 Hz Progressive Scan Production—Digital Representation.*

SMPTE 296M–2001. *Television—1280 × 720 Progressive Image Sample Structure, Analog and Digital Representation and Analog Interface.*

SMPTE EG36–1999. *Transformations Between Television Component Color Signals.*

Thomas, G. A. (1996). *A Comparison of Motion-Compensated Interlace-to-Progressive Conversion Methods.* BBC Research Department Report BBC RD1996/9.

Ultimatte®, Technical Bulletin No. 5, Ultimatte Corporation.

Watkinson, J. *The Engineer's Guide to Standards Conversion.* Snell and Wilcox Handbook Series.

Watkinson, J. *The Engineer's Guide to Motion Compensation.* Snell and Wilcox Handbook Series.

CHAPTER 6

Modulation

Ian Poole

Communications is by far the largest market for signal processing technology. Shipments of wireless handsets alone have reached an incredible one billion units per year—and this is just one of the many wireless applications we encounter every day. Add in other applications such as terrestrial radio and television, satellite radio and televison, WiFi and WiMAX, Bluetooth and RFID, and the size of this market becomes absolutely mind-boggling.

While these various applications have many important differences, all of them rely on the principles of modulation. With so many applications of modulation, there is a good chance that you will encounter it at some point in your career. That can be a scary prospect: Modulation is a complicated subject, filled with complex math and advanced DSP concepts.

In this chapter author Ian Poole accomplishes an impressive feat. He starts with a succinct, layman's definition of modulation, and without introducing complex math or advanced concepts, manages to escort us in a mere twenty-three pages to a basic understanding of how today's complex, 3G modulation schemes work.

He does this by starting with the oldest and simplest form of modulation, amplitude modulation (AM). He then works his way up in complexity (and through history), to frequency modulation (FM) and its variants, phase modulation, and finally more advanced methods such as quadrature amplitude modulation (QAM), spread spectrum techniques, and orthogonal frequency division multiplex (OFDM).

Along the way, he introduces important concepts, such as the modulation index and deviation ratio, which help you understand the theoretical limits of the various schemes. The author also provides some practical information, such as telling you what modulation schemes are used for what standards and why.

This text does assume some basic DSP theory knowledge. But the fact that the author takes that basic DSP knowledge as far as he does without more theoretical background is impressive.

—Kenton Williston

Radio signals can be used to carry information. The information, which may be audio, data or other forms, is used to modify (modulate) a single frequency known as the *carrier*. The information superimposed onto the carrier forms a radio signal which is transmitted to the receiver.

Here, the information is removed from the radio signal and reconstituted in its original format in a process known as *demodulation*. It is worth noting at this stage that the carrier itself does not convey any information.

There are many different varieties of modulation but they all fall into three basic categories, namely amplitude modulation, frequency modulation and phase modulation, although frequency and phase modulation are essentially the same. Each type has its own advantages and disadvantages. A review of all three basic types will be undertaken, although a much greater focus will be placed on those types used within phone systems. By reviewing all the techniques, a greater understanding of the advantages and disadvantages can be gained.

6.1 Radio Carrier

The basis of any radio signal or transmission is the carrier. This consists of an alternating waveform like that shown in Figure 6.1. This is generated in the transmitter, and if it is radiated in this form it carries no information—it appears at the receiver as a constant signal.

6.2 Amplitude Modulation

Possibly the most obvious method of modulating a carrier is to change its amplitude in line with the modulating signal.

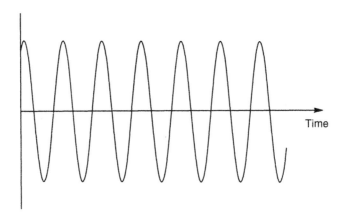

Figure 6.1: An alternating waveform

The simplest form of amplitude modulation is to employ a system known as *on-off keying (OOK)*, where the carrier is simply turned on and off. This is a very elementary form of digital modulation and was the method used to carry Morse transmissions, which were widely used especially in the early days of "wireless." Here, the length of the on and off periods defined the different characters.

More generally, the amplitude of the overall signal is varied in line with the incoming audio or other modulating signal, as shown in Figure 6.2. Here, the envelope of the carrier can be seen to change in line with the modulating signal. This is known as *amplitude modulation* (AM).

The demodulation process for AM where the radio frequency signal is converted into an audio frequency signal is very simple. It only requires a simple diode detector circuit like that shown in Figure 6.3. In this circuit the diode rectifies the signal, only allowing the one-half of the alternating radio frequency waveform through. A capacitor is used as a simple low-pass filter to remove the radio-frequency parts of the signal, leaving the audio waveform. This can be fed into an amplifier, after which it can be used to drive a loudspeaker. This form of demodulator is very cheap and easy to implement, and is still widely used in many AM receivers today.

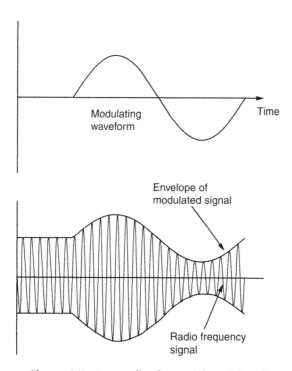

Figure 6.2: An amplitude modulated signal

The signal may also be demodulated more efficiently using a system known as *synchronous detection* (Figure 6.4). Here, the signal is mixed with a locally generated signal with the same frequency and phase as the carrier. In this way the signal is converted down to the baseband frequency. This system has the advantage of a more linear demodulation characteristic than the diode detector, and it is more resilient to various forms of distortion. There are various methods of generating the mix signal. One of the easiest is to take a feed from the signal being received and pass it through a very high-gain amplifier. This removes any modulation, leaving just the carrier with exactly the required frequency and phase. This can be mixed with the incoming signal and the result filtered to recover the original audio.

AM has the advantage of simplicity, but it is not the most efficient mode to use—both in terms of the amount of spectrum it takes up and the usage of the power. For this reason, it is rarely used for communications purposes. Its only major communications use is for VHF aircraft communications. However, it is still widely used on the long, medium, and short wave bands for broadcasting because its simplicity enables the cost of radio receivers to be kept to a minimum.

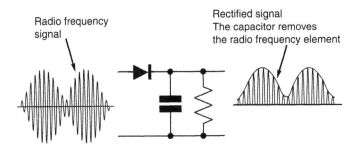

Figure 6.3: A simple diode detector circuit

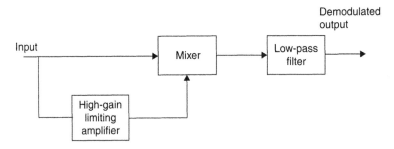

Figure 6.4: Synchronous AM demodulation

To find out why it is inefficient, it is necessary to look at a little theory behind the operation of AM. When a radio-frequency signal is modulated by an audio signal, the envelope will vary. The level of modulation can be increased to a level where the envelope falls to zero and then rises to twice the unmodulated level. Any increase above this will cause distortion because the envelope cannot fall below zero. As this is the maximum amount of modulation possible, it is called *100% modulation* (Figure 6.5).

Even with 100% modulation, the utilization of power is very poor. When the carrier is modulated, sidebands appear at either side of the carrier in its frequency spectrum. Each sideband contains the information about the audio modulation. To look at how the signal is made up and the relative powers, take the simplified case where the 1-kHz tone is modulating the carrier. In this case, two signals will be found: 1 kHz either side of the main carrier, as shown in Figure 6.6. When the carrier is fully modulated (i.e., 100%), the amplitude of the

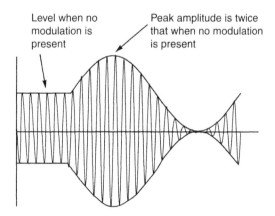

Figure 6.5: Fully modulated signal

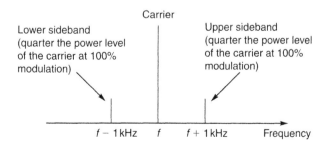

Figure 6.6: Spectrum of a signal modulated with a 1-kHz tone

modulation is equal to half that of the main carrier—that is, the sum of the powers of the sidebands is equal to half that of the carrier. This means that each sideband is just a quarter of the total power. In other words, for a transmitter with a 100 W carrier, the total sideband power will be 50 W and each individual sideband will be 25 W. During the modulation process the carrier power remains constant. It is only needed as a reference during the demodulation process. This means that the sideband power is the useful section of the signal, and this corresponds to (50/150)×100%, or only 33% of the total power transmitted.

Not only is AM wasteful in terms of power, it is also not very efficient in its use of spectrum. If the 1-kHz tone is replaced by a typical audio signal made up of a variety of sounds with different frequencies, then each frequency will be present in each sideband (Figure 6.7). Accordingly, the sidebands spread out either side of the carrier as shown and the total bandwidth used is equal to twice the top frequency that is transmitted. In the crowded conditions found on many of the short wave bands today this is a waste of space, and other modes of transmission that take up less space are often used.

To overcome the disadvantages of AM, a derivative known as *single sideband (SSB)* is often used. By removing or reducing the carrier and removing one sideband, the bandwidth can

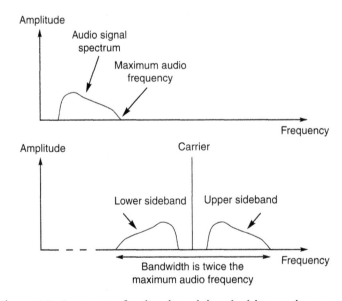

Figure 6.7: Spectrum of a signal modulated with speech or music

be halved and the efficiency improved. The carrier can be introduced by the receiver for demodulation.

Neither AM in its basic form nor SSB is used for mobile phone applications, although in some applications AM combined with phase modulation is used.

6.3 Modulation Index

It is often necessary to define the level of modulation that is applied to a signal. A factor or index known as the modulation index is used for this. When expressed as a percentage, it is the same as the depth of modulation. In other words, it can be expressed as:

$$M = \frac{\text{RMS value of modulating signal}}{\text{RMS value of unmodulated signal}}.$$

The value of the modulation index must not be allowed to exceed 1 (i.e., 100% in terms of the depth of modulation), otherwise the envelope becomes distorted and the signal will spread out either side of the wanted channel, causing interference to other users.

6.4 Frequency Modulation

While AM is the simplest form of modulation to envisage, it is also possible to vary the frequency of the signal to give frequency modulation (FM). It can be seen from Figure 6.8 that the frequency of the signal varies as the voltage of the modulating signal changes.

The amount by which the signal frequency varies is very important. This is known as the *deviation*, and is normally quoted in kilohertz. As an example, the signal may have a deviation of $\pm 3\,\text{kHz}$. In this case, the carrier is made to move up and down by $3\,\text{kHz}$.

FM is used for a number of reasons. One particular advantage is its resilience to signal-level variations and general interference. The modulation is carried only as variations in frequency, and this means that any signal-level variations will not affect the audio output provided that the signal is of a sufficient level. As a result, this makes FM ideal for mobile or portable applications where signal levels vary considerably. The other advantage of FM is its resilience to noise and interference when deviations much greater than the highest modulating frequency are used. It is for this reason that FM is used for high-quality broadcast transmissions where deviations of $\pm 75\,\text{kHz}$ are typically used to provide a high level of interference rejection. In view of these advantages, FM was chosen for use in the first-generation analog mobile phone systems.

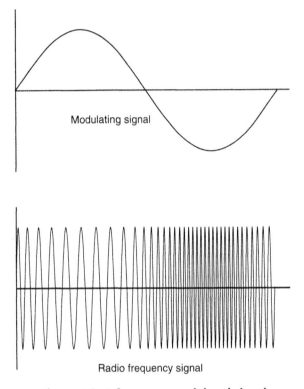

Modulating signal

Radio frequency signal

Figure 6.8: A frequency modulated signal

To demodulate an FM signal, it is necessary to convert the frequency variations into voltage variations. This is slightly more complicated than demodulating AM, but it is still relatively simple to achieve. Rather than just detecting the amplitude level using a diode, a tuned circuit has to be incorporated so that a different output voltage level is given as the signal changes its frequency. A variety of methods is used to achieve this, but one popular approach is to use a system known as a quadrature detector. It is widely used in integrated circuits, and provides a good level of linearity. It has the advantages that it requires a simple tuned circuit and it is also very easy to implement in a form that is applicable to integrated circuits.

The basic format of the quadrature detector is shown in Figure 6.9. It can be seen that the signal is split into two components. One of these passes through a network that provides a basic 90 phase-shift, plus an element of phase shift dependent upon the deviation. The original signal and the phase-shifted signal are then passed into a multiplier or mixer. The mixer output is dependent upon the phase difference between the two signals, i.e., it acts as a

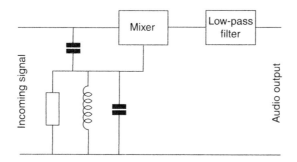

Figure 6.9: Block diagram of an FM quadrature detector

phase detector and produces a voltage output that is proportional to the phase difference and hence to the level of deviation of the signal.

6.5 Modulation Index and Deviation Ratio

In many instances a figure known as the *modulation index* is of value and is used in other calculations. The modulation index is the ratio of the frequency deviation to the modulating frequency, and will therefore vary according to the frequency that is modulating the transmitted carrier and the amount of deviation:

$$M = \frac{\text{Frequency deviation}}{\text{Modulation frequency}}.$$

However, when designing a system it is important to know the maximum permissible values. This is given by the deviation ratio, and is obtained by inserting the maximum values into the formula for the modulation index:

$$D = \frac{\text{Maximum frequency deviation}}{\text{Maximum modulation frequency}}.$$

6.6 Sidebands

Any signal that is modulated produces sidebands. In the case of an amplitude modulated signal they are easy to determine, but for frequency modulation the situation is not quite as straightforward. They are dependent upon not only the deviation, but also the level of deviation — i.e., the modulation index M. The total spectrum is an infinite series of discrete spectral components, expressed by the complex formula:

$$\text{Spectrum components} = Vc\{J_0(M)\cos\omega_c t$$
$$+ J_1(M)[\cos(\omega_c + \omega_m)t - \cos(\omega_c - \omega_m)t]$$
$$+ J_2(M)[\cos(\omega_c + 2\omega_m)t - \cos(\omega_c - 2\omega_m)t]$$
$$+ J_3(M)[\cos(\omega_c + 3\omega_m)t - \cos(\omega_c - 3\omega_m)t]$$
$$+ \ldots\}.$$

In this relationship, $J_n(M)$ are Bessel functions of the first kind, ω_c is the angular frequency of the carrier and is equal to $2\pi f$, and ω_m is the angular frequency of the modulating signal. Vc is the voltage of the carrier.

It can be seen that the total spectrum consists of the carrier plus an infinite number of sidebands spreading out on either side of the carrier at integral frequencies of the modulating frequency. The relative levels of the sidebands can be read from a table of Bessel functions, or calculated using a suitable computer program. Figure 6.10 shows the relative levels to give an indication of the way in which the levels of the various sidebands change with different values of modulation index.

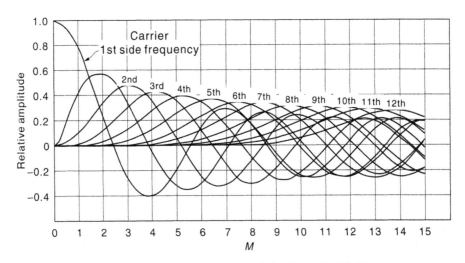

Figure 6.10: The relative amplitudes of the carrier and the first 10 side frequency components of a frequency modulated signal for different values of modulation index

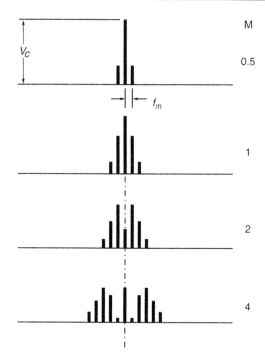

Figure 6.11: Spectra of frequency-modulated signals with various values of modulation index for a constant modulation frequency. It can be seen that for small values of the modulation index M (e.g., M = 0.5), the signal appears to consist of the carrier and two sidebands. As the modulation index increases, the number of sidebands increases and the level of the carrier can be seen to decrease for these values.

It can be gathered that for small levels of deviation (that is, what is termed *narrowband FM*) the signal consists of the carrier and the two sidebands spaced at the modulation frequency either side of the carrier. The spectrum appears the same as that of an AM signal. The major difference is that the lower sideband is out of phase by 180.

As the modulation index increases, other sidebands at twice the modulation frequency start to appear (Figure 6.11). As the index is increased, further sidebands can also be seen. It is also found that the relative levels of these sidebands change, some rising in level and others falling as the modulation index varies.

6.7 Bandwidth

It is clearly not acceptable to have a signal that occupies an infinite bandwidth. Fortunately, for low levels of modulation index all but the first two sidebands may be ignored. However, as the modulation index increases the sidebands further out increase in level, and it is often necessary to apply filtering to the signal. This should not introduce any undue distortion. To achieve this it is normally necessary to allow a bandwidth equal to twice the maximum frequency of deviation plus the maximum modulation frequency. In other words, for a VHF FM broadcast station with a deviation of $\pm 75\,kHz$ and a maximum modulation frequency of $15\,kHz$, this must be $(2\ 3 \times 75) + 15\,kHz$; i.e., $175\,kHz$. In view of this a total of $200\,kHz$ is usually allowed, enabling stations to have a small guard band and their center frequencies on integral numbers of $100\,kHz$.

6.8 Improvement in Signal-to-Noise Ratio

It has already been mentioned that FM can give a better signal-to-noise ratio than AM when wide bandwidths are used. The amplitude noise can be removed by limiting the signal. In fact, the greater the deviation, the better the noise performance. When comparing an AM signal with an FM signal, an improvement equal to $3D^2$ is obtained where D is the deviation ratio. This is true for high values of D—i.e., wideband FM.

An additional perceived improvement in signal-to-noise ratio can be achieved if the audio signal is pre-emphasized. To achieve this, the lower-level high-frequency sounds are amplified to a greater degree than the lower-frequency sounds before they are transmitted. Once at the receiver, the signals are passed through a network with the opposite effect to restore a flat frequency response.

To achieve the pre-emphasis, the signal may be passed through a capacitor-resistor (CR) network. At frequencies above the cut-off frequency, the signal increases in level by $6\,dB$ per octave. Similarly, at the receiver the response falls by the same amount.

6.9 Frequency-Shift Keying

Many signals employ a system called *frequency-shift keying (FSK)* to carry digital data (Figure 6.12). Here, the frequency of the signal is changed from one frequency to another, one frequency counting as the digital 1 (mark) and the other as a digital 0 (space). By changing the frequency of the signal between these two it is possible to send data over the radio.

There are two methods that can be employed to generate the two different frequencies needed for carrying the information. The first and most obvious is to change the frequency of the carrier. Another method is to frequency-modulate the carrier with audio tones that change in frequency, in a scheme known as *audio frequency-shift keying (AFSK)*. This second method can be of advantage when tuning accuracy is an issue.

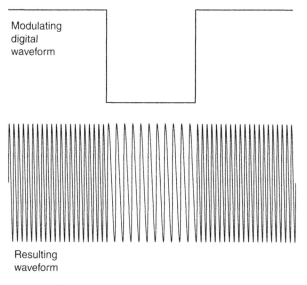

Modulating
digital
waveform

Resulting
waveform

Figure 6.12: Frequency-shift keying

6.10 Phase Modulation

Another form of modulation that is widely used, especially for data transmissions, is *phase modulation* (PM). As phase and frequency are inextricably linked (frequency being the rate of change of phase), both forms of modulation are often referred to by the common term *angle modulation*.

To explain how phase modulation works, it is first necessary to give an explanation of phase. A radio signal consists of an oscillating carrier in the form of a sine wave. The amplitude follows this curve, moving positive and then negative, and returning to the start point after one complete cycle. This can also be represented by the movement of a point around a circle, the phase at any given point being the angle between the start point and the point on the waveform as shown in Figure 6.13.

Modulating the phase of the signal changes the phase from what it would have been if no modulation were applied. In other words, the speed of rotation around the circle is modulated about the mean value. To achieve this it is necessary to change the frequency of the signal for a short time. In other words, when phase modulation is applied to a signal there are frequency changes and *vice versa*. Phase and frequency are inseparably linked, as phase is the integral of frequency. Frequency modulation can be changed to phase modulation by simply adding a CR network to the modulating signal that integrates the modulating signal. As such, the

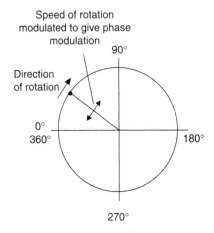

Figure 6.13: Phase modulation

information regarding sidebands, bandwidth and the like also holds true for phase modulation as it does for frequency modulation, bearing in mind their relationship.

6.11 Phase-Shift Keying

Phase modulation may be used for the transmission of data. Frequency-shift keying is robust, and has no ambiguities because one tone is higher than the other. However, phase-shift keying (PSK) has many advantages in terms of efficient use of bandwidth and is the form of modulation chosen for many cellular telecommunications applications.

The basic form of phase-shift keying is known as *binary phase-shift keying* (BPSK) or, occasionally, *phase reversal keying* (PRK). A digital signal alternating between +1 and −1 (or 1 and 0) will create phase reversals—i.e., 180° phase-shifts—as the data shifts state (Figure 6.14).

The problem with phase-shift keying is that the receiver cannot know the exact phase of the transmitted signal, to determine whether it is in a mark or space condition. This would not be possible even if the transmitter and receiver clocks were accurately linked, because the path length would determine the exact phase of the received signal. To overcome this problem, PSK systems use a differential method for encoding the data onto the carrier. This is accomplished by, for example, making a change in phase equal to a 1 and no phase change equal to a 0. Further improvements can be made upon this basic system, and a number of other types of phase-shift keying have been developed. One simple improvement can be made by making a change in phase of 90° in one direction for a 1, and 90° the other way for a 0.

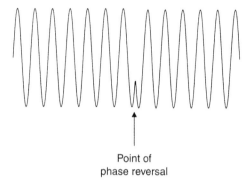

Figure 6.14: Binary phase-shift keying

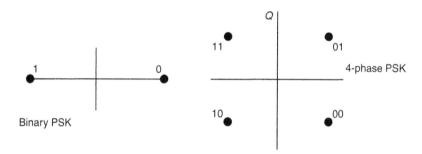

Figure 6.15: Phasor constellations for BPSK and QPSK

This retains the 180° phase reversal between the 1 and 0 states, but gives a distinct change for a 0. In a basic system not using this process it may be possible to lose synchronization if a long series of zeros is sent. This is because the phase will not change state for this occurrence.

There are many variations on the basic idea of phase-shift keying. Each has its own advantages and disadvantages, enabling system designers to choose the one most applicable for any given circumstances. Other common forms include quadrature phase-shift keying (QPSK), where four phase states are used, each at 90° to the other; 8-PSK, where there are eight states, and so forth.

It is often convenient to represent a phase-shift keyed signal, and sometimes other types of signal, using a phasor or constellation diagram (see Figure 6.15). Using this scheme, the phase of the signal is represented by the angle around the circle, and the amplitude by the distance from the origin or centre of the circle. In this way the signal can be resolved into quadrature components representing the sine or I for In-phase component, and the cosine for the quadrature component. Most phase-shift-keyed systems use a constant amplitude, and therefore points appear on one circle with a constant amplitude and the changes in state being represented by movement around the circle. For binary shift keying using phase reversals, the two points appear at opposite points on the circle. Other forms of phase-shift keying may use different points on the circle, and there can be more points on the circle.

When plotted using test equipment, errors may be seen from the ideal positions on the phase diagram. These errors may appear as the result of inaccuracies in the modulator and transmission and reception equipment, or as noise that enters the system. It can be imagined that if the position of the real measurement when compared to the ideal position becomes too large, then data errors will appear because the receiving demodulator is unable correctly to detect the intended position of the point on the circle.

Using a constellation view of the signal enables quick fault-finding in a system. If the problem is related to phase, the constellation will spread around the circle. If the problem is related to magnitude, the constellation will spread off the circle, either towards or away from the origin. These graphical techniques assist in isolating problems much faster than when using other methods.

QPSK is used for the forward link from the base station to the mobile in the IS-95 cellular system, and uses the absolute phase position to represent the symbols. There are four phase decision points, and when transitioning from one state to another it is possible to pass through the circle's origin, indicating minimum magnitude.

On the reverse link from mobile to base station, offset-quadrature phase-shift keying (O-QPSK) is used to prevent transitions through the origin. Consider the components that make up any particular vector on the constellation diagram as X and Y components. Normally, both of these components would transition simultaneously, causing the vector to move through the origin. In O-QPSK one component is delayed, so the vector will move down first and then over, thus avoiding moving through the origin, and simplifying the radio's design. A constellation diagram will show the accuracy of the modulation.

6.12 Minimum-Shift Keying

It is found that binary data consisting of sharp transitions between 1 and 0 states and vice versa potentially create signals that have sidebands extending out a long way from the carrier, and this is not ideal from many aspects. This can be overcome in part by filtering the signal, but the transitions in the data become progressively less sharp as the level of filtering is increased and the bandwidth is reduced. To overcome this, a form of modulation known as *Gaussian-filtered minimum-shift keying (GMSK)* is widely used; for example, it has been adopted for the GSM standard for mobile telecommunications. It is derived from a modulation scheme known as *minimum-shift keying (MSK)*, which is what is known as a continuous-phase scheme. Here, there are no phase discontinuities because the frequency changes occur at the carrier zero crossing points.

To illustrate this, take the example shown in Figure 6.16. Here, it can be seen that the modulating data signal changes the frequency of the signal and there are no phase discontinuities. This arises as a result of the unique factor of MSK that the frequency difference between the logical 1 and logical 0 states is always equal to half the data rate. This can be expressed in terms of the modulation index, and is always equal to 0.5.

While this method appears to be fine, in fact the bandwidth occupied by an MSK signal is too wide for many systems, where a maximum bandwidth equal to the data rate is required.

A plot of the spectrum of an MSK signal shows sidebands extending well beyond a bandwidth equal to the data rate (Figure 6.17). This can be reduced by passing the modulating signal through a low-pass filter prior to applying it to the carrier. The requirements for the filter are that it should have a sharp cut-off and a narrow bandwidth, and its impulse response should show no overshoot. The ideal filter is known as a Gaussian filter, which has a Gaussian-shaped response to an impulse and no ringing.

There are two main ways in which GMSK can be generated. The most obvious way is to filter the modulating signal using a Gaussian filter and then apply this to a frequency modulator where the modulation index is set to 0.5, as shown in Figure 6.18. While simple, this method has the drawback that the modulation index must exactly equal 0.5. In practice, this analog method is not suitable because component tolerances drift and cannot be set exactly.

A second method is more widely used. Here, what is known as a quadrature modulator is used. The term *quadrature* means that the phase of a signal is in quadrature, or 90°, to another one.

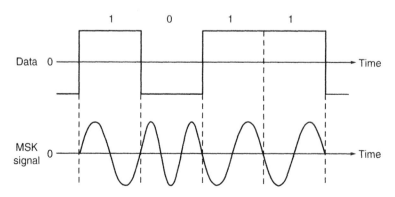

Figure 6.16: An example of an MSK signal

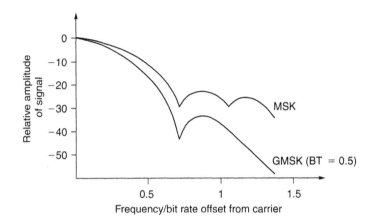

Figure 6.17: Graph of the spectral density for MSK and GMSK signals

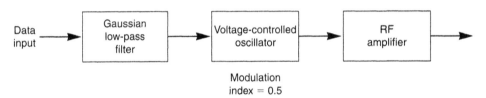

Figure 6.18: Generating GMSK using a Gaussian filter and a frequency modulator with the modulation index set to 0.5

The quadrature modulator uses one signal that is said to be in phase and another that is in quadrature to this. In view of the in-phase and quadrature elements, this type of modulator is often said to be an I-Q modulator (Figure 6.19). When using this type of modulator, the modulation index can be maintained at exactly 0.5 without the need for any settings or adjustments. This makes it much easier to use, and capable of providing the required level of performance without the need for adjustments. For demodulation, the technique can be used in reverse.

A further advantage of GMSK is that it can be amplified by a nonlinear amplifier and remain undistorted. This is because there are no elements of the signal that are carried as amplitude variations, and it is therefore more resilient to noise than some other forms of modulation.

6.13 Quadrature Amplitude Modulation

Another form of modulation that is widely used in data applications is known as *quadrature amplitude modulation (QAM)*. It is a signal in which two carriers shifted in phase by 90° are modulated, and the resultant output consists of both amplitude and phase variations. In view

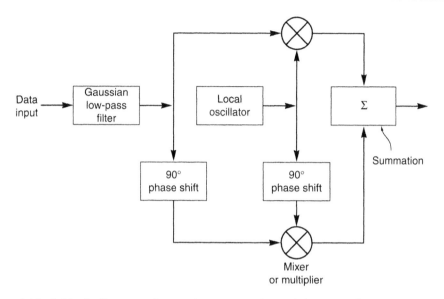

Figure 6.19: A block diagram of a quadrature or I-Q modulator used to generate GMSK

of the fact that both amplitude and phase variations are present, it may also be considered as a mixture of amplitude and phase modulation.

A continuous bit stream may be grouped into threes and represented as a sequence of eight permissible states:

Bit sequence	Amplitude	Phase (°)
000	1/2	0 (0°)
001	1	0 (0°)
010	1/2	π/2 (90°)
011	1	π/2 (90°)
100	1/2	π (180°)
101	1	π (180°)
110	1/2	3π/2 (270°)
111	1	3π/2 (270°)

Phase modulation can be considered as a special form of QAM where the amplitude remains constant and only the phase is changed. By doing this, the number of possible combinations is halved.

Although QAM appears to increase the efficiency of transmission by utilizing both amplitude and phase variations, it has a number of drawbacks. The first is that it is more susceptible to noise because the states are closer together, so that a lower level of noise is needed to move the signal to a different decision point. Receivers for use with phase or frequency modulation can both use limiting amplifiers that are able to remove any amplitude noise and thereby improve the noise reliance. This is not the case with QAM. The second limitation is also associated with the amplitude component of the signal. When a phase or frequency modulated signal is amplified in a transmitter there is no need to use linear amplifiers, whereas when using QAM that contains an amplitude component, linearity must be maintained. Unfortunately, linear amplifiers are less efficient and consume more power, and this makes them less attractive for mobile applications.

6.14 Spread Spectrum Techniques

In many instances it is necessary to keep transmissions as narrow as possible to conserve the frequency spectrum. However, under some circumstances it is advantageous to use what are known as *spread spectrum techniques*, where the transmission is spread over a wide bandwidth. There are two ways of achieving this: one is to use a technique known as *frequency hopping*, while the other involves spreading the spectrum over a wide band of frequencies so it appears as background noise. This can be done in different ways, and the two most widely used systems for this are DSSS and OFDM.

6.15 Frequency Hopping

In some instances, particularly in military applications, it is necessary to prevent any people apart from intended listeners from picking up a signal or from jamming it. Frequency hopping may also be used to reduce levels of interference. If interference is present on one channel, the hopping signal will only remain there for a short time and the effects of the interference will be short lived. Frequency hopping is a well-established principle. In this system, the signal is changed many times a second in a pseudo-random sequence from a predefined block of channels. Hop rates vary, and are dependent upon the requirements. Typically the transmission may hop a hundred times a second, although at HF this will be much less.

The transmitter will remain on each frequency for a given amount of time before moving on to the next. There is a small dead time before the signal appears on the next channel, and during this time the transmitter output is muted. This is to enable the frequency synthesizer time to settle, and to prevent interference to other channels as the signal moves.

To receive the signal, the receiver must be able to follow the hop sequence of the transmitter. To achieve this, both transmitter and receiver must know the hop sequence, and the hopping of both transmitter and receiver must be synchronized.

Frequency hopping transmissions usually use a form of digital transmission. When speech is used, this has to be digitized before being sent. The data rate over the air has to be greater than the overall throughput to allow for the dead time while the set is changing frequency.

6.16 Direct-Sequence Spread Spectrum

Direct-sequence spread spectrum (DSSS) is a form of spread spectrum modulation that is being used increasingly as it offers improvements over other systems, although this comes at the cost of greater complexity in the receiver and transmitter. It is used for some military applications, where it provides greater levels of security, and it has been chosen for many of the new cellular telecommunications systems, where it can provide an improvement in capacity. In this application it is known as *code division multiple access*, because it is a system whereby a number of different users can gain access to a receiver as a result of their different "codes." Other systems use different frequencies (frequency division multiple access—FDMA), or different times or time slots on a transmission (time division multiple access—TDMA).

Its operation is more complicated than those that have already been described. When selecting the required signal, there has to be a means by which the selection occurs. For signals such as AM and FM different frequencies are used, and the receiver can be set to a given frequency to select the required signal. Other systems use differences in time. For example, using pulse code modulation, pulses from different signals are interleaved in time, and by synchronizing the receiver and transmitter to look at the overall signal at a given time, the required signal can be selected. CDMA uses different codes to distinguish between one signal and another. To illustrate this, take the analogy of a room full of people speaking different languages. Although there is a large level of noise, it is possible to pick out the person speaking English, even when there may be people who are just as loudly (or maybe even louder) speaking a different language you may not be able to understand.

The system enables several sets of data to be placed onto a carrier and transmitted from one base station, as in the case of a cellular telecommunications base station. It also allows for individual units to send data to a receiver that can receive one of more of the signals in the presence of a large number of others. To accomplish this, the signal is spread over a given bandwidth. This is achieved by using a spreading code, which operates at a higher rate than the data. The code is sent repeatedly, each data bit being multiplied by each bit of the spreading code successively. The codes for this can be either random or orthogonal. Orthogonal codes are ones which, when multiplied together and then added up over a period of time, have a sum of zero. To illustrate this, take the example of two codes:

Code A	1	−1	−1	1
Code B	1	−1	1	−1
Product	1	1	−1	−1 summed over a period of time = 0, i.e., $1+1-1-1 = 0$

Using orthogonal codes, it is possible to transmit a large number of data channels on the same signal. To achieve this, the data are multiplied with the chip stream (Figure 6.20). This chip stream consists of the codes being sent repeatedly, so that the each data bit is multiplied with the complete code in the chip stream—in other words, if the chip stream code consists of four bits, then each data bit will be successively multiplied by four chip bits. It is also worth noting that the spread rate is the number of data bits in the chip code (i.e., the number of bits that each data bit is multiplied by). In this example the spread rate is four, because there are four bits in the chip code.

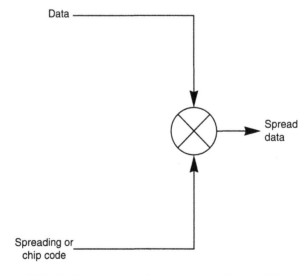

Figure 6.20: Multiplying the data stream with the chip stream

To produce the final signal that carries several data streams, the outputs from the individual multiplication processes are summed (Figure 6.21). This signal is then converted up to the transmission frequency and transmitted.

At the receiver, the reverse process is adopted. The signal is converted down to the base band frequency. Here, the signal is multiplied by the relevant chip code and the result summed over the data bit period to extract the relevant data in a process known as correlation. By multiplying by a different chip code, a different set of data will be extracted.

To see how the system operates, it is easier to refer to a diagram. In Figure 6.22, it can be seen that the waveforms (a) and (b) are the spreading codes. The spreading code streams are multiplied with their relevant data. Here, the spreading code stream (a) is multiplied by the data in (c) to give the spread data stream shown in (d). Similarly, spreading code stream (b) is multiplied by the data in (e) to give (f). The two resulting spread data streams are then added together to give the baseband signal ready to be modulated onto the carrier and transmitted.

In this case, it can be seen that chip stream (a) is repeated in waveform (h). This is multiplied by (g) to give the waveform (i). Each group of four bits (as there are four bits in the chip code used in the example) is summed, and from this the data can be reconstituted as shown in waveform (j).

When a random or, more correctly, a pseudo-random spreading code is used, a similar process is followed. Instead of using the orthogonal codes, a pseudo-random spreading sequence is used. Both the transmitter and receiver will need to be able to generate the same pseudo-random code. This is easily achieved by ensuring that both transmitter and receiver use the same algorithms to

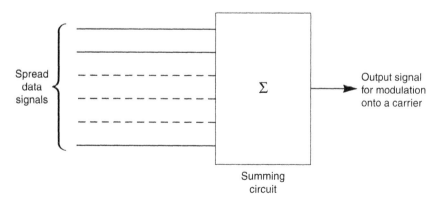

Figure 6.21: Generating a signal that carries several sets of data

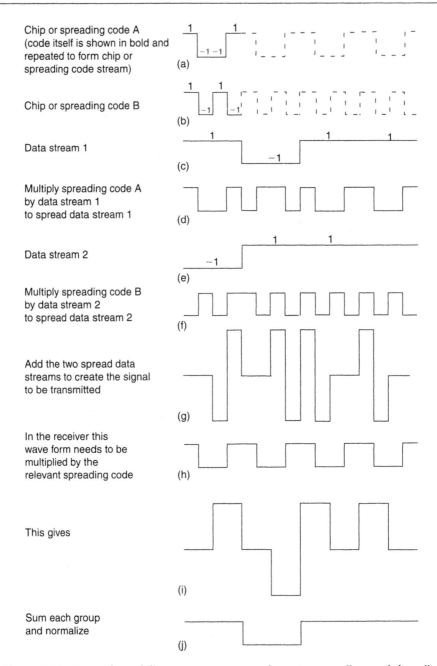

Figure 6.22: Operation of direct-sequence spread spectrum coding and decoding

generate these sequences. The drawback of using a pseudo-random code is that the codes are not orthogonal, and as a result some data errors are expected when regenerating the original data.

6.17 Orthogonal Frequency Division Multiplexing

Another form of modulation that is being used more frequently is orthogonal frequency division multiplexing (OFDM). A form of this, known as coded OFDM or COFDM, is used for many Wi-Fi applications, such as IEEE Standard 802.11 as well as digital radio (DAB), and it is likely that it will be used for the fourth-generation (4G) mobile standards to provide very high data rates.

A COFDM signal consists of a number of closely-spaced modulated carriers. When modulation of any form—voice, data, etc.—is applied to a carrier, then sidebands spread out on either side. It is necessary for a receiver to be able to receive the whole signal in order to successfully demodulate the data. As a result, when signals are transmitted close to one another they must be spaced so that the receiver can separate them using a filter, and there must be a guard band between them. This is not the case with COFDM. Although the sidebands from each carrier overlap, they can still be received without the interference that might be expected because they are orthogonal to each another. This is achieved by having the carrier spacing equal to the reciprocal of the symbol period.

Figure 6.23 shows a traditional view of receiving signals carrying modulation, and Figure 6.24 shows the spectrum of a COFDM signal.

To see how this works, we must look at the receiver. It acts as a bank of demodulators, translating each carrier down to DC. The resulting signal is integrated over the symbol period

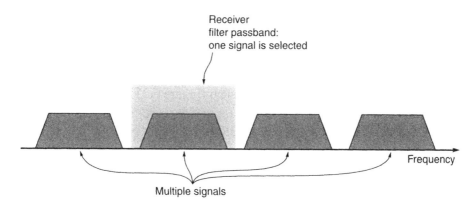

Figure 6.23: Traditional view of receiving signals carrying modulation

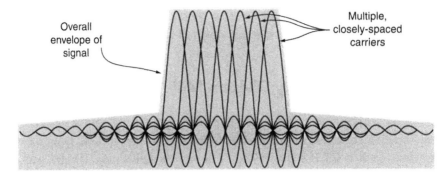

Figure 6.24: The spectrum of a COFDM signal

to regenerate the data from that carrier. The same demodulator also demodulates the other carriers. As the carrier spacing equal to the reciprocal of the symbol period means that they will have a whole number of cycles in the symbol period, their contribution will sum to zero—in other words, there is no interference contribution.

One requirement of the transmitting and receiving systems is that they must be linear, as any nonlinearity will cause interference between the carriers as a result of intermodulation distortion.

This will introduce unwanted signals that will cause interference and impair the orthogonality of the transmission.

In terms of the equipment to be used, the high peak-to-average ratio of multicarrier systems such as COFDM requires the RF final amplifier on the output of the transmitter to be able to handle the peaks while the average power is much lower, and this leads to inefficiency. In some systems, the peaks are limited. Although this introduces distortion that results in a higher level of data errors, the system can rely on the error correction to remove them.

The data to be transmitted are spread across the carriers of the signal, each carrier taking part of the payload. This reduces the data rate taken by each carrier. The lower data rate has the advantage that interference from reflections is much less critical. This is achieved by adding a guard band time (or guard interval) into the system (Figure 6.25), which ensures that the data are only sampled when the signal is stable (i.e., a sine wave) and no new delayed signals arrive that will alter the timing and phase of the signal.

The distribution of the data across a large number of carriers has some further advantages. Nulls caused by multipath effects or interference on a given frequency only affect a small number of the carriers, the remaining ones being received correctly. Using error-coding techniques, which does mean adding further data to the transmitted signal, enables many or all of the corrupted data

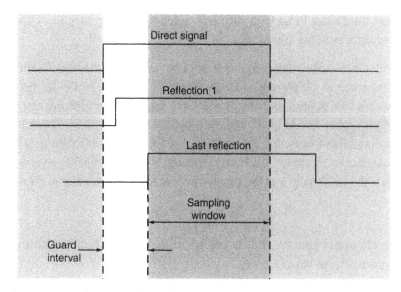

Direct signal

Reflection 1

Last reflection

Sampling
window

Guard
interval

Figure 6.25: The guard interval used to prevent intersymbol interference

to be reconstructed within the receiver. This can be done because the error correction code is transmitted in a different part of the signal. It is this error coding that is referred to in the "Coded" of COFDM.

6.18 Bandwidth and Data Capacity

One of the features that is of paramount importance in any communications system is the amount of data throughput. With users requiring more data and at faster rates, it is essential to make the optimum use of the available channels. It has been seen that there are many different types of modulation that can be used, some being more efficient than others for particular purposes. Nevertheless, there are certain laws that govern the amounts of data that can be transferred.

The bandwidth of a channel that is used is one of the major factors that influences the amount of data that can be accommodated. Bandwidth is literally the width of a band of frequencies measured in hertz (Hz). It is found simply by subtracting the lower limit of the frequencies used from the upper limit of the frequencies used.

Nyquist's theorem relates the bandwidth to the data rate by stating that a data signal with a transmission rate of 2 W can be carried by frequency of bandwidth W. The converse is also true: given a bandwidth of W, the highest signal rate that can be accommodated is 2 W. The data signal need not be encoded in binary, but if it is then the data capacity in bits per second (bps) is twice the bandwidth in hertz. Multilevel signaling can increase this capacity by transmitting more bits per data signal unit.

The problem with multilevel signaling is that it must be possible to distinguish between the different signaling levels in the presence of outside interference—in particular, noise. A law known as Shannon's Law defines the way in which this occurs. It was formulated by Claude Shannon, a mathematician who helped build the foundations for the modern computer, and it is a statement of information theory that expresses the maximum possible data speed that can be obtained in a data channel. Shannon's Law says that the highest obtainable error-free data speed is dependent upon the bandwidth and the signal-to-noise ratio. It is usually expressed in the form:

$$C = W \log_2 (1 + S/N)$$

where C is the channel capacity in bits per second, W is the bandwidth in hertz and S/N is the signal-to-noise ratio.

Theoretically, it should be possible to obtain between 2 and 12 bps/Hz, but generally this cannot be achieved and figures of between 1 and 4 bps/Hz are more reasonable. As a matter of simplicity, no attempt will be made here to provide a serious distinction between the two kinds of ways of measuring capacity and we will simply talk about "bandwidth" in terms of bits per second. However, it must be remembered that bandwidth and digital data rate are two different quantities. Bandwidth is a measure of the range of frequencies used in an analog signal, and bits per second is a measure of the digital data rate.

Error correction codes can improve the communications performance relative to un-coded transmissions, but no practical error correction coding system exists that can closely approach the theoretical performance limit given by Shannon's Law.

6.19 Summary

There are three ways in which a signal can be modulated: its amplitude, phase or frequency can be varied, although of these phase and frequency are essentially the same. However, there are a great many ways in which this can be achieved, and each type has its advantages and disadvantages. Accordingly, the choice of the correct type of modulation is critical when designing a new system.

CHAPTER 7

DSP Hardware Options

Dake Liu

One of the most confusing aspects of DSP is the fact that the acronym has two meanings: It is used as shorthand for both digital signal processing and digital signal processor. This sometimes leads to the false impression that all signal processing takes places on digital signal processors. In fact, DSP engineers have many other hardware options, and choosing the right hardware platform is an important part of the DSP development process.

In this chapter Dake Liu lays out the key DSP hardware options: general-purpose processors (GPPs), DSPs, ASIC, FPGA, and application-specific instruction-set processors (ASIPs). He also explains the implementation issues associated with each option. Liu gives an overview of a generic DSP architecture and explains how the DSP fits into an embedded system. (For a closer look at DSP architectures, check out the next chapter in this book.)

Before getting into the details of the hardware, Liu offers an overview of DSP theory and applications. This overview assumes a basic level of experience with DSP theory, applications, and implementations. If you are just getting started in DSP, you should start by reviewing the earlier chapters in this book.

The section on DSP theory is similar to what you'll find in most DSP textbooks. Here we find the usual equations and diagrams related to sampling, FIR filters, FFTs, etc. Liu deviates from these texts in the end by introducing the more advanced DSP topics of adaptive filtering and random processes.

In the DSP applications section, we're introduced to the concept of real-time processing. Liu illustrates DSP applications through two examples: a basic communications system and a multimedia processing system.

If you take only one idea away from the chapter, it should be that you have many hardware options. Choosing the wrong option can sink a project. Thinking outside the box and choosing and making an unusual choice can give you a big competitive advantage. For more on this important topic, check out this article:

http://www.dspdesignline.com/howto/199001104

—Kenton Williston

7.1 DSP Theory for Hardware Designers

Many theoretical books about DSP algorithms are available [2,3]. It is not this book's intent to review DSP algorithms; instead, DSP algorithms will be discussed from a profiling point of view in order to expose the computing cost.

The remaining part of this chapter is a collection of background knowledge. If you feel that it is insufficient, we recommend that you read a basic DSP book. However, all readers are recommended to carefully read through Section 7.3.

7.1.1 Review of DSP Theory and Fundamentals

A signal is generated by physical phenomena. It has detectable energy and it carries information; for example, variation in air pressure (sound), or variation in electromagnetic radiation (radio). In order to process these signals, they first must be converted into electrical analog signals. These signals must be further converted into digital electrical signals before any digital signal processing can take place. A digital signal is a sequence of (amplitude) quantized values. In the time domain, this sequence is in fixed order with fixed time intervals. In the frequency domain, the spectrum sequence is in fixed order with fixed frequency intervals. A continuous signal can be converted into a digital signal by sampling. A sampling circuit is called an analog-to-digital converter (ADC). Inversely, a digital signal can be converted into an analog signal by interpolation. A circuit for doing this conversion is called a digital-to-analog converter (DAC).

A system is an entity that manipulates signals to accomplish certain functions, yielding new signals. The process of digital signal manipulation therefore is called digital signal processing (DSP). A system that handles digital signal processing is called a DSP system. A DSP system handles signal processing either in the time domain or in the frequency domain. Transformation algorithms translate signals between time- and frequency domains.

DSP is a common name for the science and technology of processing digital signals using computers or digital electronic circuits (including DSP processors). Processing here stands for running algorithms based on a set of arithmetic kernels.

An arithmetic operation is specified as an atomic element of computing operations, for instance, addition, subtraction, multiplication, and division. Special DSP arithmetic operations such as guarding, saturation, truncation, and rounding will be discussed in more detail later. If it is not specially mentioned, two's complement is the default data representation in this book.

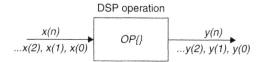

DSP operation

x(n) ...x(2), x(1), x(0) → OP{} → *y(n)* ...y(2), y(1), y(0)

Figure 7.1: A simple DSP operation or a DSP system

An algorithm is the mathematical representation of a task. An algorithm specifies a group of arithmetic operations to be performed in a certain order. However, it does not specify how the arithmetic operations involved are implemented. An algorithm can be implemented in software (SW) using a general-purpose computer or a DSP processor, or in hardware (HW) as an ASIC.

A simplified digital signal processing (DSP) system is shown in Figure 7.1. It has at least one input sequence $x(n)$ and one output sequence $y(n)$ that is generated by applying operation OP{ } to $x(n)$.

Signals $x(n)$ and $y(n)$ are actually $x(nT)$ and $y(nT)$, where T is a time interval representing the sampling period. This means that the time interval between $x(n)$ and $x(n-1)$ is T. OP{ } in Figure 7.1 can be a single operation or a group of operations. A DSP system can be as simple as pure combinational logic, or as complicated as a complete application system including several processors such as video encoders.

In Figure 7.2, fundamental knowledge is classified and depicted in the top part. Applications using fundamental knowledge are listed in the middle, and basic knowledge related to implementations is listed in the bottom part of the figure.

7.1.2 ADC and Finite-length Modeling

As mentioned earlier, a continuous electrical signal can be converted into a digital signal by an ADC. A digital signal can be converted into an analog signal by a DAC. Figure 7.3 shows a simplified and general DSP system based on a digital signal processor (DSP in this figure).

During the conversion from analog to digital and the subsequent processing, two types of errors are introduced. The first type is aliasing, which occurs if the sampling speed is close to the Nyquist rate. The second type is quantization error due to the finite word-length of the system. The ADC performs amplitude quantization of the analog input signal into binary output with finite-length precision. The maximum signal-to-quantization-noise ratio, in dB, of an ideal N-bit ADC is described in [4].

$$\text{SNR}_{Q-\max} = 6.02N + 4.77 - 3 = 6.02N + 1.77 \text{ (dB)} \tag{7.1}$$

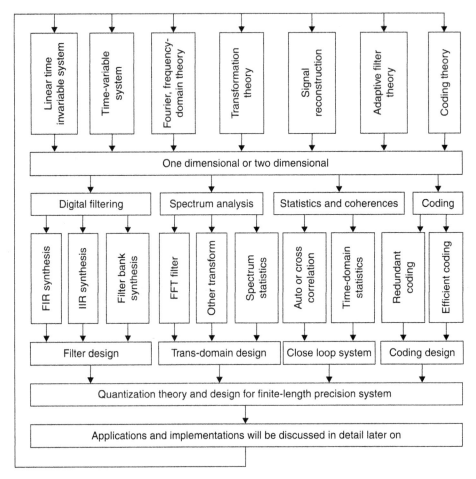

Figure 7.2: Review of DSP theory and related activities for ICT (Information and Communication Technology)

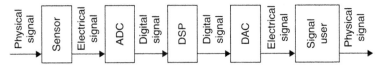

Figure 7.3: Overview of a DSP system

1.77 dB is based on a sinusoidal waveform statistic and varies for other waveforms. N represents the data word length of the ADC. The SNR_{Q-max} expression gives a commonly used rule of thumb of 6 dB/bit for the SNR of an ADC. The maximum signal-to-noise ratio of a 12 bits ADC is 74 dB. The principle of deriving the SNR_{Q-max} can be found in fundamental ADC books.

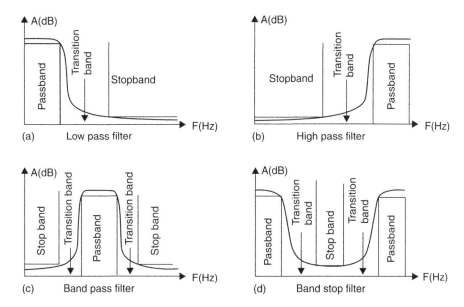

Figure 7.4: Filter specifications

7.1.3 Digital Filters

A filter attenuates certain frequency components of a signal. All frequency components of an input signal falling into the pass-band of the filter will have little or no attenuation, whereas those falling into the stop-band will have high attenuation. It is not possible to achieve an abrupt change from pass-band to stop-band, and between these bands there will be always be a transition-band. The frequency where the transition-band starts is called transition start frequency f_{pass}. The frequency where the transition-band ends is called stop frequency f_{stop}. Figure 7.4 illustrates examples of typical filters. In this figure, (a) is a low-pass filter, (b) is a high-pass filter, (c) is a band-pass filter, and (d) is a band-stop filter. All filters can be derived from the general difference equation given in the following form.[1]

$$y(n) = \sum_{k=0}^{K-1} a_k x(n-k) - \sum_{l=1}^{L-1} b_l y(n-1) \tag{7.2}$$

[1]Equations representing fundamental DSP knowledge can be found in any DSP book. We will not explain and derive equations of fundamental DSP in this book. You can find details in [2,3].

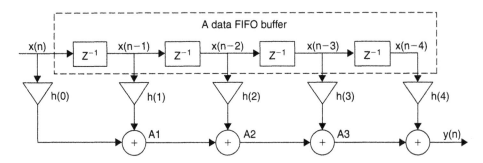

Figure 7.5: A 5-tap FIR filter with a FIFO data buffer

From Equation 7.2, the FIR (Finite Impulse Response) and the IIR (Infinite Impulse Response) filters can be defined by Equation 7.3 and Equation 7.4:

$$y(n) = \sum_{k=0}^{K-1} a_k x(n-k) \tag{7.3}$$

$$y(n) = \sum_{k=0}^{K-1} a_k x(n-k) - \sum_{l=1}^{L-1} b_l y(n-l) \tag{7.4}$$

When FIR or IIR later appears in this book, it will implicitly stand for "a filter."

Let us take a 5-tap FIR filter as an example and explore how an FIR filter works. The algorithms of the 5-tap FIR filter is $y(n) = \Sigma x(n-k)*h(k)$, $k = 0$ to 4. Here, $x(n-k)$ is the input signal, $h(k)$ is the coefficient, and k is the number of iterations. To unroll the iteration loop, the basic computing is given in the following pseudocode of a 5-tap FIR.

```
//5-tap FIR behavior code
{
    A0 = x(n)*h(0);
    A1 = A0 + x(n-1)*h(1);
    A2 = A1 + x(n-2)*h(2);
    A3 = A2 + x(n-3)*h(3);
    Y(n) = A3 + x(n-4)*h(4);
}
```

The signal flow diagram is illustrated in Figure 7.5. In this figure, a FIFO (First In First Out) buffer consists of $n-1$ memory positions keeping $n-1$ input values. Z^{-1} denotes the

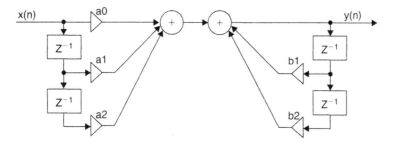

Figure 7.6: A Biquad IIR filter

delay of the input signal or the next position of the FIFO. The triangle symbol denotes the multiplication operation. If the FIFO data and the coefficients are stored in memories, FIR computing of one data sample usually consists of K multiplications, K accumulations, and $2K$ memory accesses. Here K denotes the number of taps of the filter.

Equation 7.4 illustrates an IIR filter. Because the filter output is fed back as part of the inputs for regressive computing, the filter may be unstable if it is not properly designed. The most used IIR filter is the so-called Biquad IIR. A Biquad IIR is a two-tap IIR filter that is defined by the following equation:

$$y(n) = a_0 x(n) + a_1 x(n-1) + a_2 x(n-2) + b_1 y(n-1) + b_2 y(n-2) \tag{7.5}$$

The signal flow graph for a Biquad IIR filter of Equation 7.5 is shown in Figure 7.6.

For minimizing the computing cost, b_0 is set to 1. Programming a Biquad IIR filter usually means that two short loops are unrolled. For processing one data sample, up to 19 operations are required including five multiplications, four additions, and ten memory accesses.

7.1.4 Transform

In this book, a transform is an algorithm to translate signals or elements in one style to another style without losing information. Typical transforms are FFT (fast Fourier transform), and DCT (discrete cosine transform). These two transforms translate signals between the time domain and frequency domain.

Time-domain signal processing may consume too much computing power, so signal processing in other domains might be necessary. In order to process a signal in another domain, a domain transformation of the signal is required before and after the signal processing. Digital signal processing in the frequency domain is popular because of its explicit physical meaning and lower computational cost in comparison with time-domain processing. For example, if a FIR filter is implemented in the time domain, the filtering operation is done by convolution. The computing cost of a convolution is $K * N$, where K is the number of taps of the filter and N is the number of samples. If the FIR filter is implemented in the frequency domain, the computing cost will be reduced to N because the filtering operation in the frequency domain is $Y(f) = H(f) * X(f)$. Here, Y is the output in frequency domain, H is the system transfer function in frequency domain, and X is the input signal in the frequency domain. The total cost of frequency-domain signal processing includes the computing cost in the frequency domain, and the cost of the transform and inverse transform. In general we can summarize that the condition to select frequency-domain algorithms is:

$$\text{TD}_{cost} > \text{FFT}_{cost} + \text{FD}_{cost} + \text{IFFT}_{cost} \qquad (7.6)$$

TD_{cost} is the execution time cost in the time domain, the FD_{cost} is the execution time cost in the frequency domain, and the FFT_{cost} and IFFT_{cost} are the execution time cost of FFT and inverse FFT.

The computational complexity of a direct Fourier transform (DFT) involves at least N^2 complex multiplications or at least $2N(N-1)$ arithmetic operations, without including extra addressing and memory accesses. The computing cost can be reduced if the FFT is introduced. FFT is not a new transform from the theoretical perspective; instead it is an efficient way of computing the DFT (discrete Fourier transform). If the DFT is decomposed into multiple 2-point DFTs, it is called a radix-2 algorithm. If the smallest DFT in the decomposition is a 4-point DFT, then it is called a radix-4 algorithm. Detailed discussions on FFT and DFT can be found in [2].

There are two main radix-2 based approaches, decimation in time (DIT) and decimation in frequency (DIF). The complexities of both algorithms are the same considering the number of basic operations, the so-called butterfly operations. The butterfly operations are slightly different between the DIT and DIF implementations. Both butterfly schematics (DIT and DIF) are illustrated in Figure 7.7.

X and Y in Figure 7.7 are two complex data inputs of the butterfly algorithm. W is the coefficient of the butterfly algorithm in complex data format. Computing a DIT or DIF butterfly consists of 10 operations including a complex data multiplication (equivalent to six integer operations), two complex data additions (equivalent to four integer operations), and five memory accesses of complex data (two data load, one coefficient load, and two data

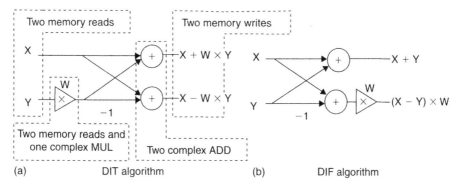

Figure 7.7: DIT butterfly and DIF butterfly of Radix-2 FFT

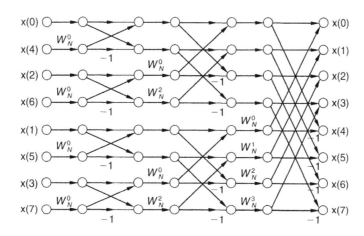

Figure 7.8: Signal flow of an 8-point Radix-2 DIF FFT

store, equivalent to 10 integer data accesses). For the DIT algorithm, the input data should be in bit-reversed order. Bit-reversal addressing is given in Figure 7.8 for an 8-point FFT. The bit-reversal computing cost is at least N.

An 8-point FFT signal flow is given as an example in Figure 7.8.

It can be seen in Figure 7.8 that the computation consists of $\log_2 N$ computing layers. Each layer consists of $N/2$ butterfly subroutines. The total computing cost is at least $N + 0.5 N \times$ (the cost of a butterfly) $\times \log_2 N$. Therefore, the computing cost of a 256-point FFT is at least $0.5 \times 256 \times 10 \times \log_2 256 = 10240$ arithmetic operations and 10240 basic memory accesses (not including bit-reversal addressing, illustrated in Table 7.1).

Table 7.1: Bit-reversal addressing

Sequential Order		Bit-reversed Order	
Sample order	Binary	Binary	Sample order
$X[0]$	000	000	$X[0]$
$X[1]$	001	100	$X[4]$
$X[2]$	010	010	$X[2]$
$X[3]$	011	110	$X[6]$
$X[4]$	100	001	$X[1]$
$X[5]$	101	101	$X[5]$
$X[6]$	110	011	$X[3]$
$X[7]$	111	111	$X[7]$

7.1.5 Adaptive Filter and Signal Enhancement

In order to adapt to the dynamic environment, such as dynamic noise, dynamic echo, or dynamic radio channel of a mobile radio transceiver, some filter behaviors should be updated according to the change of the environment.

An adaptive filter is a kind of filter where the coefficients can be updated by an adaptive algorithm in order to improve or optimize the filter's response to a desired performance criterion. An adaptive filter consists of two basic parts: the filter, which applies the required processing to the incoming signal, and the adaptive algorithm, which adjusts the coefficients of the filter to improve its performance.

The structure of an adaptive filter is depicted in Figure 7.9. The input signal, $x(n)$, is filtered (or weighted) in a digital filter, which provides the output $y(n)$. The adaptive algorithm will continuously adjust the coefficients $c(n)$ in the filter in order to minimize the error $e(n)$. The error is the difference between the filtered output $y(n)$, and the desired response of the filter $d(n)$.

A room acoustic echo canceller, shown in Figure 7.10, is a typical application of an adaptive filter. The purpose of the filter is to cancel the echo sampled by the microphone.

In this acoustic echo canceller, the new coefficient is calculated by the following adaptive algorithm:

$$c(n)_{new} = Kx(n) - c(n)_{old} \tag{7.7}$$

Here $K = f(e(n))$ is the convergence factor. The filter is usually a FIR filter operating in the time domain using convolution. A convolution of a data sample consists of N multiplications,

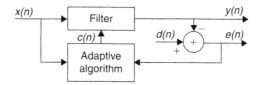

Figure 7.9: General form of an adaptive filter

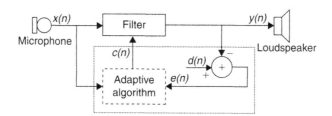

Figure 7.10: A room acoustic echo canceller

N accumulations, and $2N$ memory read operations. Similar to convolution, coefficient adaptation during one data sample consists of N multiplication and N accumulator operations. There is no extra memory load because the access of the old coefficient and the data were counted for by the computing of the filter. However, the new coefficient should be updated and written back to the coefficient memory. Thus there will be N memory write operations. To conclude, the processing cost of an adaptive filter during one data sampling period could be up to $4N$ computations and $3N$ memory accesses if the coefficients are updated continuously. The computing cost and the memory accesses are at least twice that of a fixed FIR filter.

7.1.6 Random Process and Autocorrelation

Statistics and probability theories are used in DSP to characterize signals and processes. One purpose of DSP is to reduce different types of interference such as noise and other undesirable components in the acquired data. Signals are always impaired by noise during data acquisition, or noise is induced as an unavoidable byproduct of finite-length DSP operations (quantization). The theories of statistics and probability allow these disruptive features to be measured and classified.

Mean and standard deviation are basic measurements for statistics. The *mean*, μ, is the average value of a signal. The mean is simply calculated by dividing the sum of N samples by N, the number of samples. The standard variance is achieved by squaring each of the deviations before taking the average, and the standard deviation σ is the square root of the standard variation.

$$\sigma^2 = \frac{1}{N-1}\sum_{i=0}^{N-1}(x-\mu)^2 \tag{7.8}$$

The mean indicates the statistics behavior of a signal and is an estimation of the expected signal value. The standard deviation gives a creditable measurement of the estimation. The computing cost of mean is $2N+1$ (N additions, N memory accesses, and the final step of $1/N$). Similarly, the computing cost of standard deviation is $4N+2$ (three arithmetic and one memory access operations in each tap).

Two other algorithms, autocorrelation and cross-correlation, are useful for signal detection. Autocorrelation is used for finding regularities or periodical features of a signal. Autocorrelation is defined as:

$$a(k) = \sum_{i=0}^{N-1}x(i)x(i-k) \tag{7.9}$$

The computing cost of autocorrelation is similar to the cost of FIR. The main difference is that autocorrelation uses two variables from the same data array.

Cross-correlation is used for measuring the similarity of a signal with a known signal pattern. Cross-correlation is defined as:

$$y(k) = \sum_{i=0}^{N-1}c(i)x(i-k) \tag{7.10}$$

The computing cost is exactly the same as the cost of a FIR.

7.2 Theory, Applications, and Implementations

The scope of DSP is huge, and DSP as an abbreviation has been used by different people with different meanings. In this book, DSP is divided into three categories: DSP theory, DSP applications, and DSP implementations.

DSP theory is the mathematical description of signals and systems using discrete-time or discrete-frequency methods. Most DSP books have DSP theory as the main subject, and this area is today a mature science. On the other hand, applications and implementations based on DSP theory have become major challenges in today's DSP academia and industry. The three DSP categories are summarized in Figure 7.11.

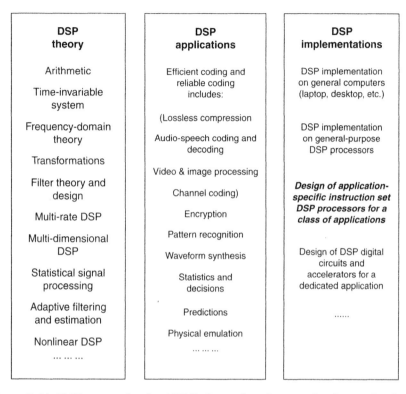

Figure 7.11: DSP categories for ICT (Information Communication Technology)

The theory of digital signal processing (DSP) is the mathematical study of signals in a digital representation and the processing methods of these signals. By using DSP theory, a system behavior is modeled and represented mathematically in a discrete-time domain or discrete-frequency domain. The scope of DSP theory is huge, and the main activities (related to this chapter) are selectively listed in Figure 7.11.

DSP has turned out to be one of the most important technologies in the development of communications systems and other electronics. After Marconi and Armstrong's invention of basic radio communication and transceivers, users were soon dissatisfied with the poor communication quality. The noise was high and the bandwidth utility was inefficient. Parameters changed due to variations in temperature, and the signals could not be reproduced exactly. In order to improve the quality, analog radio communication had to give way to digital radio communication.

DSP applications in this chapter are processes through which systems with specific purposes can be modeled using the knowledge of DSP theory. DSP applications based on established

DSP theory can be found everywhere in daily life. For example, coding for communications is a kind of DSP application. Coding for communications can be further divided into reliable coding (for error detection and error correction) and efficient coding (compression). Other DSP applications will be discussed.

DSP applications can be divided into two categories: applications following standards and those not following standards. When a system must communicate with other systems, or access data from certain storage media, standards are needed. Standards regulate the data format and the transmission protocol between the transmitter and the receiver or between the data source and the data consumer. There are several committees and organizations working with standards, such as IEEE (Institute of Electronics and Electrical Engineering), ISO (International Standard Organization), ITU (International Telecom Union), and ETSI (European Telecommunications Standard Institute) [5]. A standard gives requirements, regulations, and descriptions of functions, performance, and constraints.

Standards do not describe or define implementation details. For example, a physical layer standard of WLAN (Wireless Local Area Network) specifies the format and quality requirement of a radio channel and regulations for radio signal coding and transmission. However, this standard does not regulate the implementation. A semiconductor manufacturer can implement the radio and the baseband part in an integrated or a distributed way, and into a programmable or a dedicated nonprogrammable device. Efficient implementations make successful stories, and all companies have fair chances to compete in the market.

DSP implementation is about realizing DSP algorithms in a programmable device or into an ASIC. DSP implementation can be divided into concept development, system development, software development, hardware development, product development, and fabrication. In this book, focus will be on development and implementation of hardware systems.

7.3 DSP Applications

In this section, DSP applications are introduced briefly through examples. The motivation is to provide examples of real-time systems where digital signal processing is intensively used, because real-time applications form the implicit scope throughout this book. For entry-level readers and students, it might be a bit hard to fully understand every application case presented here. However, this will not prevent you from understanding other parts of the book. From an engineering point of view, the examples selected in this section are simpler than those real-world designs in industrial products. Simplification is necessary because understanding a real-world design might take a very long time and may even be confusing sometimes.

As ASIP designers, you should know more about the methods to analyze the cost of applications instead of a deep understanding of the theory behind them. You should understand where the cost is from, and how to analyze and estimate it. To obtain more details, visit the related company websites [6].

7.3.1 Real-Time Concepts

There are two types of systems: real-time and non-real-time [7,8]. A real-time system is the simplification of a real-time computing system. The system processes data in a time-predictable fashion so that the results are available in time to influence the process being monitored, played, or controlled. The definition of a real-time system is not absolutely unambiguous; however, it is generally accepted that a real-time DSP subsystem is a system that processes periodical signals. A real-time system in this chapter, therefore, creates the output signal samples at the same rate at which the input signals arrive. It means that the processing capacity must be enough for processing one sample within the time interval of two consecutively arriving samples. For example, digital signal processing for voice communication in a mobile phone must be real-time. If the decoding of a voice data packet cannot be finished before the arrival of the next data packet, the computing resource has to abort the current computation and information will be lost. To formalize, a real-time system processes tasks within a time interval and finishes the processing before a certain deadline.

Non-real-time digital signal processing has no strict requirements on the execution time. For instance, image encoding in a digital camera does not necessarily have to be real-time because the slow execution will not introduce any system error (though it may annoy the camera user).

7.3.2 Communication Systems

A communication system is usually a real-time system, which transmits a message stream from the source to the destination through a noisy channel. The message source is called a transmitter, and the destination is called a receiver. A simplified communication system is given in Figure 7.12 [9].

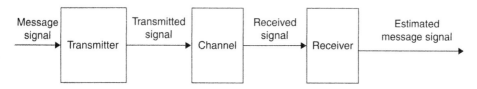

Figure 7.12: A communication system

The transmitter and the receiver are usually at different physical locations. The channel is the physical medium between them, which can be either the air interface of a radio system (e.g., a mobile phone and a radio base station) or a telephone wire.

The received signal that has propagated through the channel is interfered with by the external environment and distorted internally by the receiver itself due to the physical characteristics of the channel. Thus, the signal received is actually different from the signal sent by the transmitter. For example, a receiver may get continuous interference from other neighboring radio systems and interference such as burst glitches from surrounding equipment. The receiver may receive both the wave in the direct path with normal delay and waves from paths experiencing multiple reflections with longer delays. The signal attenuation may change quickly according to the change of the relative position between the transmitter and the receiver.

The function of the receiver in Figure 7.13 is to recover the transmitted data by estimating the received signal and dynamically modeling the channel behavior. The function of the transmitter is not only to send data, but also to code it in order to enable and facilitate the estimation of the channel impulse response at the receiver side. This enables the receiver to compare the received signal with a known training data and to find the difference between the expected signal and the signal actually received, in order to finally calculate the approximation of the channel impulse response. As soon as the channel is correctly modeled, data can be received and recovered by symbol detection with sufficient accuracy. All these heavy computations in the transmitter and the receiver are called radio baseband signal processing, and include coding, channel model estimation, signal reception, error detection, and error correction. All radio baseband signal processing is a class of DSP applications based on well-known DSP theory.

The computing cost of baseband signal processing varies significantly according to the channel condition and the coding algorithms. An advanced radio receiver requires several

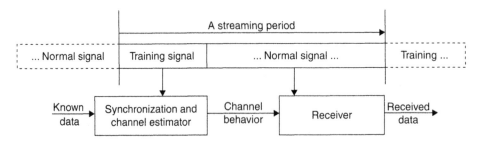

Figure 7.13: Streaming signal between a transceiver pair

hundred to several thousand operations (including memory accesses) for recovering one data bit received from a radio channel. Thus, receiving one megabit per second might require several giga-operations per second. The principle of typical radio baseband signal processing is illustrated in Figure 7.13. The receiver gets a training signal packet first, which is the known signal used for detecting the current channel model as the transfer function $H(f) = Y(f)/X(f)$ by comparing the known training signal stored in the receiver (X) to the received signal (Y). Note that this computation must be conducted in a very short time in order to process the normal signal (payload) in time [10]. The estimated channel model H is used for recovering the normal signal during data reception.

7.3.3 Multimedia Signal Processing Systems

Multimedia signal processing is an important class of DSP applications. Here, the concept of multimedia covers various information formats such as voice, audio, image, and video. Data can be stored in different ways—for example, on CD, hard disk, or memory card, and transmitted via fixed-line or radio channel. Both transmission bandwidth and storage volume will directly affect the end cost. Therefore, data must be compressed for transmission and storage. The compressed data needs to be decompressed before being presented [11].

There are two efficient coding techniques for compression and decompression of data: lossy compression and lossless compression. Lossy compression can be used for voice-, audio-, and image-based applications because the imperfect human hearing and visual capability allows certain types of information to be removed. Lossless compression is required when all information in the data has to be fully recovered.

7.3.3.1 Lossless Compression

Lossless compression is a method of reducing the size of a data file without losing any information, which means that less storage space or transmission bandwidth is needed after the data compression. However, the file after compression and decompression must exactly match the original information. The principle of lossless compression is to find and remove any redundant information in the data. For example, when encoding the characters in a computer system, the length of the code assigned to each character is the same. However, some characters may appear more often than the others, thus making it more efficient to use shorter codes for representing these characters.

A common lossless compression scheme is Huffman coding, which has the following properties:

- Codes for more probable symbols are shorter than those for less probable symbols.
- Each code can be uniquely decoded.

A Huffman tree is used in Huffman coding. This tree is built based on statistical measurements of the data to be encoded. As an example, the frequencies of the different symbols in the sequence ABAACDAAAB are calculated and listed in Table 7.2.

The Huffman tree is illustrated in Figure 7.14.

In this case, there are four different symbols (A, B, C, and D), and at least two bits per symbol are needed. Thus $10 \times 2 = 20$ bits are required for encoding the string ABAACDAAAB. If the Huffman codes in Table 7.4 are used, only $6 \times 1 + 2 \times 2 + 3 + 3 = 16$ bits are needed. Thus four bits are saved. Once the Huffman codes have been decided, the code of each symbol can be found from a simple lookup table.

Decoding can be illustrated by the same example. Assume that the bit stream 01000110111 was generated from the Huffman codes in Table 7.2. This binary code will be translated in the following way: $0 \rightarrow A$, $10 \rightarrow B$, $0 \rightarrow A$, $0 \rightarrow A$, $110 \rightarrow C$, and $111 \rightarrow D$. Obviously the Huffman tree must be known to the decoder.

Without special hardware acceleration, the performance of Huffman coding and decoding is usually much lower than one bit per instruction. With special hardware acceleration, two bits per instruction can be achieved. In reference [12], the throughput of Huffman encoding was enhanced from 0.083 bit per clock cycle to more than 2 bits per clock cycle by using a special hardware architecture. Huffman coding and decoding are typical applications for a FSM (finite state machine).

Table 7.2: Symbol frequencies and Huffman codes[2]

Symbol	Frequency	Normal code	Huffman code
A	6	00	0
B	2	01	10
C	1	10	110
D	1	11	111

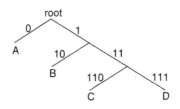

Figure 7.14: Huffman tree

---

[2]Power point files are available at the web page of the book.

7.3.3.2 Voice Compression

In order to reach a high compression ratio, lossy compression can be used. In this case, data cannot be completely recovered. However, if the lost information is not important to the user or is negligible for other reasons, lossy compression can be very useful for some applications [13]. For example, modern voice compression techniques can compress 104 kbits/s voice data (8 kHz sampling rate with 13 bits resolution) into a new stream with a data rate of 1.2 kbits/s with reasonable (limited) distortion using an eMELP voice codec. The compression ratio here is 86:1 [14].

Voice (speech) compression has been thoroughly investigated, and advanced voice compression today is based on voice synthesis techniques. A simplified example of a voice encoder can be found in Figure 7.15.

The voice encoder depicted in Figure 7.15 [13] synthesizes both the vocal model and the voice features within a period of time. A typical period is between 5 and 20 milliseconds. The vocal mode is modeled using a 10-tap IIR filter emulating the shape of the mouth. The voice features are described by four parameters: gain (volume of the voice), pitch (fundamental frequency of the voice providing the vocal feature), attenuation (describes the change in voice volume), and noise pattern (describes the consonants of the voice).

Instead of transferring or storing the original voice, the vocal model and voice patterns will be transferred or stored. Taking the voice codec in Figure 7.15 as an example, the data size of the original voice during 20 milliseconds is $104\,\text{kb} \times 0.02 = 2.08\,\text{kb}$. After compression, the vocal and voice patterns can be coded using $10 \times 8 + 4 \times 8 = 112$ bits. This corresponds to a compression ratio of 18:1. For implementing the encoder and the decoder, many DSP algorithms will be used such as autocorrelation, Fast Fourier transform, adaptive filtering, quantization, and waveform generation. The computing cost of a complete voice encoder is

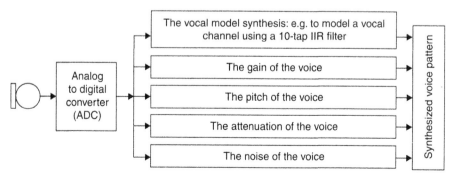

Figure 7.15: A voice encoder based on voice synthesis technique

around 10 to 50 million operations per second for compressing 104 kbits per second using the voice synthesis technique. This is equivalent to about 100 to 500 operations per voice sample.

A voice codec (coder and decoder) is a real-time system, which means that the complete coding and decoding must be finished before the next data arrives (in other words, the data processing speed must be faster than the data rate).

7.3.3.3 Image and Video Compression

Generally, compression techniques for image and video [15] are based on two-dimensional signal processing. Both lossy and lossless compression techniques are used for these applications. Image and video compression techniques are illustrated in Figure 7.16.

The complete algorithm flow in Figure 7.16 is a simplified video compression flow, and the shaded subset is a simplified image compression flow. The first step in image and video compression is color transformation. The three original color planes R, G, B (R = red, G = green, B = blue) are translated to the Y, U, V color planes (Y = luminance, U and V = chrominance). Because the human sensitivity to chrominance (color) is lower than the sensitivity to luminance (brightness), the U and V planes can be down-sampled to a quarter of their original size. Therefore, a frame with a size factor of 3 (3 color planes) is down-sampled to a frame with the size factor of $1 + 1/4 + 1/4 = 1.5$. The compression ratio here is 2.

Frequency-domain compression is executed after the RGB to YUV transformation. The DCT is used for transferring the image from the time domain to the frequency domain. Because the human sensitivity to spatially high-frequency details is low, the information located in the higher frequency parts can be reduced by quantization (the information is represented with lower resolution).

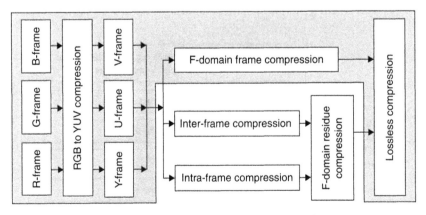

Figure 7.16: Image and video compression

After the frequency-domain compression, a lossless compression such as Huffman coding will finally be carried out.

The classical JPEG (Joint Picture Expert Group) compression for still images can reach 20:1 compression on average. Including the memory access cost, JPEG encoding consumes roughly 200 operations and decoding roughly 150 operations, for one RGB pixel (including three color components). The processing of Huffman coding and decoding can vary a lot between different images, so the JPEG computing cost of a complete image cannot be accurately estimated.

Video compression is an extension of image compression. Video compression is utilizing two types of redundancies: spatial and temporal. Compression of the first video frame, the reference frame, is similar to compression of a still image frame. The reference frame is used for compressing later frames, the inter-frames.

If the difference between a reference frame and a neighbor frame is calculated, the result (motion vectors) can be used for representing this frame. The size of this data is usually very small compared to the size of the original frame. Often there is little difference between the corresponding pixels in consecutive frames except for the movement of some objects. The data transferred will consist of a reference frame and a number of consecutive frames represented by motion vectors. The video stream can therefore be significantly compressed.

The classical MPEG2 (Moving Picture Expert Group) video compression standard can reach a 50:1 compression ratio on average. The advanced video codec H.264/AVC standard can increase this ratio to more than 100:1. The computing cost of a video encoder is very dependent on the complexity of the video stream and the motion estimation algorithm used. Including the cost of memory accesses, an H.264 encoder may consume as much as 4000 operations and its decoder about 500 operations for a pixel on average. As an example, for encoding a video stream with QCIF size (176×144) and 30 frames per second, the encoder requires about 3×10^9 operations per second. The corresponding decoder requires about 4×10^8 operations per second.

7.3.4 Review on Applications

You may have realized that the application examples discussed in this section are applicable to a high-end mobile phone with a video camera. The total digital signal processing cost, including 3G (third-generation) baseband, voice codec, digital camera, and video camera, is actually much more than the capacity of a Pentium running at 2 GHz!

How can a DSP subsystem in a mobile phone supply so much computing capacity while keeping the power consumption low? The answer is the use of application-specific processors! Both the radio baseband and media (voice, audio, and video) processors are application-specific instruction set processors (ASIP) or ASIC modules. They are designed for a class of applications and optimized for low power and low silicon cost.

For designing an ASIP, the computing cost of the target applications is an essential design input. In this section, the cost is measured as the unit cost, the cost per sample (a voice sample, a pixel in a picture, or a bit of recovered data from radio channel). We strongly recommend that you use this way of counting the computing cost.

7.4 DSP Implementations

A DSP application can be implemented in a variety of ways. One way is to implement the application algorithms using a general-purpose computer, like a personal computer or a workstation. There are two reasons for implementing a DSP application on a general-purpose computer:

1. To quickly supply the application to the final user within the shortest possible time.

2. To use this implementation as a reference model for the design of an embedded system.

The discussion of the DSP implementation using a general computer in this section follows the first reason. Many DSP applications are implemented using a general-purpose DSP (off-the-shelf processor). Here, general-purpose DSP stands for a DSP available from a semiconductor supplier and not targeted for a specific class of DSP applications. For example, the TMS320C55X processor from Texas Instruments [6] is available on the market and can be used for many DSP applications requiring a computing performance of less than 500 million arithmetic operations per second.

DSP applications can also be implemented using an ASIP DSP. An ASIP is designed for a class of applications such as radio baseband processing in a multimode mobile phone, audio processing in an MP3 player, or image processing in a digital camera.

Another alternative for implementing DSP applications is the nonprogrammable ASIC (Application Specific Integrated Circuit). Many DSP applications were implemented this way in the 1980s due to limitations in silicon area and performance. Recently the nonprogrammable DSP ASIC has been taken over gradually by the DSP ASIP in accordance with increasing requirements on flexibility.

An FPGA (Field-Programmable Gate Array) is also an alternative as an intermediate or the final implementation of DSP applications.

7.4.1 DSP Implementation on GPP

Many DSP applications, with or without real-time requirements, can be implemented on a general-purpose processor (GPP). Recently, applications for media entertainment have become popular in personal computers, for example, audio and video players. A video player is a video decoder that is decompressing data according to international standards (e.g., ISO/IEC MPEG-4) or proprietary standards (e.g., Windows Media Video from Microsoft).

The video player must be able to decode the video stream in real-time using the operating system (OS). However, most operating systems for desktops are not originally designed for real-time applications. When designing a high quality media player using such an OS, attention must be paid to the real-time features by correctly setting the priorities of the running tasks.

Another type of DSP application, without real-time requirements, is high performance computing. Typical examples are analysis of the stock market, weather forecast, or earthquakes. This type of analysis often is implemented in software running on a personal or supercomputer. The execution time of such a software program is not strictly regulated and this software therefore is defined as general software instead of real-time DSP software.

7.4.2 DSP Implementation on GP DSP Processors

A general-purpose DSP can be bought off-the-shelf from a number of different DSP suppliers (e.g., Texas Instruments, Analog Devices, or Freescale) [6]. A general-purpose DSP has a general assembly instruction set that provides good flexibility for many applications. However, high flexibility usually means fewer application-specific features or less acceleration of both arithmetic and control operations. Therefore, a general-purpose DSP is not suitable for applications with very high performance requirements. High flexibility also means that the chip area will be large. A general-purpose DSP processor can be used for initializing a product because the system design time will be short. When the volume has gone up, a DSP ASIP could replace the general-purpose processor in order to reduce the component cost.

No general-purpose DSP is 100% general. Most general-purpose DSP processors are actually designed with different target applications in mind. General-purpose DSP processors can be divided into processors targeted for either low power or high performance. For example, TMS320C2X of Texas Instruments is designed for low-cost applications, whereas TMS320C55 of Texas Instruments is designed for applications with medium performance and medium power. TMS320C6X of Texas Instruments is designed for applications requiring high performance. General-purpose DSP processors can also be divided into floating-point processors and fixed-point processors.

"General-purpose" implies only that the DSP is designed neither for a specific task nor for a class of specific tasks. It is available on the device market for all possible applications. The instruction set and the architecture must be general, meaning that the instruction set covers all basic arithmetic functions and sufficient control functions. The instruction set is not designed for accelerating a specific group of algorithms. At the same time, peripherals and interfaces must be comprehensive in order to be able to connect to various microcontrollers, memories, and peripheral devices.

Since the processor is off-the-shelf, the main development of a DSP product will be software design. The hardware development is limited to peripheral design, which means connecting the DSP to the surrounding components, including the MCU (microcontroller), main memory, and other input–output components. The best way of designing peripherals for an available DSP is to find a reference design from the DSP supplier such as a debug board [6].

7.4.3 DSP Implementation on ASIP

A DSP ASIP has an instruction set optimized for a single application or a class of applications. On the one hand, a DSP ASIP is a programmable machine with a certain level of flexibility, which allows it to run different software programs. On the other hand, its instruction set is designed based on specific application requirements, making the processor very suitable for these applications. Low power consumption, high performance, and low cost by manufacturing in high volume can be achieved. In case the processor is used for applications for which it was not intended, poor performance can be expected. For example, using a video or image DSP for radio baseband applications will result in catastrophically poor performance. DSP ASIPs are suitable in volume products such as mobile phones, digital cameras, video camcorders, and audio players (MP3 player).

An ASIP DSP has a dedicated instruction set and dedicated data types. These two features will be discussed and implemented throughout this book. You will find later in this book that one instruction of an ASIP DSP could be equivalent to a kernel subroutine or part of a kernel subroutine running on a general-purpose DSP. For supporting algorithm-level acceleration, special functions will be implemented using specific hardware. This yields better performance, lower power consumption, and higher performance/silicon ratio compared to a general-purpose DSP. At the same time, the range of application is limited due to simplification of the instruction set.

An ASIP can be an in-house product or a commercial off-the-shelf (COTS) component available on the market. In-house means a dedicated design for a specific product within a company. COTS means that the processor is designed as an ASSP (application-specific standard product).

7.4.4 DSP Implementation on ASIC

There are two cases when an ASIC is needed for digital signal processing. The first is to meet extreme performance requirements. In this case, a programmable device would not be able to handle the processing load. The second case is to meet ultra-low power or ultra-low silicon area, when the algorithm is stable and simple. In this case, there is no requirement on flexibility, and a programmable solution is not needed.

A typical ASIC application example is decimation for synthetic aperture radar baseband processing. The requirement for this application is up to 1011 MAC (multiplication and accumulation) operations per second with relative low power consumption. Nonprogrammable devices can give such performance (as of 2006). Another typical application is a hearing aid device that includes a band-pass filter bank (synthesizer) and an echo canceller. About 40 MIPS (million instructions per second), high data dynamic range, and the power consumption below 1 mW are required. A programmable device will not likely meet these requirements (as of 2006).

ASIC implementation is to map algorithms directly to an integrated circuit [16]. Comparing a programmable device supplying the flexibility at every clock cycle, an ASIC has very limited flexibility. It can be configurable to some extent in order to accommodate very similar algorithms, but typically it cannot be updated in every clock cycle.

When designing an ASIP DSP, functions are mapped to subroutines consisting of assembly instructions. When designing an ASIC, the algorithms are directly mapped to circuits. However, most DSP applications are so complicated that mapping functions to circuits is becoming increasingly difficult. On the other hand, mapping DSP functions to an instruction set is becoming more popular because the challenge of complexity is handled in both software and hardware, and conquered separately.

Example 7.1 exposes the way to map functions directly to a circuit.

Finally the circuit is shown in Figure 7.17.

In order to avoid accumulation-induced overflow in this example, the data width should be increased after each accumulation. Because the result $y(n)$ shall have the same data type as the input $x(n)$, a rounding operation is necessary before truncating the lower 16 bits. Overflow might happen, so saturation is necessary before using the result.

We can see in Example 7.1 that mapping an FIR to hardware is simple. However, when algorithms or applications are complicated, especially when the algorithm details cannot be decided during the system design, this method cannot be used. In this case mapping applications to an instruction set is the only solution.

Example 7.1

Mapping a 5-tap FIR filter to a hardware circuit: The algorithm of the 5-tap FIR filter is $y(n) = \Sigma x(n - k) * h(k)$. Here, k is the number of iterations from $k = 0$ to $k = 4$.

The algorithm can be mapped to hardware after unrolling the iteration. The pseudocode of a 5-tap FIR becomes:

```
//5-tap FIR behavior
{
    A1 = x(n)*h0 + x(n-1)*h1;
    A2 = A1 + x(n-2)*h2;
    A3 = A2 + x(n-3)*h3;
    Y(n) = A3 + x(n-4)*h4;
}
```

The pseudo HDL code of a 5-tap FIR becomes:

```
//5-tap FIR HDL
{
    A0[32:0] >= x(n)[15:0]*h0[15:0]
    A1[32:0] >= A0[32:0] + x(n-1)[15:0]*h1[15:0];
    A2[33:0] >= A1[32:0] + x(n-2)[15:0]*h2[15:0];
    A3[34:0] >= A2[33:0] + x(n-3)[15:0]*h3[15:0];
    A4[35:0] >= A3[34:0] + x(n-2)[15:0]*h4[15:0];
    Y(n)[15:0] >= saturation (round (A4[35:0]));
}

clk = '1' and clk'event
{
    X(n-4)>= x(n-3); x(n-3)>= x(n-2); X(n-2)>= x(n-1);
    x(n-1)>= x(n);
}
```

7.4.5 Trade-off and Decision of Implementations

In Figure 7.18, the power consumption (based on a 90 nm digital silicon process) is shown as a function of the MOPS (million operations per second) figure for different DSP implementations. The power consumption of the memory is very dependent on the application-specific hardware configuration, so the memory cost is not included in this figure.

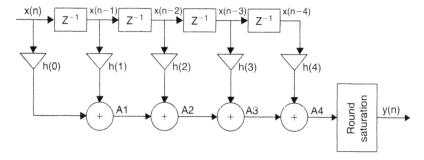

Figure 7.17: A 5-tap FIR filter

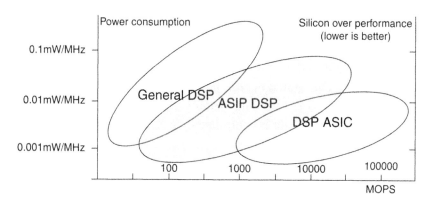

Figure 7.18: Comparing three types of DSP implementations

The general DSP processor typically is used for applications with not more than a thousand MOPS. The DSP ASIC can support applications with very low power consumption (less than 0.01 mW/MHz) and high performance but without requirements on flexibility. The DSP ASIP, however, is used mostly when a trade-off between silicon/power cost, performance, and development effort is required.

7.5 Review of Processors and Systems

Most DSP applications are implemented using either general-purpose or ASIP DSP processors. A digital signal processor is a programmable integrated circuit for data manipulation. A DSP processor is designed for performing arithmetic functions like add, subtract, multiply, and shift as well as logic functions.

7.5.1 DSP Processor Architecture

Learning processor design is an iterative process. In this chapter, you will get a bird's-eye view of the methodology of DSP processor design and the processor architecture.

7.5.1.1 What Is Inside a DSP?

Similar to other types of processors, a DSP contains five key components:

- Program memory (PM): PM is used for storing programs (in binary machine code). PM is part of the control path.

- Programmable FSM: It is a programmable finite state machine consisting of a program counter (PC) and an instruction decoder (ID). It supplies addresses to the program memory for fetching instructions. Meanwhile, it also performs instruction decoding and supplies control signals to the data processing unit and data addressing unit.

- Data memory and data memory addressing: DM stores information to be processed. Three types of data are stored in DM: input/output data, intermediate data in a computing buffer (a part of the data memory), and parameters or coefficients. The data memory addressing unit is controlled by programmable FSM and supplies addresses to data memories.

- Data processing unit (DU): The data processing unit, or datapath, performs arithmetic and logic computing. A DU includes at least a register file (RF), a multiplication and accumulation unit (MAC), and an arithmetic logic unit (ALU). A data processing unit may also include some special or accelerated functions.

- Input/output unit (I/O): I/O serves as an interface for functional units connected to the outside world. I/O also handles the synchronization of external signals. Memory buses and peripherals are also included.

A simplified block diagram is given in Figure 7.19.

7.5.2 DSP Firmware

Before introducing firmware in detail, you should simply accept that firmware is fixed software running in an electronic product. Only real-time firmware will be discussed in this book. Real-time DSP firmware usually consists of an infinite loop that processes real-time signals continuously and periodically. Figure 7.20 depicts a typical top-level infinite loop in a real-time system. One data unit (one data packet or one data sample) is processed in the figure through one complete execution of the infinite loop.

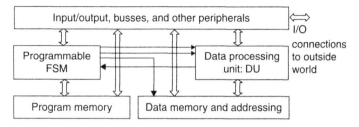

Figure 7.19: DSP processor architecture

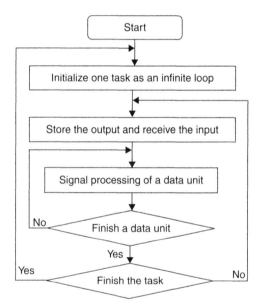

Figure 7.20: An infinite loop in a DSP processor

The executable binary code is developed in four steps:

1. **Design the behavior source code.** The behavior source code is the original description that models an application or an algorithm.

2. **Design the hardware dependent source code.** The behavior source code must be modified or rewritten in order to adapt it to the hardware. Hardware adaptation includes modifying hardware constraints and utilizing hardware features.

3. **Design the assembly code.** The hardware-dependent source code is translated to assembly code, or the assembly code is written using the hardware-dependent source code as a reference.

4. **Generate and debug the binary machine code.** Finally, the assembly code is assembled and linked to executable binary machine code. The binary machine code is verified by executing it on an assembly instruction set simulator.

Most DSP applications do not require ultra-high data precision. Therefore, fixed-point processors can be used in order to reduce the silicon cost. For a fixed-point DSP the silicon cost can be less than half compared to a standard floating-point DSP. As fixed-point DSP processors are dominant on the DSP market, the discussions in this book will be focused on this type of processor.

Quantization (noise) error and offset of frequency response are unwanted behaviors when using a fixed-point processor. These errors are due to the finite data length, and handling them will increase firmware complexity. The firmware must maintain the quality of input and intermediate data in the computing buffers. Data quality here stands for a measure of the quantization error (which should be minimized) and the dynamic range (which should be maximized). The final firmware for a fixed-point DSP will contain both functional firmware and firmware for data quality control. The firmware for data quality control includes data quality measurement and scaling (to be discussed with Figure 7.22). Data quality can be optimized by data scaling. Normally, the execution of the data quality control firmware does not occur very often, and it is only needed when the amplitude of the input data is changed or the algorithm in the firmware is changed.

The program flow is shown in Figure 7.21 and further illustrated in Figure 7.22. The complete firmware running on a fixed-point processor can be divided into three flows. The main flow is shown with gray background, the measurement flow is shown with a background of vertical filling lines, and the scaling flow is shown with a background of horizontal filling lines.

In Figure 7.21 and Figure 7.22, the main flow is executed once for each input streaming data. The data quality control flows are not executed very often. In most cases, the quality control part is skipped (via the default path in Figure 7.21) until it is needed. The measurement flow monitors inputs and intermediate results. In the measurement flow, parameters such as signal-to-noise ratio, signal level on average, and maximum/minimum values will be measured. Other parameters may also be observed, for example, event counting of overflow and saturation. Counters are used for collecting statistics of certain events. The results from measurements will be used in the scaling flow. Finally, scaling factors for input, output, and internal data are modified.

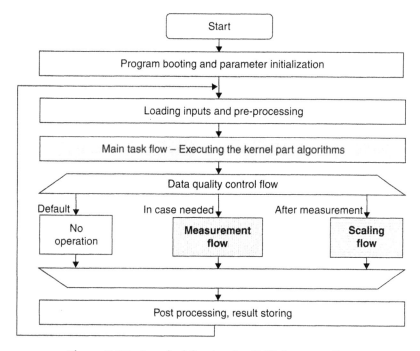

Figure 7.21: A typical fixed-point DSP firmware flow

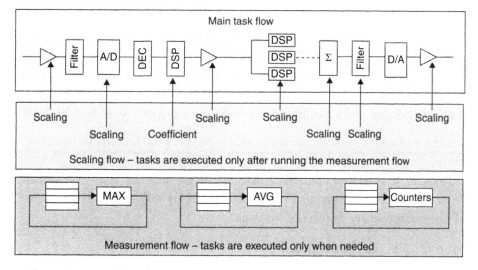

Figure 7.22: Relation between function firmware and quality control firmware

7.5.3 Embedded System Overview

An embedded system is a system that is inside a product system or a product [18]. An embedded system is a special-purpose computer system designed to perform one or a class of dedicated functions. In contrast, a general-purpose computer, such as a personal computer, can do many different tasks, depending on programming. The product user may not even be aware of the existence of the embedded system, although it may play an important role in the function of the product. An embedded system could be a component of a personal computer such as a keyboard controller, mouse controller, or a wireless modem. An embedded system could also be a digital subsystem inside a mobile phone, a digital camera, a digital TV, or in medical equipment. Except for general computers, most microelectronic systems are embedded systems.

A general computer system is not designed for any specific purpose. A desktop computer can be a general-purpose computing engine, a home electronics system, a documentation editing system, a media terminal, or a network terminal. An embedded system is an application-specific system, different from a general computer system. Within the specific application domain, the embedded system may have much higher performance or much lower power consumption compared to a general computer system. An embedded system, such as a radar signal processing system or a computer tomography (CT) processing system, can have teraflops performance (~1000 times more than the performance of a Pentium). Ultra low power consumption in the range 30 to 50 MIPS/mW is possible for hearing aid embedded systems.

However, it will be impossible to use an embedded processor for applications outside the specific domain. For example, a processor designed for controlling a washing machine can never be used as a DSP for video applications. On the other hand, a desktop computer can handle both these tasks. But flexibility has a price tag. A processor for a washing machine can cost less than one dollar, and a Pentium 4 in a desktop costs much more than $100 (in 2005).

To summarize: embedded systems are application-specific. Product cost, design cost, performance, power consumption, and lifetime are all application-specific. In general, embedded system design covers almost all activities in the area of electrical engineering. Thus, the number of different types of embedded systems is very big. In this book, we discuss only DSP subsystems inside embedded systems.

7.5.4 DSP in an Embedded System

DSP processors are essential components in many embedded systems. One or several DSP processors consist of a DSP subsystem in an embedded system. A general embedded system,

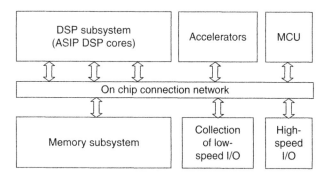

Figure 7.23: DSP processor in an embedded system

including a DSP subsystem, is shown in Figure 7.23. Such a system is also called a system on a chip (SoC) platform for embedded applications.

The system in Figure 7.23 can be divided into four parts. The first part is the microcontroller (MCU), which is the master of the chip or the system. The MCU is responsible for handling miscellaneous tasks, except computing for real-time algorithms. Typical miscellaneous tasks are operating system, connection protocols, Java programs, human-machine-interface, and hardware management.

The second part is the ASIP DSP subsystem including accelerators, which is the main computing engine of the system. All heavy computing tasks should be allocated to this subsystem. The DSP subsystem could include a single processor with accelerators or a multicore processor cluster. For example, in a 3G mobile phone, the DSP subsystem usually has two processors: a baseband processor and an application processor.

The third part is the memory subsystem, which supports data and program storage for the DSP subsystem and the MCU. A SoC usually has multiple levels of memories. Within the MCU core and the DSP core, local memories or level-1 caches can be found. At SoC level, a level-2 cache or an on-chip main memory can be found. Level three in the memory hierarchy is the off-chip main memory.

The fourth part consists of peripherals including high-speed and low-speed I/Os. Analog circuits could be part of the low-speed I/O.

7.5.5 Fundamentals of Embedded Computing

Within a DSP subsystem, embedded computing can be divided into three parts: computing using ASIC, computing using hardware/software (HW/SW) (processors + accelerators), and computing using SW. Here, the focus is to discuss the computing using HW/SW and SW

Example 7.2

Classify the following systems. Which is a real-time system, and which is not a real-time system?

- A mobile phone receiving and sending voice data.
- A mobile phone receiving and sending short message packets.
- A person calculating the statistics on the quality of stored data.
- A computer that analyzes stock information.

Answers

- *Mobile phone sending and receiving voice data.* It is a real-time system because the decoding of the received voice packet must be finished before the arrival of the next voice data packet.

- *Mobile phone receiving and sending short messages.* The baseband part is a real-time system because a symbol must be processed before the arrival of the next symbol. The application part for message display is not a real-time system, because the display of the short message can be delayed for a while without affecting the functionality of the system.

- *Statistics on stored data.* It is not a real-time system. Your boss might ask you to speed up the process. Nevertheless, the arriving new data will not be lost if your processing speed is low.

- *Stock analysis.* Even though the processing result must be available in time, it is not a real-time system because the new arriving data will not be lost when the processing is slow.

---

instead of using ASIC. Embedded SW can be divided further into three categories: operating system, SW for real-time computing, and SW for best-effort embedded computing. In this book, the focus is the SW for real-time computing.

SW for real-time computing must be executed on a specific DSP platform based on real-time scheduling. The platform must supply enough computational capacity, and the real-time scheduling guarantees that the execution will consume less time than the time interval between arriving input data.

In an embedded system, the complexity or dependency is relatively known before runtime. Static scheduling can therefore be used to enhance the utilization of the time and hardware resources. In a general computing system, on the contrary, the management of running applications is dynamic, and the issue of a task is based on a dynamic priority table.

Another special feature of embedded computing is the application-specific precision. Both the precision of operands and the precision of computing are specified according to the application. The precision of audio processing can be rather high (around 24 bits), whereas the precision of video can be as low as 8 to 12 bits.

7.6 Design Flow

The system development process from conceptualization to manufacturing includes modeling, implementation, and verification. This is known as the design flow. Conducting designs using the design flow is called the design methodology. In this section, methodologies for designing hardware, including processors, will be briefly reviewed.

7.6.1 Hardware Design Flow in General

The design of an embedded system includes implementation of complete and correct functions with a specified performance (not necessarily the highest), affordable cost, reasonable reliability, and within a limited amount of design time. In most cases, the product lifetime is also a design parameter. In order to design a system, with complex functions optimized and allocated to both hardware and software, an efficient and reliable design flow or methodology is required.

A design consists of several transformation steps from a high-level to a low-level description. More hardware, control information, and constraints are inserted during each transformation to a lower level. A transformation from one level to the next lower level can be executed via description or synthesis. Two basic types of transformations, described as two design methodologies, are introduced in most methodology books: the capture-and-simulate methodology and the describe-and-synthesize methodology [19].

Capture-and-simulate has been the dominant methodology since the 1960s. According to this method, the system is described at every level and each description is proved by simulation. This method is not efficient, because it consumes a lot of design time and designs cannot be sufficiently optimized. However, this is still the only proven way (as of 2007) of high-level design from system specification down to RTL coding.

The describe-and-synthesize methodology was introduced during the late 1980s by the success of logic synthesis. This methodology can be further divided into two levels of abstraction. At the higher level, behavioral synthesis translates the behavioral description into a structural RTL description, including the data flow at bit level and the control details at cycle and bit level. The main tasks for behavioral synthesis are allocation, scheduling, and binding. At the lower level, RTL synthesis translates the RTL code to a gate level description, the net list.

Behavioral synthesis is currently (in 2007) far from mature. In most cases, behavioral synthesis of an ASIP is impossible today. There is hardly any automatic method for translating an assembly instruction set to a good HW architecture. One reason for this is the extremely large design space that makes it difficult to optimize the design toward HW multiplexing, high performance, and power efficiency.

A well-known methodology for embedded system design is to divide the design activities using the famous *Y*-chart, shown in Figure 7.24, which was proposed by Professor Daniel Gajski [19].

In using the *Y*-chart, an assumption is made that each design can be modeled in three basic ways, emphasizing different properties, no matter how complex the design is. The *Y*-chart has three axes representing design behavior (function, specification), design structure (net list, block diagram), and physical design (layout, boards, packages).

Behavior design of a system means that the design is represented as a black box with specified input and output data relations. The black box behavior does not indicate in any way how it is implemented or how its internal structure looks.

Structure design gives a description of the hardware partition and the relation (interconnection) between the black boxes. Each black box should be further refined by a

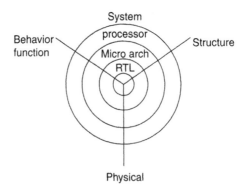

Figure 7.24: *Y*-chart

more detailed structure description including functionality, timing cost, and interconnects until the design reaches the register transfer level (RTL).

Physical design implements functions based on logic gates or transistors and interconnects. Physical features include the size and the position of each component, the wire routing between components, the placement, physical delay, power consumption, and thermo behavior. Physical parameters are taken into account, such as parasitic capacitance and resistance of interconnects.

7.6.2 ASIP Hardware Design Flow

Processor design is a complicated process. Without an advanced design flow, a processor cannot be designed in time and the quality of the design will not be high. The design flow is therefore essential for complicated systems such as ASIP.

The ASIP design flow is introduced briefly here. The ASIP design flow is divided into three parts: architecture design, design of programming tools, and firmware design, as depicted in Figure 7.25. In this section, focus will be on architecture design, and in particular on system architecture and hardware development flow. Other topics such as design of programming tools and design of application firmware will be addressed briefly.

The first and most important step in the design of a processor is the instruction set design. This design step is complicated, and no one can really claim that a certain instruction set is the best. The instruction set design is a trade-off among a multitude of parameters including performance, functional coverage, flexibility, power consumption, silicon cost, and design

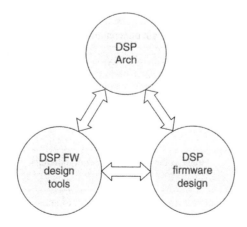

Figure 7.25: Knowledge required in ASIP design

time. In Figure 7.26, a simplified design flow is described, including the basic flow for the design of an instruction set architecture.

The starting point of the design of an ASIP is the application analysis. Application coverage should be specified first and then translated to functional (algorithm) coverage. Application coverage is the process of reading and understanding specifications and standards of the relevant applications. Functional coverage of an ASIP is decided based on both the current standard specifications and carefully collected knowledge (e.g., books and research publications) in order to add extra features for future usage. Performance and cost should also be specified as design constraints.

After the functional coverage is determined, the partitioning of hardware and software should be decided through profiling of the source code. Hardware/software partitioning for an ASIP is to meet the performance constraint by defining what functions should be accelerated by application-specific instructions and what functions should be implemented as software routines using conventional instructions. This is an important design step of an instruction set, which is called the 10%-90% code locality. The locality rule means that 10% of the

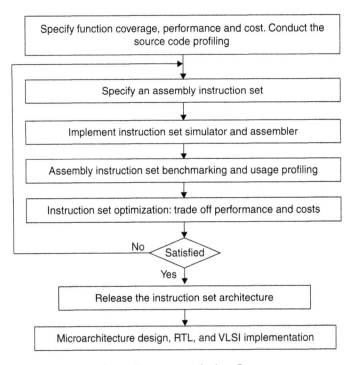

Figure 7.26: ASIP design flow

instructions run 90% of the time and 90% of the instructions appear only 10% of the time during execution. In other words, ASIP design is to find the best instruction set architecture optimized for the 10% most frequently used instructions and to select those among the 90% of the not often used instructions in order to guarantee the functional coverage.

During the process of hardware and software partitioning, the instruction set of the ASIP is gradually specified. The next design step is to implement the instruction set, which includes instruction coding, design of the instruction set simulator, and benchmarking. The coding of the instruction set includes the design of the assembly syntax and the design of the binary machine codes. The instruction set simulator must be implemented after the instruction set has been coded. Finally, the instruction set must be evaluated by benchmarking. The performance of the instruction set and the usage of each instruction will be exposed as inputs for further optimization.

The ASIP architecture can be specified when the assembly instruction set is released. The microarchitecture design is a refinement of the architecture design including fine-grained function allocation and hardware pipeline scheduling, specifying hardware modules, and interconnections between modules.

The ASIP design flow starts from the requirement specification and is complete after the microarchitecture design. The design of an ASIP is based mostly on experience, and it is essential to minimize the cost of design iteration. The implementation of the microarchitecture, which involves RTL coding, is not the focus of this section. You can find more information in references [17] and [22].

7.6.3 ASIP Design Automation

This subsection is written for researchers and project managers. Custom design of an ASIP DSP Processor is based on experience and is error prone. Design automation has been investigated and can be used to replace the custom ASIP design. ASIP design automation tools is summarized in Figure 7.27.

Figure 7.27 presents the research on ASIP design automation, which can be divided into three steps. The first step is the architecture exploration (selecting or generating an architecture and assembly instruction set according to the application analyses). Different profilers were designed by researchers, but the constraint specification tool has not been investigated. The tool to merge multiple CFG has not been extensively investigated. (CFG in this step stands for *control flow graph*.)

The second step is to specify an ADL (Architecture Description Language) to model the instruction set and architecture. This step can be very difficult. The language must be easy enough that ASIP designers can use it in modeling the design. However, if an ADL is easy, it

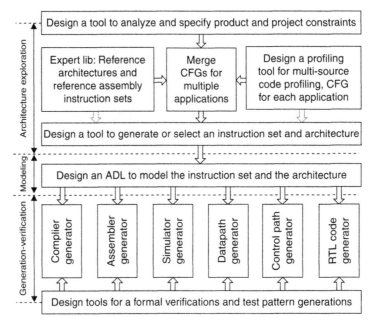

Figure 7.27: Automatic ASIP design flow (tool researcher's view)

cannot carry sufficient information to generate all tools and architectures. For example, tool generation requires sufficient modeling of the instruction set, and the hardware (datapath, control path) generation requires sufficient modeling of the microarchitecture (for example, structure of hardware multiplexing) and its function. If the ADL carries sufficient information for generating tools and architectures, the ADL will not be readable and cannot be used by ASIP designers. Details are beyond the scope of this chapter and can be found in our NoGAP research [38,39].

The third step involves generations and verifications. Enormous research on generation of tools and architectures can be found in Tensilica [32,33] and the famous LISA project [34,35]. As of 2007, there have been few research contributions to the ASIP formal verification.

It is good for designers to know the basic concepts behind design automation tools. Designers' interests actually focus on how to use the tools to generate instruction set, architecture, and assembly programming tools, as well as support for design verifications. This is summarized in Figure 7.28.

Architecture and assembly instruction set exploration according to application profiling is the first step of processor design automation. This is actually the most difficult part, because the

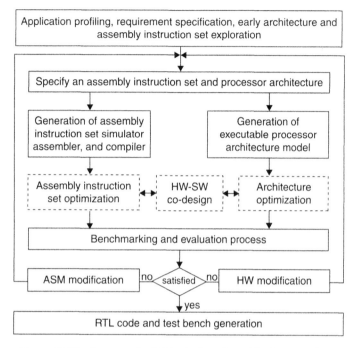

Figure 7.28: Automatic ASIP design flow (designer's view)

distance between CFGs of multiple applications and an ASM (assembly instruction set) is very large, and there are too many choices to make in selecting different instruction set architectures. To manage the large gap, another design step (constraint specification) might be needed.

Optimum ways to make decisions concerning tools have not yet been thoroughly investigated. Tools to generate accelerator instructions are available, but the tool to generate a complete processor instruction set does not exist.

Instruction set architecture of a processor is proposed and decided by designers, and the instruction set and the architecture selected will be inputs of processor modeling. The processor model will be used for generating the instruction set simulator, the compiler, assembler, and the architecture behavior model. After benchmarking of the instruction set and architecture, RTL code will finally be generated by the ASIP automation design tools. In Table 7.3, selected ASIP design tools are briefly discussed [25–39].

The last column in the table shows a feature of ASIP design automation tools; the instruction set and architecture selection may or may not be limited by a built-in architecture. Limited by an architecture means that the automation tool offers instruction extension based on a built-in

Table 7.3: Review of ASIP design automation tools

Tool or solution	Profiling and architecture exploring	Compiler generator	Assembler generator	Simulator generator	Cycle accurate architecture behavior model generator	Instruction set and architecture Optimizer	Datapath RTL code generator	Control path RTL code generator	Limited by an architecture
MIMOLA				yes			yes	yes	no
Cathedral-II	yes			yes	yes		yes	yes	no
Target	yes	yes	yes	yes	yes		yes	yes	no
ARC		yes	yes	yes	yes		yes	yes	limited
Tensilica	yes	yes	yes	yes	yes	yes	yes	yes	limited
LISA	yes	yes	yes	yes	yes		yes	yes	limited
MESCAL	yes		yes	yes	yes		yes	yes	no
PEAS-III			yes	yes	yes		yes	yes	limited
NoGAP	yes		yes	yes	yes		yes	yes	no
...									

processor instruction set architecture. The drawback is obviously that the built-in processor may not be suitable for all applications. Therefore, "no" in that column is superior to "limited".

MIMOLA is possibly the first processor design automation tool. It was proposed by Professor Zimmermann in 1976 and was the research product of Professor Zimmermann and Professor Marwedel of Kiel University and TU Dortmund in Germany. Cathedral-I and II proposed by Professor Rabaey and Professor De Man of IMEC in Belgium was possibly the first tool successfully used by industry (on the Mentor DSP station). Target was a successful research spin-off of Dr. Gert Goossens of IMEC, Belgium. ARC and Tensilica are successful companies that supply programmable hardware acceleration based on their processor core as the master of the platform.

LISA of CoWare was the research spin-off of Professor Meyr and Professor Leupers of ISS Aachen University, Germany. MESCAL is the research project of GigaScale of UC, Berkeley, California. NoGAP is the research project of Linköping University, Sweden.

7.7 Conclusions

The scope of DSP is huge, and includes DSP theory, DSP-related standards and applications, and DSP implementations. The intention of this chapter is to make a good partitioning of

DSP concepts into these three categories, and to focus on the hardware implementation of application-specific instruction set DSP processors.

In this chapter, knowledge required by the ASIP designer was briefly reviewed. DSP implementation was discussed based on four platforms—the general computer, the general-purpose DSP, the ASIP DSP, and the ASIC. DSP architectures were briefly introduced. Furthermore, the concepts of embedded systems and embedded computing also have been introduced, because the DSP subsystem is an essential part of most embedded systems. ASIP design flow and methodologies of ASIP design automation were briefly introduced. All topics covered in this chapter are only introductions.

We strongly recommend that you study the listed reference books [2,17,21,22] if you are not acquainted with the background knowledge discussed in this chapter.

References

1. http://www.da.isy.liu.se/.

2. Sanjit V, Mitra K. *Digital Signal Processing, A Computer-Based Approach*: McGraw-Hill; 1998.

3. Madisetti VK, Williams DB. *The Digital Signal Processing Handbook*: CRC Press; 1997 IEEE Press.

4. Norsworthy SR, Schreier R, Gabor Temes C, editors. *Delta-Sigma Data Converters, Theory, Design, and Simulation*: IEEE Press; 1997.

5. For example, www.ieee.org; www.iso.org; www.itu.org; and www.etsi.org.

6. For example, www.ti.com, www.adi.com, www.ceva.com, and www.freescale.com.

7. Krishna CM, Shin KG. *Real-Time Systems*: McGraw-Hill; 1997.

8. Laplante PA. *Real-Time Systems Design and Analysis*. 3rd ed.: IEEE Press and Wiley; 2004.

9. Gibson JD, editor. *The Mobile Communications Handbook*. 2nd ed.: CRC Press and IEEE Press; 1999.

10. Liu D, Tell E. *Low Power Baseband Processors for Communications*, Chapter 23. In: Piguet C, editor. *Low Power Electronics Design*: CRC Press; 2005.

11. Salomon D. *Data Compression, the Complete Reference*. 3rd ed.: Springer; 2006.

12. Kumaki T, et al. Multi-port CAM based VLSI architecture for Huffman coding with real-time optimized code word table. *48th Midwest Symposium on Circuits and Systems* 2005; **1**:55–58.

13. Goldberg R, Riek L. *A Practical Handbook of Speech Coders*: CRC Press; 2000.

14. Collura JS, Brandt DF, Rahikka DJ. The 1.2 Kbps/2.4 Kbps MELP Speech Coding Suite with Integrated Noise Pre-Processing. *IEEE MILCOM 2* 1999:1449–53.

15. www.mpeg.org and www.jpeg.org.

16. Wanhammar L. *DSP Integrated Circuits*: Academic Press; 1999.

17. Smith MJS. *Application-Specific Integrated Circuits (ASIC)*: Addison-Wesley VLSI Systems Series; 1997.

18. Wolf W. *Computers as Components, Principles of Embedded Computing System Design*: Morgan Kaufmann; 2001.

19. Gajski DD, Vahid F, Narayan S, Gong J. *Specification and Design of Embedded Systems, Technique or Embedded Systems*: Prentice Hall; 1995.

20. Lapsley P, Bier J, Shoham A, Lee EA. *DSP Processor Fundamentals, Architectures and Features*: IEEE Press; 1997.

21. Patterson DA, Hennessy JL. *Computer Organization & Design, The Hardware/Software Interface*: Morgan Kaufmann; 2004.

22. Keiting M, Bricaud P. *Reuse Methodology Manual for System On Chip Designs*. 3rd ed.: KAP; 2007.

23. Kuo SM, Gan W-S. *Digital Signal Processors, Architectures, Implementations, and Applications*: Prentice Hall; 2005.

24. Fettweis GP. DSP Cores for Mobile Communications: Where are we going?. *IEEE International Conference on Acoustics, Speech, and Signal Processing (ICASSP'97)* pp. 279–82, 21–24, April 1997.

25. Zimmermann G. The MIMOLA design system, a computer aided digital processor design method. *16th Design Automation Conference*; 1979.

26. Marwedel P. The MIMOLA design system: Tools for the design of digital processors. *21st Design Automation Conference*; 1984.

27. Krüger G. Entwurf einer Rechnerzentraleinheit für den Maschinenbefehlssatz des SIEMENS Systems 7.000 mit dem MIMOLA-Rechnerentwurfssystem. *Diploma Thesis. Kiel: University of Kiel*; 1980.

28. De Man H, Rabaey J, Six P, Clesen L. Cathedral-II, A Silicon Compiler for Digital Signal Processing. *IEEE J. of Design and Test of Computers*; December 1986.

29. http://www.retarget.com/.

30. Goossens G, Lanneer D, Geurts W, Van Praet J. Design of ASIPs in multi-processor SoCs using the Chess/Checker retargetable tool suit. *IEEE International Symposium on System-on-Chip*, 2006.

31. http://www.arc.com/

32. http://www.tensilica.com/

33. Rowen C. *Engineering the Complex SOC*. PTR 2004.

34. Hoffmann A, Meyr H, Leupers R. Architecture Exploration for Embedded Processors with LISA: *Kluwer Academic Publishers*; Dec. 2002.

35. http://www.coware.com/PDF/products/ProcessorDesigner.pdf.

36. Mihal A, Kulkarni C, Moskewicz M, Tsai M, Shah N, Weber S, Jin Yujia, Keutzer K, Vissers K, Sauer C, Malik S. Developing architectural platforms: a disciplined approach, *Design & Test of Computers. IEEE* 2002; **19**(6):6–16. Nov.–Dec. 2002.

37. Itoh M, Higaki S, Sato J, Shiomi A, Takeuchi Y, Kitajima A, Imai M. Peas-iii: an asip design environment, In: *Computer Design, 2000. Proceedings, 2000 International Conference on*. 2000. p. 430–436.

38. Karlström P, Liu D. NoGAP, a Micro Architecture Construction Framework. *19th IEEE International Conference on Application-specific Systems, Architectures and Processors*; July 2008; Belgium.

39. http://www.da.isy.liu.se/research/nogap/

CHAPTER 8

DSP Processors and Fixed-Point Arithmetic

Li Tan

This chapter presents the basic concepts of DSP hardware and software. And by basic, I mean it starts by referencing von Neumann's seminal 1946 paper introducing the von Neumann architecture, and contrasts it with the almost-as-old Harvard architecture favored by most DSPs. It then delves into the operation of basic execution units such as the multiply-accumulate (MAC) unit, shifter, and address generation unit (AGU).

You may be wondering, "Why do I need to know how a shifter works? Why do I need to know the difference between an opcode and operand?" The reason is that DSP algorithms are computationally intensive, and generally require high levels of optimization to meet performance, power, and cost targets. This optimization often entails working in assembly or another low-level language. To work in these low-level languages, you must know the underlying hardware.

On the software side, Tan gives us fundamental concepts that every DSP engineer needs. He starts with an explanation of fixed-point and floating-point data formats and then dives into the problems of underflow and overflow—the main reasons fixed-point implementations take so long. Then, he gets practical and lays out the IEEE standards for fixed and floating-point operations.

In the end of the chapter, Tan illustrates the above concepts with real hardware and software. He implements FIR and IIR filters—two of the most common DSP functions—on Texas Instruments' fixed-point C54x DSP and floating-point C3x DSP. If you want to tinker with your own filters, you are in luck: Tan provides sample C and MATLAB code for each example.

—Kenton Williston

8.1 Digital Signal Processor Architecture

Unlike microprocessors and microcontrollers, digital signal (DS) processors have special features that require operations such as fast Fourier transform (FFT), filtering, convolution and correlation, and real-time sample-based and block-based processing. Therefore, DS processors use a different dedicated hardware architecture.

We first compare the architecture of the general microprocessor with that of the DS processor. The design of general microprocessors and microcontrollers is based on the *von Neumann architecture,* which was developed from a research paper written by John von Neumann and others in 1946. Von Neumann suggested that computer instructions, as we shall discuss, be numerical codes instead of special wiring. Figure 8.1 shows the von Neumann architecture.

As shown in Figure 8.1, a von Neumann processor contains a single, shared memory for programs and data, a single bus for memory access, an arithmetic unit, and a program control unit. The processor proceeds in a serial fashion in terms of fetching and execution cycles. This means that the central processing unit (CPU) fetches an instruction from memory and decodes it to figure out what operation to do, then executes the instruction. The instruction (in machine code) has two parts: the *opcode* and the *operand*. The opcode specifies what the operation is, that is, tells the CPU what to do. The operand informs the CPU what data to operate on. These instructions will modify memory, or input and output (I/O). After an instruction is completed, the cycles will resume for the next instruction. One an instruction or piece of data can be retrieved at a time. Since the processor proceeds in a serial fashion, it causes most units to stay in a wait state.

As noted, the von Neumann architecture operates the cycles of fetching and execution by fetching an instruction from memory, decoding it via the program control unit, and finally executing the instruction. When execution requires data movement—that is, data to be read from or written to memory—the next instruction will be fetched after the current instruction

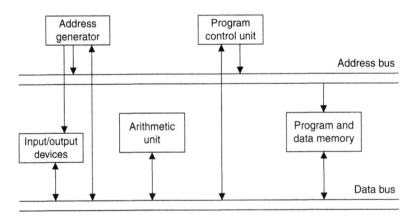

Figure 8.1: General microprocessor based on von Neumann architecture

is completed. The von Neumann-based processor has this bottleneck mainly due to the use of a single, shared memory for both program instructions and data. Increasing the speed of the bus, memory, and computational units can improve speed, but not significantly.

To accelerate the execution speed of digital signal processing, DS processors are designed based on the *Harvard architecture,* which originated from the Mark 1 relay-based computers built by IBM in 1944 at Harvard University. This computer stored its instructions on punched tape and data using relay latches. Figure 8.2 shows today's Harvard architecture. As depicted, the DS processor has two separate memory spaces. One is dedicated to the program code, while the other is employed for data. Hence, to accommodate two memory spaces, two corresponding address buses and two data buses are used. In this way, the program memory and data memory have their own connections to the program memory bus and data memory bus, respectively. This means that the Harvard processor can fetch the program instruction and data in parallel at the same time, the former via the program memory bus and the latter via the data memory bus. There is an additional unit called a *multiplier and accumulator* (MAC), which is the dedicated hardware used for the digital filtering operation. The last additional unit, the shift unit, is used for the scaling operation for fixed-point implementation when the processor performs digital filtering.

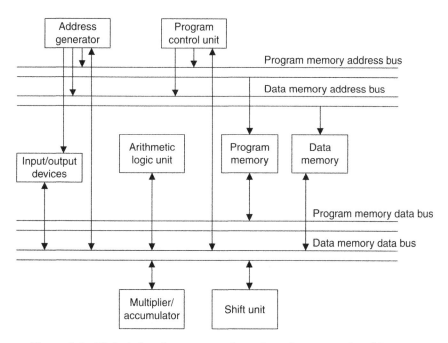

Figure 8.2: Digital signal processors based on the Harvard architecture

Let us compare the executions of the two architectures. The von Neumann architecture generally has the execution cycles described in Figure 8.3. The fetch cycle obtains the opcode from the memory, and the control unit will decode the instruction to determine the operation. Next is the execute cycle. Based on the decoded information, execution will modify the content of the register or the memory. Once this is completed, the process will fetch the next instruction and continue. The processor operates one instruction at a time in a serial fashion.

To improve the speed of the processor operation, the Harvard architecture takes advantage of a common DS processor, in which one register holds the filter coefficient while the other register holds the data to be processed, as depicted in Figure 8.4.

As shown in Figure 8.4, the execute and fetch cycles are overlapped. We call this the *pipelining* operation. The DS processor performs one execution cycle while also fetching the next instruction to be executed. Hence, the processing speed is dramatically increased.

The Harvard architecture is preferred for all DS processors due to the requirements of most DSP algorithms, such as filtering, convolution, and FFT, which need repetitive arithmetic operations, including multiplications, additions, memory access, and heavy data flow through the CPU.

For other applications, such as those dependent on simple microcontrollers with less of a timing requirement, the von Neumann architecture may be a better choice, since it offers much less silicon area and is thus less expensive.

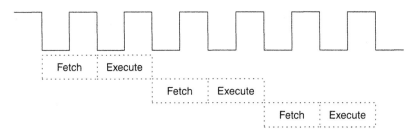

Figure 8.3: Execution cycle based on the von Neumann architecture

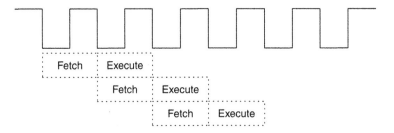

Figure 8.4: Execution cycle based on the Harvard architecture

8.2 Digital Signal Processor Hardware Units

In this section, we will briefly discuss special DS processor hardware units.

8.2.1 Multiplier and Accumulator

As compared with the general microprocessors based on the von Neumann architecture, the DS processor uses the MAC, a special hardware unit for enhancing the speed of digital filtering. This is dedicated hardware, and the corresponding instruction is generally referred to as a MAC operation. The basic structure of the MAC is shown in Figure 8.5.

As shown in Figure 8.5, in a typical hardware MAC, the multiplier has a pair of input registers, each holding the 16-bit input to the multiplier. The result of the multiplication is accumulated in a 32-bit accumulator unit. The result register holds the double precision data from the accumulator.

8.2.2 Shifters

In digital filtering, to prevent overflow, a scaling operation is required. A simple scaling-down operation shifts data to the right, while a scaling-up operation shifts data to the left. Shifting data to the right is the same as dividing the data by 2 and truncating the fraction part; shifting data to the left is equivalent to multiplying the data by 2. As an example, for a 3-bit data word $011_2 = 3_{10}$,

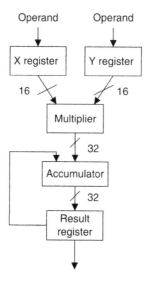

Figure 8.5: The multiplier and accumulator (MAC) dedicated to DSP

shifting 011 to the right gives $001_2 = 1$; that is, $3/2 = 1.5$, and truncating 1.5 results in 1. Shifting the same number to the left, we have $110_2 = 6_{10}$; that is, $3 \times 2 = 6$. The DS processor often shifts data by several bits for each data word. To speed up such operations, the special hardware shift unit is designed to accommodate the scaling operation, as depicted in Figure 8.2.

8.2.3 Address Generators

The DS processor generates the addresses for each datum on the data buffer to be processed. A special hardware unit for circular buffering is used (see the address generator in Figure 8.2). Figure 8.6 describes the basic mechanism of circular buffering for a buffer having eight data samples.

In circular buffering, a pointer is used and always points to the newest data sample, as shown in the figure. After the next sample is obtained from analog-to-digital conversion (ADC), the data will be placed at the location of $x(n - 7)$, and the oldest sample is pushed out. Thus, the location for $x(n - 7)$ becomes the location for the current sample. The original location for $x(n)$ becomes a location for the past sample of $x(n - 1)$. The process continues according to the mechanism just described. For each new data sample, only one location on the circular buffer needs to be updated.

The circular buffer acts like a first-in/first-out (FIFO) buffer, but each datum on the buffer does not have to be moved. Figure 8.7 gives a simple illustration of the 2-bit circular buffer. In the figure, there is data flow to the ADC ($a, b, c, d, e, f, g, \ldots$) and a circular buffer initially containing $a, b, c,$ and d. The pointer specifies the current data of d, and the equivalent FIFO buffer is shown on the right side with a current data of d at the top of the memory. When e

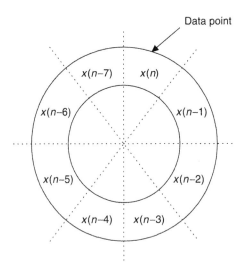

Figure 8.6: Illustration of circular buffering

comes in, as shown in the middle drawing in Figure 8.7, the circular buffer will change the pointer to the next position and update old *a* with a new datum *e*. It costs the pointer only one movement to update one datum in one step. However, on the right side, the FIFO has to move each of the other data down to let in the new datum *e* at the top. For this FIFO, it takes four data movements. In the bottom drawing in Figure 8.7, the incoming datum *f* for both the circular buffer and the FIFO buffer continues to confirm our observations.

Like finite impulse response (FIR) filtering, the data buffer size can reach several hundreds. Hence, using the circular buffer will significantly enhance the processing speed.

8.3 Digital Signal Processors and Manufacturers

DS processors are classified for general DSP and special DSP. The general-DSP processor is designed and optimized for applications such as digital filtering, correlation, convolution,

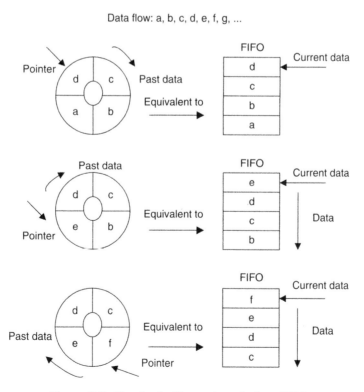

Figure 8.7: Circular buffer and equivalent FIFO

and FFT. In addition to these applications, the special DSP processor has features that are optimized for unique applications such as audio processing, compression, echo cancellation, and adaptive filtering. Here, we will focus on the general-DSP processor.

The major manufacturers in the DSP industry are Texas Instruments (TI), Analog Devices, and Motorola. TI and Analog Devices offer both fixed-point DSP families and floating-point DSP families, while Motorola offers fixed-point DSP families. We will concentrate on TI families, review their architectures, and study real-time implementation using the fixed-and floating-point formats.

8.4 Fixed-Point and Floating-Point Formats

In order to process real-world data, we need to select an appropriate DS processor, as well as a DSP algorithm or algorithms for a certain application. Whether a DS processor uses a fixed- or floating-point method depends on how the processor's CPU performs arithmetic. A fixed-point DS processor represents data in *2's complement integer format* and manipulates data using integer arithmetic, while a floating-point processor represents numbers using a mantissa (fractional part) and an exponent in addition to the integer format and operates data using floating-point arithmetic (discussed in a later section).

Since the fixed-point DS processor operates using the integer format, which represents only a very narrow dynamic range of the integer number, a problem such as overflow of data manipulation may occur. Hence, we need to spend much more coding effort to deal with such a problem. As we shall see, we may use floating-point DS processors, which offer a wider dynamic range of data, so that coding becomes much easier. However, the floating-point DS processor contains more hardware units to handle the integer arithmetic and the floating-point arithmetic, hence is more expensive and slower than fixed-point processors in terms of instruction cycles. It is usually a choice for prototyping or proof-of-concept development.

When it is time to make the DSP an application-specific integrated circuit (ASIC), a chip designed for a particular application, a dedicated hand-coded fixed-point implementation can be the best choice in terms of performance and small silica area.

The formats used by DSP implementation can be classified as fixed or floating point.

Table 8.1: A 3-bit 2's complement number representation

Decimal Number	2's Complement
3	011
2	010
1	001
0	000
−1	111
−2	110
−3	101
−4	100

8.4.1 Fixed-Point Format

We begin with 2's complement representation. Considering a 3-bit 2's complement, we can represent all the decimal numbers shown in Table 8.1.

Let us review the 2's complement number system using Table 8.1. Converting a decimal number to its 2's complement requires the following steps:

1. Convert the magnitude in the decimal to its binary number using the required number of bits.

2. If the decimal number is positive, its binary number is its 2's complement representation; if the decimal number is negative, perform the 2's complement operation, where we negate the binary number by changing the logic 1's to logic 0's and logic 0's to logic 1's and then add a logic 1 to the data. For example, a decimal number of 3 is converted to its 3-bit 2's complement as 011; however, for converting a decimal number of −3, we first get a 3-bit binary number for the magnitude in the decimal, that is, 011. Next, negating the binary number 011 yields the binary number 100. Finally, adding a binary logic 1 achieves the 3-bit 2's complement representation of −3, that is, 100 + 1 = 101, as shown in Table 8.1.

As we see, a 3-bit 2's complement number system has a dynamic range from −4 to 3, which is very narrow. Since the basic DSP operations include multiplications and additions, results of operation can cause overflow problems. Let us examine the multiplications in Example 8.1.

Example 8.1

Given:

1. $2 \times (-1)$
2. $2 \times (-3)$,

a. Operate each using its 2's complement.

a. **1.**
```
        010
     ×  001
     ------
        010
       000
     + 000
     ------
      00010
```

and 2's complement of $00010 = 11110$. Removing two extended sign bits gives 110.

The answer is 110 (-2), which is within the system.

   2.
```
        010
     ×  011
     ------
        010
        010
     + 000
     ------
      00110
```

and 2's complement of $00110 = 11010$. Removing two extended sign bits achieves 010.

Since the binary number 010 is 2, which is not (-6) as we expect, overflow occurs; that is, the result of the multiplication (-6) is out of our dynamic range $(-4$ to $3)$.

Let us design a system treating all the decimal values as fractional numbers, so that we obtain the fractional binary 2's complement system shown in Table 8.2.

To become familiar with the fractional binary 2's complement system, let us convert a positive fraction number $\frac{3}{4}$ and a negative fraction number $-\frac{1}{4}$ in decimals to their 2's complements. Since

$$\frac{3}{4} = 0 \times 2^0 + 1 \times 2^{-1} + 1 \times 2^{-2},$$

Table 8.2: A 3-bit 2's complement system using fractional representation

Decimal Number	Decimal Fraction	2's Complement
3	3/4	0.11
2	2/4	0.10
1	1/4	0.01
0	0	0.00
−1	−1/4	1.11
−2	−2/4	1.10
−3	−3/4	1.01
−4	−4/4 = −1	1.00

its 2's complement is 011. Note that we did not mark the binary point for clarity. Again, since

$$\frac{1}{4} = 0 \times 2^0 + 0 \times 2^{-1} + 1 \times 2^{-2},$$

its positive-number 2's complement is 001. For the negative number, applying the 2's complement to the binary number 001 leads to $110 + 1 = 111$, as we see in Table 8.2.

Now let us focus on the fractional binary 2's complement system. The data are normalized to the fractional range from –1 to $1 - 2^{-2} = \frac{3}{4}$. When we carry out multiplications with two fractions, the result should be a fraction, so that multiplication overflow can be prevented. Let us verify the multiplication $(010) \times (101)$, which is the overflow case in Example 8.1:

$$
\begin{array}{r}
0.10 \\
\times\, 0.11 \\
\hline
010 \\
010 \\
+\ 000 \\
\hline
0.0110
\end{array}
$$

2's complement of $0.0110 = 1.1010$.

The answer in decimal form should be:

$$1.1010 = (-1) \times (0.0110)_2 = -(0 \times (2)^{-1} + 1 \times (2)^{-2} + 1 \times (2)^{-3} + 0 \times (2)^{-4})$$
$$= -\frac{3}{8}.$$

This number is correct, as we can verify from Table 8.2, that is, $(\frac{2}{4} \times (-\frac{3}{4})) = -\frac{3}{8}$.

If we truncate the last two least-significant bits to keep the 3-bit binary number, we have an approximated answer as:

$$1.10 = (-1) \times (0.10)_2 = -(1 \times (2)^{-1} + 0 \times (2)^{-2}) = -\frac{1}{2}.$$

The truncation error occurs. The error should be bounded by $2^{-2} = \frac{1}{4}$. We can verify that:

$$|-1/2 - (-3/8)| = 1/8 < 1/4.$$

To use such a scheme, we can avoid the overflow due to multiplications but cannot prevent the additional overflow. In the following addition example,

$$
\begin{array}{r}
0.11 \\
+\,0.01 \\
\hline
1.00
\end{array}
$$

where the result 1.00 is a negative number.

Adding two positive fractional numbers yields a negative number. Hence, overflow occurs. We see that this signed fractional number scheme partially solves the overflow in multiplications. Such fractional number format is called the signed Q-2 format, where there are 2 magnitude bits plus one sign bit. The additional overflow will be tackled using a scaling method discussed in a later section.

Q-format number representation is the most common one used in fixed-point DSP implementation. It is defined in Figure 8.8.

As indicated in Figure 8.8, Q-15 means that the data are in a sign magnitude form in which there are 15 bits for magnitude and one bit for sign. Note that after the sign bit, the dot shown in Figure 8.8 implies the binary point. The number is normalized to the fractional range from -1 to 1. The range is divided into 2^{16} intervals, each with a size of 2^{-15}. The most negative number is -1, while the most positive number is $1-2^{-15}$. Any result from multiplication is within the fractional range of -1 to 1. Let us study the following examples to become familiar with Q-format number representation.

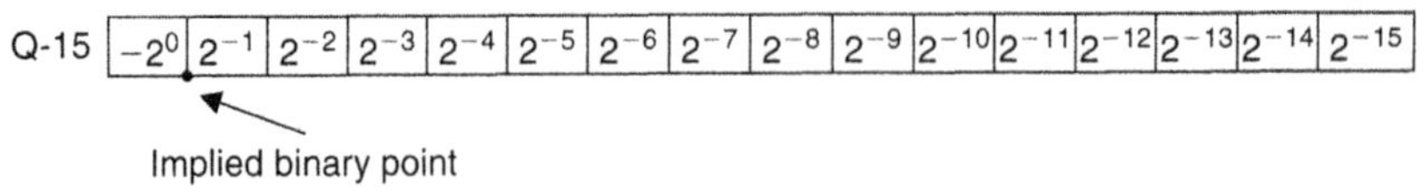

Figure 8.8: Q-15 (fixed-point) format

Example 8.2

a. Find the signed Q-15 representation for the decimal number 0.560123.

Solution

a. The conversion process is illustrated using Table 8.3. For a positive fractional number, we multiply the number by 2 if the product is larger than 1, carry bit 1 as a most-significant bit (MSB), and copy the fractional part to the next line for the next multiplication by 2; if the product is less than 1, we carry bit 0 to MSB. The procedure continues to collect all 15 magnitude bits.

We yield the Q-15 format representation as

$$0.100011110110010.$$

Since we use only 16 bits to represent the number, we may lose accuracy after conversion. Like quantization, the truncation error is introduced. However, this error should be less than the interval size, in this case, $2^{-15} = 0.000030517$. We shall verify this in

Table 8.3: Conversion process of Q-15 representation

Number	Product	Carry
0:560123 × 2	1.120246	1 (MSB)
0:120246 × 2	0.240492	0
0:240492 × 2	0.480984	0
0:480984 × 2	0.961968	0
0:961968 × 2	1.923936	1
0:923936 × 2	1.847872	1
0:847872 × 2	1.695744	1
0:695744 × 2	1.391488	1
0:391488 × 2	0.782976	0
0:782976 × 2	1.565952	1
0:565952 × 2	1.131904	1
0:131904 × 2	0.263808	0
0:263808 × 2	0.527616	0
0:527616 × 2	1.055232	1
0:055232 × 2	0.110464	0 (LSB)

MSB, most-significant bit; LSB, least-significant bit.

Example 8.5. An alternative way of conversion is to convert a fraction, let's say ¾, to Q-2 format, multiply it by 2^2, and then convert the truncated integer to its binary, that is,

$$(3/4) \times 2^2 = 3 = 011_2.$$

In this way, it follows that:

$$(0.560123) \times 2^{15} = 18354.$$

Converting 18354 to its binary representation will achieve the same answer. The next example illustrates the signed Q-15 representation for a negative number.

Example 8.3

a. Find the signed Q-15 representation for the decimal number -0.160123.

Solution

a. Converting the Q-15 format for the corresponding positive number with the same magnitude using the procedure described in Example 8.2, we have:

$$0.160123 = 0.0001010001111110.$$

Then, after applying 2's complement, the Q-15 format becomes:

$$-0.160123 = 1.110101110000010.$$

Alternative way: Since $(-0.160123) \times 2^{15} = -5246.9$, converting the truncated number -5246 to its 16-bit 2's complement yields 1110101110000010.

Example 8.4

a. Convert the Q-15 signed number 1.110101110000010 to the decimal number.

Solution

a. Since the number is negative, applying the 2's complement yields:

$$0.001010001111110.$$

Then the decimal number is:

$$-(2^{-3} + 2^{-5} + 2^{-9} + 2^{-10} + 2^{-11} + 2^{-12} + 2^{-13} + 2^{-14}) = -0.160095.$$

Example 8.5

a. Convert the Q-15 signed number 0.100011110110010 to the decimal number.

Solution

a. The decimal number is:

$$2^{-1} + 2^{-5} + 2^{-6} + 2^{-7} + 2^{-8} + 2^{-10} + 2^{-11} + 2^{-14} = 0.560120.$$

As we know, the truncation error in Example 8.2 is less than $2^{-15}=0:000030517$. We verify that the truncation error is bounded by:

$$|0.560120 - 0.560123| = 0.000003 < 0.000030517.$$

Note that the larger the number of bits used, the smaller the round-off error that may accompany it.

Examples 8.6 and 8.7 are devoted to illustrating data manipulations in the Q-15 format.

Example 8.6

a. Add the two numbers in Examples 8.4 and 8.5 in Q-15 format.

Solution

a. Binary addition is carried out as follows:

$$
\begin{array}{r}
1.110101110000010 \\
+\ 0.100011110110010 \\
\hline
10.011001100110100
\end{array}
$$

Then the result is:

$$0.011001100110100.$$

This number in the decimal form can be found to be:

$$2^{-2} + 2^{-3} + 2^{-6} + 2^{-7} + 2^{-10} + 2^{-11} + 2^{-13} = 0.400024.$$

Example 8.7

This is a simple illustration of fixed-point multiplication.

a. Determine the fixed-point multiplication of 0.25 and 0.5 in Q-3 fixed-point 2's complement format.

Solution

a. Since $0.25 = 0.010$ and $0.5 = 0.100$, we carry out binary multiplication as follows:

$$
\begin{array}{r}
0.010 \\
\times\,0.100 \\
\hline
0000 \\
0000 \\
0010 \\
+\,0000 \\
\hline
0.001000
\end{array}
$$

Truncating the least-significant bits to convert the result to Q-3 format, we have:

$$0.010 \times 0.100 = 0.001.$$

Note that $0.001 = 2^{-3} = 0.125$. We can also verify that $0.25 \times 0.5 = 0.125$.

As a result, the Q-format number representation is a better choice than the 2's complement integer representation. But we need to be concerned with the following problems.

1. Converting a decimal number to its Q-N format, where N denotes the number of magnitude bits, we may lose accuracy due to the truncation error, which is bounded by the size of the interval, that is, 2^{-N}.

2. Addition and subtraction may cause overflow, where adding two positive numbers leads to a negative number, or adding two negative numbers yields a positive number; similarly, subtracting a positive number from a negative number gives a positive number, while subtracting a negative number from a positive number results in a negative number.

3. Multiplying two numbers in Q-15 format will lead to a Q-30 format, which has 31 bits in total. As in Example 8.7, the multiplication of Q-3 yields a Q-6 format, that is, 6 magnitude bits and a sign bit. In practice, it is common for a DS processor to hold the multiplication result using a double word size such as MAC operation, as shown in Figure 8.9 for multiplying two numbers in Q-15 format. In Q-30 format, there is one

sign-extended bit. We may get rid of it by shifting left by one bit to obtain Q-31 format and maintaining the Q-31 format for each MAC operation. Sometimes, the number in Q-31 format needs to be converted to Q-15; for example, the 32-bit data in the accumulator needs to be sent for 16-bit digital-to-analog conversion (DAC), where the upper most-significant 16 bits in the Q-31 format must be used to maintain accuracy. We can shift the number in Q-30 to the right by 15 bits or shift the Q-31 number to the right by 16 bits. The useful result is stored in the lower 16-bit memory location. Note that after truncation, the maximum error is bounded by the interval size of 2^{-15}, which satisfies most applications. In using the Q-format in the fixed-point DS processor, it is costive to maintain the accuracy of data manipulation.

4. Underflow can happen when the result of multiplication is too small to be represented in the Q-format. As an example, in the Q-2 system shown in Table 8.2, multiplying 0.01×0.01 leads to 0.0001. To keep the result in Q-2, we truncate the last two bits of 0.0001 to achieve 0.00, which is zero. Hence, underflow occurs.

8.4.2 Floating-Point Format

To increase the dynamic range of number representation, a floating-point format, which is similar to scientific notation, is used. The general format for floating-point number representation is given by:

$$x = M \cdot 2^{E}, \qquad (8.1)$$

where M is the mantissa, or fractional part, in Q format, and E is the exponent. The mantissa and exponent are signed numbers. If we assign 12 bits for the mantissa and 4 bits for the exponent, the format looks like Figure 8.10.

Since the 12-bit mantissa has limits between −1 and +1, the dynamic range is controlled by the number of bits assigned to the exponent. The bigger the number of bits assigned to

Figure 8.9: Sign bit extended Q-30 format

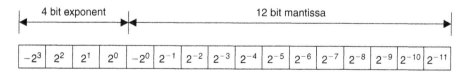

Figure 8.10: Floating-point format

the exponent, the larger the dynamic range. The number of bits for the mantissa defines the interval in the normalized range; as shown in Figure 8.10, the interval size is 2^{-11} in the normalized range, which is smaller than the Q-15. However, when more mantissa bits are used, the smaller interval size will be achieved. Using the format in Figure 8.10, we can determine the most negative and most positive numbers as:

Most negative number $= (1.00000000000)_2 \cdot 2^{0111_2} = (-1) \times 2^7 = -128.0$

Most positive number $= (0.11111111111)_2 \cdot 2^{0111_2} = (1 - 2^{-11}) \times 2^7 = 127.9375.$

The smallest positive number is given by

Smallest positive number $= (0.000000000001)_2 \cdot 2^{1000_2} = (2^{-11}) \times 2^{-8} = 2^{-19}.$

As we can see, the exponent acts like a scale factor to increase the dynamic range of the number representation. We study the floating-point format in the following example.

Example 8.8

Convert each of the following decimal numbers to the floating-point number using the format specified in Figure 8.10.

1. 0.1601230
2. −20.430527

Solution

a. 1. We first scale the number 0.1601230 to $0.160123/2^{-2} = 0:640492$ with an exponent of −2 (other choices could be 0 or −1) to get $0.160123 = 0.640492 \times 2^{-2}$. Using 2's complement, we have −2 = 1110. Now we convert the value 0.640492 using Q-11 format to get 010100011111. Cascading the exponent bits and the mantissa bits yields:

$$1110010100011111.$$

2. Since $-20.430527/2^5 = -0.638454$, we can convert it into the fractional part and exponent part as $-20.430527 = -0.638454 \times 2^5$.

Note that this conversion is not particularly unique; the forms $-20.430527 = -0.319227 \times 2^6$ and $-20.430527 = -0.1596135 \times 2^7 \dots$ are still valid choices. Let us keep what we have now. Therefore, the exponent bits should be 0101. Converting the number 0.638454 using Q-11 format gives:

$$010100011011.$$

Using 2's complement, we obtain the representation for the decimal number -0.638454 as:

$$101011100101.$$

Cascading the exponent bits and mantissa bits, we achieve:

$$0101101011100101.$$

The floating arithmetic is more complicated. We must obey the rules for manipulating two floating-point numbers. Rules for arithmetic addition are given as:

$$x_1 = M_1 2^{E_1}$$
$$x_2 = M_2 2^{E_2}.$$

The floating-point sum is performed as follows:

$$x_1 + x_2 = \begin{cases} (M_1 + M_2 \times 2^{-(E_1 - E_2)}) \times 2^{E_1}, & \text{if } E_1 \geq E_2 \\ (M_1 \times 2^{-(E_2 - E_1)} + M_2) \times 2^{E_2} & \text{if } E_1 < E_2 \end{cases}$$

As a multiplication rule, given two properly normalized floating-point numbers:

$$x_1 = M_1 2^{E_1}$$
$$x_2 = M_2 2^{E_2},$$

where $0.5 \leq |M_1| < 1$ and $0.5 \leq |M_2| < 1$. Then multiplication can be performed as follows:

$$x_1 \times x_2 = (M_1 \times M_2) \times 2^{E_1 + E_2} = M \times 2^{E}.$$

That is, the mantissas are multiplied while the exponents are added:

$$M = M_1 \times M_2$$
$$E = E_1 + E_2.$$

Examples 8.9 and 8.10 serve to illustrate manipulators.

Next, we examine overflow and underflow in the floating-point number system.

8.4.2.1 Overflow

During operation, overflow will occur when a number is too large to be represented in the floating-point number system. Adding two mantissa numbers may lead to a number larger than

Example 8.9

a. Add two floating-point numbers achieved in Example 8.8:

$$1110\ 010100011111 = 0.640136718 \times 2^{-2}$$
$$0101\ 101011100101 = -0.638183593 \times 2^{5}.$$

Solution

a. Before addition, we change the first number to have the same exponent as the second number, that is,

$$0101\ 000000001010 = 0.005001068 \times 2^{5}.$$

Then we add two mantissa numbers:

$$0.00000001010$$
$$+ \quad 1.01011100101$$
$$\overline{ 1.01011101111}$$

and we get the floating number as:

$$0101\ 101011101111.$$

We can verify the result by the following:

$$0101\ 101011101111 = -(2^{-1} + 2^{-3} + 2^{-7} + 2^{-11}) \times 2^{5}$$
$$= -0.633300781 \times 2^{5} = -20.265625.$$

Example 8.10

a. Multiply two floating-point numbers achieved in Example 8.8:

$$1110\ 010100011111 = 0.640136718 \times 2^{-2}$$
$$0101\ 101011100101 = -0.638183593 \times 2^{5}.$$

Solution

a. From the results in Example 8.8, we have the bit patterns for these two numbers as:

$$E_1 = 1110,\ E_2 = 0101,\ M_1 = 010100011111,\ M_2 = 101011100101.$$

Adding two exponents in 2's complement form leads to:

$$E = E_1 + E_2 = 1110 + 0101 = 0011,$$

which is $+3$, as we expected, since in decimal domain $(-2) + 5 = 3$.

As previously shown in the multiplication rule, when multiplying two mantissas, we need to apply their corresponding positive values. If the sign for the final value is negative, then we convert it to its 2's complement form. In our example, $M_1 = 010100011111$ is a positive mantissa. However, $M_2 = 101011100101$ is a negative mantissa, since the MSB is 1. To perform multiplication, we use 2's complement to convert M_2 to its positive value, 010100011011, and note that the multiplication result is negative. We multiply two positive mantissas and truncate the result to 12 bits to give:

$$010100011111 \times 010100011011 = 001101000100.$$

Now we need to add a negative sign to the multiplication result with 2's complement operation. Taking 2's complement, we have:

$$M = 110010111100.$$

Hence, the product is achieved by cascading the 4-bit exponent and 12-bit mantissa as:

$$0011\ 110010111100.$$

Converting this number back to the decimal number, we verify the result to be $0.408203125 \times 2^3 = -3.265625$.

1 or less than –1; and multiplying two numbers causes the addition of their two exponents, so that the sum of the two exponents could overflow. Consider the following overflow cases.

Case 1. Add the following two floating-point numbers:

$$0111\ 011000000000 + 0111\ 010000000000.$$

Note that two exponents are the same and they are the biggest positive number in 4-bit 2's complement representation. We add two positive mantissa numbers as:

$$
\begin{array}{r}
0.11000000000 \\
+ \quad 0.10000000000 \\
\hline
1.01000000000.
\end{array}
$$

The result for adding mantissa numbers is negative. Hence, the overflow occurs.

Case 2. Multiply the following two numbers:

$$0111\ 011000000000 + 0111\ 011000000000.$$

Adding two positive exponents gives:

$$0111 + 0111 = 1000 \text{ (negative; the overflow occurs)}.$$

Multiplying two mantissa numbers gives:

$$0.11000000000 \times 0.11000000000 = 0.10010000000 \text{ (OK!)}.$$

8.4.2.2 Underflow

As we discussed before, underflow will occur when a number is too small to be represented in the number system. Let us divide the following two floating-point numbers:

$$1001\ 001000000000 \div 0111\ 010000000000.$$

First, subtracting two exponents leads to:

$$1001 \text{ (negative)} - 0111 \text{ (positive)} = 1001 + 1001$$
$$= 0010 \text{ (positive; the underflow occurs)}$$

Then, dividing two mantissa numbers, it follows that:

$$0.01000000000 \div 0.10000000000 = 0.10000000000 \text{ (OK!)}.$$

However, in this case, the expected resulting exponent is -14 in decimal, which is too small to be presented in the 4-bit 2's complement system. Hence the underflow occurs.

Understanding basic principles of the floating-point formats, we can next examine two floating-point formats of the Institute of Electrical and Electronics Engineers (IEEE).

8.4.3 IEEE Floating-Point Formats

8.4.3.1 Single Precision Format

IEEE floating-point formats are widely used in many modern DS processors. There are two types of IEEE floating-point formats (IEEE 754 standard). One is the IEEE single precision format, and the other is the IEEE double precision format. The single precision format is described in Figure 8.11.

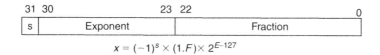

$$x = (-1)^s \times (1.F) \times 2^{E-127}$$

Figure 8.11: IEEE single precision floating-point format

The format of IEEE single precision floating-point standard representation requires 23 fraction bits F, 8 exponent bits E, and 1 sign bit S, with a total of 32 bits for each word. F is the mantissa in 2's complement positive binary fraction represented from bit 0 to bit 22. The mantissa is within the normalized range limits between $+1$ and $+2$. The sign bit S is employed to indicate the sign of the number, where when $S = 1$ the number is negative, and when $S = 0$ the number is positive. The exponent E is in excess 127 form. The value of 127 is the offset from the 8-bit exponent range from 0 to 255, so that E-127 will have a range from -127 to $+128$. The formula shown in Figure 8.11 can be applied to convert the IEEE 754 standard (single precision) to the decimal number. The following simple examples also illustrate this conversion:

$$0\ 1000000\ 00000000000000000000000 = (-1)^0 \times (1.0_2) \times 2^{128-127} = 2.0$$
$$0\ 1000000\ 11010000000000000000000 = (-1)^0 \times (1.101_2) \times 2^{129-127} = 6.51$$
$$1\ 1000000\ 11010000000000000000000 = (-1)^1 \times (1.101_2) \times 2^{129-127} = -6.5.$$

Let us look at Example 8.11 for more explanation.

8.4.3.2 Double Precision Format

The IEEE double precision format is described in Figure 8.12.

The IEEE double precision floating-point standard representation requires a 64-bit word, which may be numbered from 0 to 63, left to right. The first bit is the sign bit S, the next eleven bits are the exponent bits E, and the final 52 bits are the fraction bits F. The IEEE floating-point format in double precision significantly increases the dynamic range of number representation, since there are eleven exponent bits; the double precision format also reduces the interval size in the mantissa normalized range of $+1$ to $+2$, since there are 52 mantissa bits as compared with the single precision case of 23 bits. Applying the conversion formula shown in Figure 8.12 is similar to the single precision case.

Example 8.11

a. Convert the following number in the IEEE single precision format to the decimal format:

$$1\ 10000000.010\ldots0000.$$

Solution

a. From the bit pattern in Figure 8.11, we can identify the sign bit, exponent, and fractional as:

$$s = 1, E = 2^7 = 128$$

$$1.F = 1.01_2 = (2)^0 + (2)^{-2} = 1.25.$$

Then, applying the conversion formula leads to:

$$x = (-1)^1(1.25) \times 2^{128-127} = -1.25 \times 2^1 = -2.5.$$

In conclusion, the value x represented by the word can be determined based on the following rules, including all the exceptional cases:

- If $E = 255$ and F is nonzero, then $x = NaN$ ("Not a number").
- If $E = 255$, F is zero, and S is 1, then $x = -$Infinity.
- If $E = 255$, F is zero, and S is 0, then $x = +$Infinity.
- If $0 < E < 255$, then $x = (-1)^s \times (1.F) \times 2^{E-127}$, where $1.F$ represents the binary number created by prefixing F with an implicit leading 1 and a binary point.
- If $E = 0$ and F is nonzero, then $x = (-1)^s \times (0.F) \times 2^{-126}$. This is an "unnormalized" value.
- If $E = 0$, F is zero, and S is 1, then $x = -0$.
- If $E = 0$, F is zero, and S is 0, then $x = 0$.

Typical and exceptional examples are shown as follows:

$$0\ 00000000\ 00000000000000000000000 = 0$$
$$1\ 00000000\ 00000000000000000000000 = -0$$
$$0\ 11111111\ 00000000000000000000000 = \text{Infinity}$$
$$1\ 11111111\ 00000000000000000000000 = -\text{Infinity}$$
$$0\ 11111111\ 00000100000000000000000 = \text{NaN}$$
$$1\ 11111111\ 00100010001001010101010 = \text{NaN}$$
$$0\ 00000001\ 00000000000000000000000 = (-1)^0 \times (1.0_2) \times 2^{1-127} = 2^{-126}$$
$$0\ 00000000\ 10000000000000000000000 = (-1)^0 \times (0.1_2) \times 2^{0-126} = 2^{-127}$$
$$0\ 00000000\ 00000000000000000000001 = (-1)^0 \times (0.00000000000000000000001_2) \times 2^{0-126}$$
$$= 2^{-149}\ (\text{smallest positive value})$$

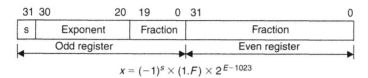

Figure 8.12: IEEE double precision floating-point format

Example 8.12

a. Convert the following number in IEEE double precision format to the decimal format:

$$001000\ldots0.110\ldots0000$$

Solution

b. Using the bit pattern in Figure 8.12, we have:

$$s = 0, E = 2^9 = 512 \text{ and}$$

$$1.F = 1.11_2 = (2)^0 + (2)^{-1} + (2)^{-2} = 1.75.$$

Then, applying the double precision formula yields:

$$x = (-1)^0(1.75) \times 2^{512-1023} = 1.75 \times 2^{-511} = 2.6104 \times 10^{-154}.$$

For purposes of completeness, rules for determining the value x represented by the double precision word are listed as follows:

- If $E = 2047$ and F is nonzero, then $x = NaN$ ("Not a number").

- If $E = 2047$, F is zero, and S is 1, then $x = -$Infinity.

- If $E = 2047$, F is zero, and S is 0, then $x = +$Infinity.

- If $0 < E < 2047$, then $x = (-1)^s \times (1.F) \times 2^{E-1023}$, where $1.F$ is intended to represent the binary number created by prefixing F with an implicit leading 1 and a binary point.

- If $E = 0$ and F is nonzero, then $x = (-1)^s \times (0.F) \times 2^{-1022}$. This is an "unnormalized" value.

- If $E = 0$, F is zero, and S is 1, then $x = -0$.

- If $E = 0$, F is zero, and S is 0, then $x = 0$.

8.4.4 Fixed-Point Digital Signal Processors

Analog Devices, Texas Instruments, and Motorola all manufacture fixed-point DS processors. Analog Devices offers a fixed-point DSP family such as ADSP21xx. Texas Instruments provides various generations of fixed-point DSP processors based on historical development, architectural features, and computational performance. Some of the most common ones are TMS320C1x (first generation), TMS320C2x, TMS320C5x, and TMS320C62x. Motorola manufactures varieties of fixed-point processors, such as the DSP5600x family. The new families of fixed-point DS processors are expected to continue to grow. Since they share some basic common features such as program memory and data memory with associated address buses, arithmetic logic units (ALUs), program control units, MACs, shift units, and address generators, here we focus on an overview of the TMS320C54x processor. The typical TMS320C54x fixed-point DSP architecture appears in Figure 8.13.

The fixed-point TMS320C54x families supporting 16-bit data have on-chip program memory and data memory in various sizes and configurations. They include data RAM (random access memory) and program ROM (read-only memory) used for program code, instruction, and data. Four data buses and four address buses are accommodated to work with the data memories and program memories. The program memory address bus and program memory data bus are responsible for fetching the program instruction. As shown in Figure 8.13, the C and D data memory address buses and the C and D data memory data buses deal with fetching data from the data memory, while the E data memory address bus and the E data memory data bus are dedicated to moving data into data memory. In addition, the E memory data bus can access the I/O devices.

Computational units consist of an ALU, an MAC, and a shift unit. For TMS320C54x families, the ALU can fetch data from the C, D, and program memory data buses and access the E memory data bus. It has two independent 40-bit accumulators, which are able to operate 40-bit addition. The multiplier, which can fetch data from C and D memory data buses and write data via the E memory data bus, is capable of operating 17-bit×17-bit multiplications. The 40-bit shifter has the same capability of bus access as the MAC, allowing all possible shifts for scaling and fractional arithmetic such as we have discussed for the Q-format.

The program control unit fetches instructions via the program memory data bus. Again, in order to speed up memory access, there are two address generators available: one responsible for program addresses and one for data addresses.

Advanced Harvard architecture is employed, where several instructions operate at the same time for a given single instruction cycle. Processing performance offers 40 MIPS (million

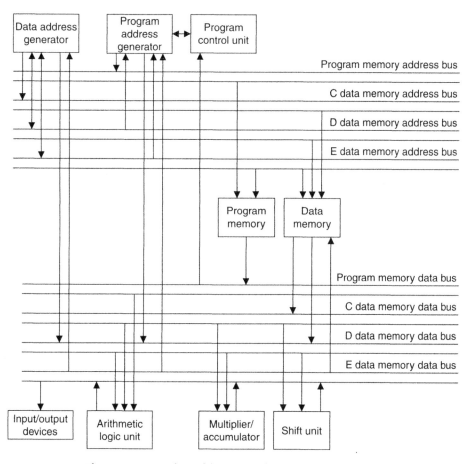

Figure 8.13: Basic architecture of TMS320C54x family

instruction sets per second). To further explore this subject, the reader is referred to Dahnoun (2000), Embree (1995), Ifeachor and Jervis (2002), and Van der Vegte (2002), as well as the website for Texas Instruments (www.ti.com).

8.4.5 Floating-Point Processors

Floating-point DS processors perform DSP operations using floating-point arithmetic, as we discussed before. The advantages of using the floating-point processor include getting rid of finite word length effects such as overflows, round-off errors, truncation errors, and coefficient quantization errors. Hence, in terms of coding, we do not need to do scaling input samples to avoid overflow, shift the accumulator result to fit the DAC word size, scale the filter coefficients, or apply Q-format arithmetic. The floating-point DS processor with high-performance speed and calculation precision facilitates a friendly environment to develop and implement DSP algorithms.

Analog Devices provides floating-point DSP families such as ADSP210xx and TigerSHARC. Texas Instruments offers a wide range of floating-point DSP families, in which the TMS320C3x is the first generation, followed by the TMS320C4x and TMS320C67x families. Since the first generation of a floating-point DS processor is less complicated than later generations but still has the common basic features, we overview the first-generation architecture first.

Figure 8.14 shows the typical architecture of Texas Instruments' TMS320C3x families. We discuss some key features briefly. Further detail can be found in the TMS320C3x User's Guide (Texas Instruments, 1991), the TMS320C6x CPU and Instruction Set Reference Guide (Texas Instruments, 1998), and other studies (Dahnoun, 2000; Embree, 1995; Ifeachor and Jervis, 2002; Kehtarnavaz and Simsek, 2000; Sorensen and Chen, 1997; van der Vegte, 2002). The TMS320C3x family consists of 32-bit single chip floating-point processors that support both integer and floating-point operations.

The processor has a large memory space and is equipped with dual-access on-chip memories. A program cache is employed to enhance the execution of commonly used codes. Similar to

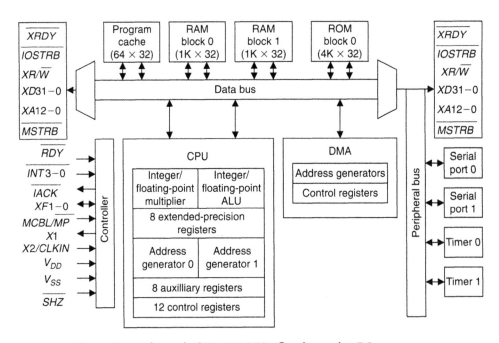

Figure 8.14: The typical TMS320C3x floating-point DS processor

the fixed-point processor, it uses the Harvard architecture, in which there are separate buses used for program and data so that instructions can be fetched at the same time that data are being accessed. There also exist memory buses and data buses for direct-memory access (DMA) for concurrent I/O and CPU operations, and peripheral access such as serial ports, I/O ports, memory expansion, and an external clock.

The C3x CPU contains the floating-point/integer multiplier; an ALU, which is capable of operating both integer and floating-point arithmetic; a 32-bit barrel shifter; internal buses; a CPU register file; and dedicated auxiliary register arithmetic units (ARAUs). The multiplier operates single-cycle multiplications on 24-bit integers and on 32-bit floating-point values. Using parallel instructions to perform a multiplication, an ALU will cost a single cycle, which means that a multiplication and an addition are equally fast. The ARAUs support addressing modes, in which some of them are specific to DSP such as circular buffering and bit-reversal addressing (digital filtering and FFT operations). The CPU register file offers 28 registers, which can be operated on by the multiplier and ALU. The special functions of the registers include eight extended 40-bit precision registers for maintaining accuracy of the floating-point results. Eight auxiliary registers can be used for addressing and for integer arithmetic. These registers provide internal temporary storage of internal variables instead of external memory storage, to allow performance of arithmetic between registers. In this way, program efficiency is greatly increased.

The prominent feature of C3x is its floating-point capability, allowing operation of numbers with a very large dynamic range. It offers implementation of the DSP algorithm without worrying about problems such as overflows and coefficient quantization. Three floating-point formats are supported. A short 16-bit floating-point format has 4 exponent bits, 1 sign bit, and 11 mantissa bits. A 32-bit single precision format has 8 exponent bits, 1 sign bit, and 23 fraction bits. A 40-bit extended precision format contains 8 exponent bits, 1 sign bit, and 31 fraction bits. Although the formats are slightly different from the IEEE 754 standard, conversions are available between these formats.

The TMS320C30 offers high-speed performance with 60-nanosecond single-cycle instruction execution time, which is equivalent to 16.7 MIPS. For speech-quality applications with an 8 kHz sampling rate, it can handle over 2,000 single-cycle instructions between two samples (125 microseconds). With instruction enhancement such as pipelines executing each instruction in a single cycle (four cycles required from fetch to execution by the instruction itself) and a multiple interrupt structure, this high-speed processor validates implementation of real-time applications in floating-point arithmetic.

8.5 Finite Impulse Response and Infinite Impulse Response Filter Implementations in Fixed-Point Systems

With knowledge of the IEEE formats and of filter realization structures such as the direct form I, direct form II, and parallel and cascade forms, we can study digital filter implementation in the fixed-point processor. In the fixed-point system, where only integer arithmetic is used, we prefer input data, filter coefficients, and processed output data to be in the Q-format. In this way, we avoid overflow due to multiplications and can prevent overflow due to addition by scaling input data. When the filter coefficients are out of the Q-format range, coefficient scaling must be taken into account to maintain the Q-format. We develop FIR filter implementation in Q-format first, and then infinite impulse response (IIR) filter implementation next. In addition, we assume that with a given input range in Q-format, the filter output is always in Q-format even if the filter passband gain is larger than 1.

First, to avoid overflow for an adder, we can scale the input down by a scale factor S, which can be safely determined by the equation:

$$S = I_{max} \cdot \sum_{k=0}^{\infty} |h(k)| = I_{max} \cdot \left(|h(0)| + |h(1)| + |h(2)| + \cdots \right), \tag{8.2}$$

where $h(k)$ is the impulse response of the adder output and I_{max} the maximum amplitude of the input in Q-format. Note that this is not an optimal factor in terms of reduced signal-to-noise ratio. However, it shall prevent the overflow. Equation (8.2) means that the adder output can actually be expressed as a convolution output:

$$\text{adder output} = h(0)x(n) + h(1)x(n-1) + h(2)x(n-2) + \cdots$$

Assuming the worst condition, that is, that all the inputs $x(n)$ reach a maximum value of I_{max} and all the impulse coefficients are positive, the sum of the adder gives the most conservative scale factor, as shown in Equation (8.2). Hence, scaling down of the input by a factor of S will guarantee that the output of the adder is in Q-format.

When some of the FIR coefficients are larger than 1, which is beyond the range of Q-format representation, coefficient scaling is required. The idea is that scaling down the coefficients will make them less than 1, and later the filtered output will be scaled up by the same amount before it is sent to DAC. Figure 8.15 describes the implementation.

In the figure, the scale factor B makes the coefficients b_k/B convertible to the Q-format. The scale factors of S and B are usually chosen to be a power of 2, so the simple shift operation can be used in the coding process. Let us implement an FIR filter containing filter coefficients larger than 1 in the fixed-point implementation.

Example 8.13

Given the FIR filter:

$$y(n) = 0.9x(n) + 3x(n-1) - 0.9x(n-2),$$

with a passband gain of 4, and assuming that the input range occupies only 1/4 of the full range for a particular application,

a. Develop the DSP implementation equations in the Q-15 fixed-point system.

Solution

a. The adder may cause overflow if the input data exists for ¼ of a full dynamic range. The scale factor is determined using the impulse response, which consists of the FIR filter coefficients.

$$S = \frac{1}{4}\left(|h(0)| + |h(1)| + |h(2)|\right) = \frac{1}{4}(0.9 + 3 + 0.9) = 1.2.$$

Overflow may occur. Hence, we select $S = 2$ (a power of 2). We choose $B = 4$ to scale all the coefficients to be less than 1, so the Q-15 format can be used. According to Figure 8.15, the developed difference equations are given by:

$$x_s(n) = \frac{x(n)}{2}$$
$$y_s(n) = 0.225x_s(n) + 0.75x_s(n-1) - 0.225x_s(n-2)$$
$$y(n) = 8y_s(n)$$

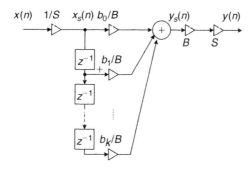

Figure 8.15: Direct-form I implementation of the FIR filter

Next, the direct-form I implementation of the IIR filter is illustrated in Figure 8.16.

As shown in Figure 8.16, the purpose of a scale factor C is to scale down the original filter coefficients to the Q-format. The factor C is usually chosen to be a power of 2 for using a simple shift operation in DSP.

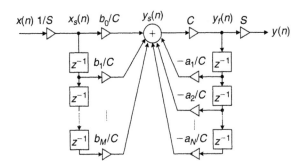

Figure 8.16: Direct-form I implementation of the IIR filter

Example 8.14

The following IIR filter,

$$y(n) = 2x(n) + 0.5\,y(n-1),$$

uses the direct form I, and for a particular application, the maximum input is $I_{max} = 0.010\ldots0_2 = 0.25$.

a.   Develop the DSP implementation equations in the Q-15 fixed-point system.

Solution

a.   This is an IIR filter whose transfer function is:

$$H(z) = \frac{2}{1 - 0.5z^{-1}} = \frac{2z}{z - 0.5}.$$

Applying the inverse z-transform, we have the impulse response:

$$h(n) = 2 \times (0.5)^n u(n).$$

To prevent overflow in the adder, we can compute the S factor with the help of the Maclaurin series or approximate Eq. (9.2) numerically. We get:

$$S = 0.25 \times (2(0.5)^0 + 2(0.5)^1 + 2(0.5)^2 + \ldots) = \frac{0.25 \times 2 \times 1}{1 - 0.5} = 1.$$

MATLAB function **impz()** can also be applied to find the impulse response and the S factor:

```
» h = impz(2,[1 - 0.5]); % Find the impulse response
» sf = 0.25* sum(abs(h)) % Determine the sum of absolute
values of h(k)

  sf = 1
```

Hence, we do not need to perform input scaling. However, we need to scale down all the coefficients to use the Q-15 format. A factor of $C = 4$ is selected. From Figure 8.16, we get the difference equations as:

$$x_s(n) = x(n)$$
$$y_s(n) = 0.5x_s(n) + 0.125y_f(n-1)$$
$$y_f(n) = 4y_s(n)$$
$$y(n) = y_f(n).$$

We can develop these equations directly. First, we divide the original difference equation by a factor of 4 to scale down all the coefficients to be less than 1, that is,

$$\frac{1}{4}y_f(n) = \frac{1}{4} \times 2x_s(n) + \frac{1}{4} \times 0.5y_f(n-1),$$

and define a scaled output:

$$y_s(n) = \frac{1}{4}y_f(n).$$

Finally, substituting $y_s(n)$ to the left side of the scaled equation and rescaling up the filter output as $y_f(n) = 4y_s(n)$ we have the same results we got before.

The fixed-point implementation for the direct form II is more complicated. The developed direct-form II implementation of the IIR filter is illustrated in Figure 8.17.

As shown in Figure 8.17, two scale factors A and B are designated to scale denominator coefficients and numerator coefficients to their Q-format representations, respectively. Here

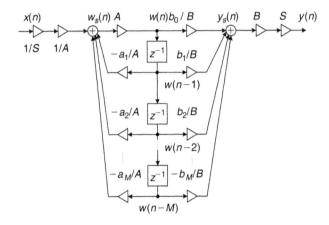

Figure 8.17: Direct-form II implementation of the IIR filter

S is a special factor to scale down the input sample so that the numerical overflow in the first sum in Figure 8.17 can be prevented. The difference equations are listed here:

$$w(n) = x(n) - a_1 w(n - 1) - a_2 w(n - 2) - \cdots - a_M w(n - M)$$
$$y(n) = b_0 w(n) + b_1 w(n - 1) + \cdots + b_M w(n - M).$$

The first equation is scaled down by the factor A to ensure that all the denominator coefficients are less than 1, that is,

$$w_s(n) = \frac{1}{A}w(n) = \frac{1}{A}x(n) - \frac{1}{A}a_1 w(n - 1) - \frac{1}{A}a_2 w(n - 2) - \cdots - \frac{1}{A}a_M w(n - M)$$
$$w(n) = A \times w_s(n).$$

Similarly, scaling the second equation yields:

$$y_s(n) = \frac{1}{B}y(n) = \frac{1}{B}b_0 w(n) + \frac{1}{B}b_1 w(n - 1) + \cdots + \frac{1}{B}b_M w(n - M)$$

and

$$y(n) = B \times y_s(n)$$

To avoid the first adder overflow (first equation), the scale factor S can be safely determined by Eq. (8.3):

$$S = I_{max}(|h(0)| + |h(1)| + |h(2)| + \cdots), \tag{8.3}$$

where $h(k)$ is the impulse response of the filter whose transfer function is the reciprocal of the denominator polynomial, where the poles can cause a larger value to the first sum:

$$h(n) = Z^{-1}\left(\frac{1}{1 + a_1 z^{-1} + \cdots + az^{-M}}\right). \tag{8.4}$$

All the scale factors A, B, and S are usually chosen to be a power of 2, respectively, so that the shift operations can be used in the coding process. Example 8.15 serves for illustration.

The implementation for cascading the second-order section filters can be found in Ifeachor and Jervis (2002).

A practical example will be presented in the next section. Note that if a floating-point DS processor is used, all the scaling concerns should be ignored, since the floating-point format offers a large dynamic range, so that overflow hardly ever happens.

Example 8.15

Given the following IIR filter:

$$y(n) = 0.75x(n) + 1.49x(n-1) + 0.75x(n-2) - 1.52y(n-1) - 0.64y(n-2),$$

with a passband gain of 1 and a full range of input,

a. Use the direct-form II implementation to develop the DSP implementation equations in the Q-15 fixed-point system.

Solution

a. The difference equations without scaling in the direct-form II implementation are given by:

$$w(n) = x(n) - 1.52w(n-1) - 0.64w(n-2)$$
$$y(n) = 0.75w(n) + 1.49w(n-1) + 0.75w(n-2).$$

To prevent overflow in the first adder, we have the reciprocal of the denominator polynomial as:

$$A(z) = \frac{1}{1 + 1.52z^{-1} + 0.64z^{-2}}.$$

Using the MATLAB function leads to:

```
» h = impz(1, [1 1:52 0.64]);
» sf = sum(abs(h))
sf = 10.4093.
```

We choose the S factor as $S = 16$ and we choose $A = 2$ to scale down the denominator coefficients by half. Since the second adder output after scaling is:

$$y_s(n) = \frac{0.75}{B}w(n) + \frac{1.49}{B}w(n-1) + \frac{0.75}{B}w(n-2),$$

we have to ensure that each coefficient is less than 1, as well as the sum of the absolute values:

$$\frac{0.75}{B} + \frac{1.49}{B} + \frac{0.75}{B} < 1$$

to avoid second adder overflow. Hence $B = 4$ is selected. We develop the DSP equations as:

$$x_s(n) = x(n)/16$$
$$w_s(n) = 0.5x_s(n) - 0.76w(n-1) - 0.32w(n-2)$$
$$w(n) = 2w_s(n)$$
$$y_s(n) = 0.1875w(n) + 0.3725w(n-1) + 0.1875w(n-2)$$
$$y(n) = (B \times S)y_s(n) = 64\,y_s(n)$$

8.6 Digital Signal Processing Programming Examples

In this section, we first review the TMS320C67x DSK (DSP Starter Kit), which offers floating-point and fixed-point arithmetic. We will then investigate real-time implementation of digital filters.

8.6.1 Overview of TMS320C67x DSK

In this section, a Texas Instruments TMS320C67x DSK is chosen for demonstration in Figure 8.18. This DSK board (Kehtarnavaz and Simsek, 2001; Texas Instruments, 1998) consists of the TMS320C67x chip, SDRAM (synchronous dynamic random access memory) and ROM

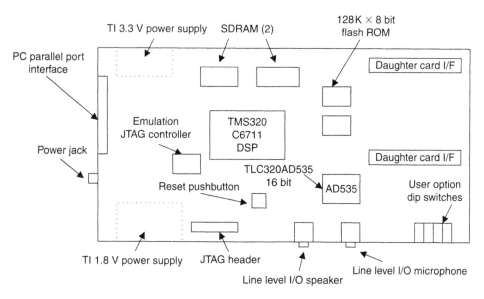

Figure 8.18: C6711 DSK board

for storing program code and data, and an ADC535 chip performing 16-bit ADC and DAC operations. The gain of the ADC channel is programmable to provide microphone or other line inputs, such as from the function generator or other sensors. The DAC channel is also programmable to deliver the power gain to drive a speaker or other devices. The ADC535 chip sets a fixed sampling rate of 8 kHz for speech applications. The on-board daughter card connections facilitate the external units for advanced applications. For example, a daughter card designed using PCM3001/3 offers a variable high sampling rate, such as 44.1 kHz, and its own programmable ADC and DAC for CD-quality audio applications. The parallel port is used for connection between the DSK board and the host computer, where the user program is developed, compiled, and downloaded to the DSK for real-time applications using the user-friendly software called the Code Composer Studio, which we shall discuss later.

The TMS320C67x operates at a high clock rate of 300 MHz. Combining with high speed and multiple units operating at the same time has pushed its performance up to 2,400 MIPS at 300 MHz. Using this number, the C67x can handle 0.3 MIPS between two speech samples at a sampling rate of 8 kHz and can handle over 54,000 instructions between two audio samples with a sampling rate of 44.1 kHz. Hence, the C67x offers great flexibility for real-time applications with a high-level C language.

Figure 8.19 shows a C67x architecture overview, while Figure 8.20 displays a more detailed block diagram. C67x contains three main parts, which are the CPU, the memories, and

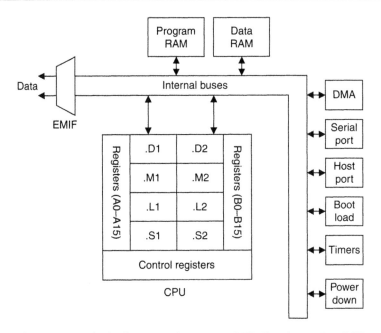

Figure 8.19: Block diagram of TMS320C67x floating-point DSP

the peripherals. As shown in Figure 8.19, these three main parts are joined by an external memory interface (EMIF) interconnected by internal buses to facilitate interface with common memory devices; DMA; a serial port; and a host port interface (HPI).

Since this section is devoted to showing DSP coding examples, C67x key features and references are briefly listed here:

1. Architecture: The system uses Texas Instruments Veloci™ architecture, which is an enhancement of the VLIW (very long instruction word architecture) (Dahnoun, 2000; Ifeachor and Jervis, 2002; Kehtarnavaz and Simsek, 2000).

2. CPU: As shown in Figure 8.20, the CPU has eight functional units divided into two sides *A* and *B*, each consisting of units .D, .M, .L, and .S. For each side, an .M unit is used for multiplication operations, an .L unit is used for logical and arithmetic operations, and a .D unit is used for loading/storing and arithmetic operations. Each side of the C67x CPU has sixteen 32-bit registers that the CPU must go through for interface. More detail can be found in Kehtarnavaz and Simsek (2000) and Texas Instruments (1998).

3. Memory and internal buses: Memory space is divided into internal program memory, internal data memory, and internal peripheral and external memory space. The internal buses include a 32-bit program address bus, a 256-bit program data bus to carry out eight

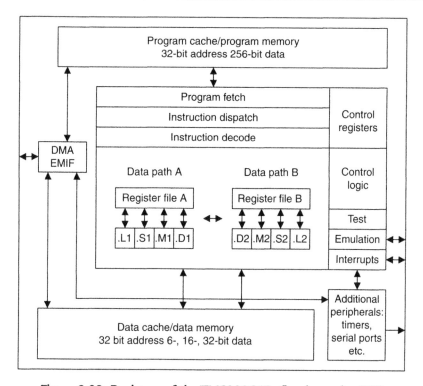

Figure 8.20: Registers of the TMS320C67x floating-point DSP

32-bit instructions (VLIW), two 32-bit data address buses, two 64-bit load data buses, two 64-bit store data buses, two 32-bit DMA buses, and two 32-bit DMA address buses responsible for reading and writing. There also exist a 22-bit address bus and a 32-bit data bus for accessing off-chip or external memory.

4. Peripherals:

- EMIF, which provides the required timing for accessing external memory.

- DMA, which moves data from one memory location to another without interfering with the CPU operations.

- Multichannel buffered serial port (McBSP) with a high-speed multichannel serial communication link.

- HPI, which lets a host access internal memory.

- Boot loader for loading code from off-chip memory or the HPI to internal memory.

- Timers (two 32-bit counters).

- Power-down units for saving power for periods when the CPU is inactive.

The software tool for the C67x is the Code Composer Studio (CCS) provided by TI. It allows the user to build and debug programs from a user-friendly graphical user interface (GUI) and extends the capabilities of code development tools to include real-time analysis. Installation, tutorial, coding, and debugging can be found in the *CCS Getting Started Guide* (Texas Instruments, 2001) and in Kehtarnavaz and Simsek (2000).

8.6.2 Concept of Real-Time Processing

We illustrate real-time implementation in Figure 8.21, where the sampling rate is 8,000 samples per second; that is, the sampling period $T = 1/f_s = 125$ microseconds, which is the time between two samples.

As shown in Figure 8.21, the required timing includes an input sample clock and an output sample clock. The input sample clock maintains the accuracy of sampling time for each ADC operation, while the output sample clock keeps the accuracy of time instant for each DAC operation. Time between the input sample clock n and the output sample clock n consists of the ADC operation, algorithm processing, and the wait for the next ADC operation. The numbers of instructions for ADC and DSP algorithms must be estimated and verified to ensure that all instructions have been completed before the DAC begins. Similarly, the number of instructions for DAC must be verified so that DAC instructions will be finished

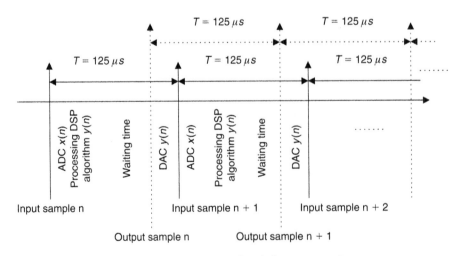

Figure 8.21: Concept of real-time processing

between the output sample clock n and the next input sample clock $n+1$. Timing usually is set up using the DSP interrupts (we will not pursue the interrupt setup here).

Next, we focus on the implementation of the DSP algorithm in the floating-point system for simplicity.

8.6.3 Linear Buffering

During DSP such as digital filtering, past inputs and past outputs are required to be buffered and updated for processing the next input sample. Let us first study the FIR filter implementation.

8.6.3.1 Finite Impulse Response Filtering

Consider implementation for the following 3-tap FIR filter:

$$y(n) = 0.5x(n) + 0.2x(n-1) + 0.5x(n-2).$$

The buffer requirements are shown in Figure 8.22. The coefficient buffer b[3] contains 3 FIR coefficients, and the coefficient buffer is fixed during the process. The input buffer x[3], which holds the current and past inputs, is required to be updated. The FIFO update is adopted here with the segment of codes shown in Figure 8.22. For each input sample, we update the input buffer using FIFO, which begins at the end of the data buffer; the oldest sampled is kicked out first from the buffer and updated with the value from the upper location. When the FIFO completes, the first memory location x[0] will be free to be used to store the current input sample. The segment of code in Figure 8.22 explains implementation.

Note that in the code segment, x[0] holds the current input sample x(n), while b[0] is the corresponding coefficient; x[1] and x[2] hold the past input samples $x(n-1)$ and $x(n-2)$, respectively; similarly, b[1] and b[2] are the corresponding coefficients.

Again, note that using the array and loop structures in the code segment is for simplicity in notations and assumes that the reader is not familiar with the C pointers in C-language. This concern for simplicity has to do mainly with the DSP algorithm. More coding efficiency can be achieved using the C pointers and circular buffer. The DSP-oriented coding implementation can be found in Kehtarnavaz and Simsek (2000).

8.6.3.2 Infinite Impulse Response Filtering

Similarly, we can implement an IIR filter. It requires an input buffer, which holds the current and past inputs; an output buffer, which holds the past outputs; a numerator

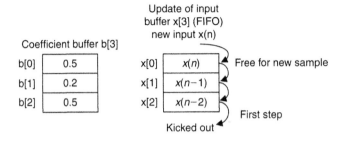

```
volatile int sample;
float x[3]={0.0, 0.0, 0.0};
float b[3]={0.5, 0.2, 0.5};
float y[1]={0.0};
interrupt void AtoD()
{
int i;
  sample = mcbsp0_read(); /*ADC*/
  for(i = 2; i > 0; i--)/* Update the input buffer x[3] */
  {
  x[i] = x[i - 1];
  }
  x[0]=(float) sample;
  y[0]=0;
  for(i = 0; i < 3; i++)
  {
  y[0] = y[0] +b[i]*x[i];
  }
  sample = (int) y[0]; /* The processed sample will be sent to DAC */
}
```

Figure 8.22: Example of FIR filtering with linear buffer update

coefficient buffer; and a denominator coefficient buffer. Considering the following IIR filter for implementation,

$$y(n) = 0.5x(n) + 0.7x(n - 1) - 0.5x(n - 2) - 0.4y(n - 1) + 0.6y(n - 2),$$

we accommodate the numerator coefficient buffer b[3], the denominator coefficient buffer a[3], the input buffer x[3], and the output buffer y[3] shown in Figure 8.23. The buffer updates for input x[3] and output y[3] are FIFO. The implementation is illustrated in the segment of code listed in Figure 8.23.

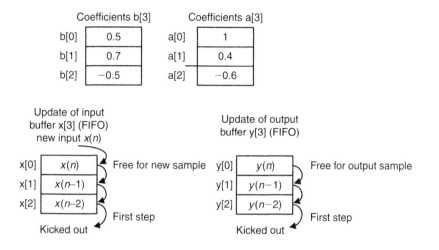

```
volatile int sample;
float b[3]={0.5, 0.7, −0.5};
float a[3]={1, 0.4, −0.6};
float x[3]={0.0, 0.0, 0.0};
float y[3]={0.0, 0.0, 0.0};
interrupt void AtoD()
{
  int i;
  sample=mcbsp0_read(); /* ADC */
  for(i = 2; i > 0; i--)   /* Update the input buffer */
  {
  x[i] = x[i − 1];
  }
  x[0]= (float) sample;
  for (i = 2;i > 0;i--)/* Update the output buffer */
  {
  y[i]=y[i − 1];
  }
  y[0] = b[0]*x[0] +b[1]*x[1] +b[2]*x[2]-a[1]*y[1]-a[2]*y[2];
  sample= (int) y[0]; /* the processed sample will be sent to DAC */
}
```

Figure 8.23: Example of IIR filtering using linear buffer update

Again, note that in the code segment, x[0] holds the current input sample, while y[0] holds the current processed output, which will be sent to the DAC unit for conversion. The coefficient a[0] is never modified in the code. We keep that for a purpose of notation simplicity and consistency during the programming process.

8.6.3.3 Digital Oscillation with Infinite Impulse Response Filtering

The principle for generating digital oscillation is where the input to the digital filter is the impulse sequence, and the transfer function is obtained by applying the z-transform of the digital sinusoidal function. Applications can be found in dual-tone multifrequency (DTMF) tone generation, digital carrier generation for communications, and so on. Hence, we can modify the implementation of IIR filtering for tone generation with the input generated internally instead of by using the ADC channel.

Let us generate an 800 Hz tone with a digital amplitude of 5,000. The transfer function, difference equation, and impulse input sequence are found to be, respectively,

$$H(z) = \frac{0.587785z^{-1}}{1 - 1.618034z^{-1} + z^{-2}}$$
$$y(n) = 0.587785x(n-1) + 1.618034y(n-1) - y(n-2)$$
$$x(n) = 5000\delta(n).$$

We define the numerator coefficient buffer b[2], the denominator coefficient buffer a[3], the input buffer x[2], and the output buffer y[3], shown in Figure 8.24, which also shows the modified implementation for the tone generation.

Initially, we set x[0] = 5000. Then it will be updated with x[0] = 0 for each current processed output sample y[0].

8.6.4 Sample C Programs

8.6.4.1 Floating-Point Implementation Example

Real-time DSP implementation using the floating-point processor is easy to program. The overflow problem hardly ever occurs. Therefore, we do not need to consider scaling factors, as described in the last section. The code segment shown in Figure 8.25 demonstrates the simplicity of coding the floating-point IIR filter using the direct-form I structure.

8.6.5 Fixed-Point Implementation Example

Where execution time is critical, fixed-point implementation is preferred in a floating-point processor. We implement the following IIR filter with a unit passband gain in direct form II:

$$H(z) = \frac{0.0201 - 0.0402z^{-2} + 0.0201z^{-4}}{1 - 2.1192z^{-1} + 2.6952z^{-2} - 1.6924z^{-3} + 0.6414z^{-4}}$$
$$w(n) = x(n) + 2.1192w(n-1) - 2.6952w(n-2) + 1.6924w(n-3) - 0.6414w(n-4)$$
$$y(n) = 0.0201w(n) - 0.0402w(n-2) + 0.0201w(n-4).$$

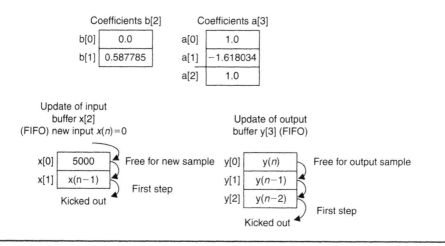

```
volatile int sample;
float b[2]={0.0,0.587785};
float a[3]={1, -1.618034,1};
float x[2]={5000, 0.0};  /*Set up the input as an impulse function */
float y[3]={0.0, 0.0, 0.0};
interrupt void AtoD()
{
  int i;
  sample = mcbsp0_read(); /*ADC */
  y[0] = b[0]*x[0] + b[1]*x[1]-a[1]*y[1]-a[2]*y[2];
  sample= (int) y[0]; /*The processed sample will be sent to DAC */
  for(i = 1; i > 0; i--)/*Update the input buffer */
  {
  x[i] = x[i-1];
  }
  x[0] = 0;
  for(i = 2 > 0;i--)/* Update the output buffer */
  {
  y[i] = y[i - 1];
  }
}
```

Figure 8.24: Example of IIR filtering using linear buffer update and the impulse sequence input

Using MATLAB to calculate the scale factor *S*, it follows that:

$$\gg \quad h = impz([1],[1 - 2.1192 \ 2.6952 - 1.6924 \ 0.6414]);$$

$$\gg \quad sf = sum(abs(h))$$

$$sf = 28.2196$$

```
volatile int sample;
float a[5] = {1.00, -2.1192, 2.6952, -1.6924, 0.6414};
float b[5] = {0.0201, 0.00, -0.0402, 0.00, 0.0201};
float x[5] = {0.0, 0.0, 0.0, 0.0, 0.0};
float y[5] = {0.0, 0.0, 0.0, 0.0, 0.0};
/*****************************************************************/
/*AtoD() Interrupt Service Routine (ISR) ->interrupt 12 defined in IST
of vectors.asm (read McBSP)*/
/*****************************************************************/
interrupt void AtoD()
{
int i;
float temp, sum;
sample = mcbsp0_read(); /*ADC*/
//Insert DSP Algorithm Here ()
temp = (float) sample;
for (i = 4; i > 0; i--)
{
  x[i] = x[i-1];
}
x[0] = temp;
for (i = 4; i > 0; i--)
{
y[i] = y[i -1];
}
sum = b[0]*x[0] +b[1]*x[1] +b[2]*x[2] +b[3]*x[3] +b[4]*x[4] -a[1]*y[1] -a[2]*
y[2] -a[3]*y[3] -a[4]*y[4];

y[0] = sum;
sample = sum;
}
/*****************************************************************/
/*DtoA() Interrupt Service Routine (ISR) ->interrupt 11 defined in IST
of vectors.asm (write to McBSP)*/
/*****************************************************************/
interrupt void DtoA()
{
sample = sample & 0xfffe; /*set LSB to 0 for primary communication*/
mcbsp0_write(sample); /*DAC*/
}
```

Figure 8.25: Sample C code for IIR filtering (float-point implementation)

Table 8.4: Filter coefficients in Q-15 format

IIR Filter	Filter Coefficients	Q-15 Format (Hex)
$-a_1$	0.5298	0x43D0
$-a_2$	−0.6738	0xA9C1
$-a_3$	0.4230	0x3628
$-a_4$	−0.16035	0xEB7A
b_0	0.0201	0×0293
b_1	0.0000	0×0000
b_2	−0.0402	$0 \times FADB$
b_3	0.0000	0×000
b_4	0.0201	0×0293

Hence we choose $S = 32$. To scale the filter coefficients in the Q-15 format, we use the factors $A = 4$ and $B = 1$. Then the developed DSP equations are:

$$x_s(n) = x(n)/32$$
$$w_s(n) = 0.25x_s(n) + 0.5298w_s(n-1) - 0.6738w_s(n-2) + 0.4231w_s(n-3)$$
$$+ 0.4231w_s(n-3) + 0.16035w_s(n-4)$$
$$w(n) = 4w_s(n)$$
$$y_s(n) = 0.0201w(n) - 0.0402w(n-2) + 0.0201w(n-4)$$
$$y(n) = 32y_s(n).$$

We can convert filter coefficients into the Q-15 format; each coefficient is listed in Table 8.4.

The list of codes for the fixed-point implementation is displayed in Figure 8.26, and some coding notations are given in Figure 8.27.

8.7 Summary

1. The von Neumann architecture consists of a single, shared memory for programs and data, a single bus for memory access, an arithmetic unit, and a program control unit. The von Neumann processor operates fetching and execution cycles in series.

2. The Harvard architecture has two separate memory spaces dedicated to program code and to data, respectively, two corresponding address buses, and two data buses for accessing two memory spaces. The Harvard processor offers fetching and execution cycles in parallel.

```
volatile in tsample;
/*float a[5]={1.00,-2.1192,2.6952,-1.6924,0.6414};
float b[5]={0.0201,0.00,-0.0402,0.00,0.0201};*/
short a[5]={0×2000,0×43D0,0×A9C1,0×3628,0×EB7A};/*coefficients in Q-15 forma
short b[5]={0×0293,0×0000,0×FADB,0×0000,0×0293};
int w[5]={0,0,0,0,0};
interrupt void AtoD()
{int i,sum=0;
sample=mcbsp0_read();/*ADC*/
//Insert DSP Algorithm Here()
sample=(sample≪16);/*Move to high 16 bits*/
sample=(sample≫5);/*Scaled down by 32 to avoid overflow*/
for(i=4;i>0;i--)
{
w[i]=w[i-1];
}
sum=(sample≫2);/*Scaled down by 4 to use Q-15*/
for(i=1;i<5;i++)
{
sum +=(_mpyhl(w[i],a[i]))≪1;
}
sum=(sum≪2);/*scaled up by 4*/
w[0]=sum;
sum=0;
for(i=0;i<5;i++)
{
sum +=(_mpyhl(w[i],b[i]))≪1;
}
sum=(sum≪5);/*Scaled up by 32 to get y(n)*/
sample=sum≫16);/*Move to low 16 bits*/
}


/*****************************************************************************/

/*DtoA()Interrupt Service Routine (ISR)->interrupt 11 defined in IST of
vectors.asm(write to McBSP)*/
/*****************************************************************************/

interrupt void DtoA()
{
sample=sample & 0xfffe;/*set LSB to 0 for primary communication*/
mcbsp0_write(sample);/*DAC*/
}
```

Figure 8.26: Sample C code for IIR filtering (fixed-point implementation)

---

short coefficient;   declaration of 16 bit signed integer

int sample, result;   declaration of 32 bit signed integer

MPYHL   assembly instruction (signed multiply high low 16 MSB ×16 LSB)

      result = (_mpyhl(sample, coefficient)) <<1;

sample must be shifted left by 16 bits to be stored in the high 16 MSB.

coefficient is the 16 bit data to be stored in the low 16 LSB.

result is shifted left by one bit to get rid of the extended sign bit, and high 16

MSB's are designated for the processed data.

Final **result** will be shifted down to right by 16 bits before DAC conversion.

      sample = (result>>16);

Figure 8.27: Some coding notations for the Q-15 fixed-point implementation

3. The DSP special hardware units include an MAC dedicated to DSP filtering operations, a shifter unit for scaling, and address generators for circular buffering.

4. The fixed-point DS processor uses integer arithmetic. The data format Q-15 for the fixed-point system is preferred to avoid overflows.

5. The floating-point processor uses floating-point arithmetic. The standard floating-point formats include the IEEE single precision and double precision formats.

6. The architectures and features of fixed-point processors and floating-point processors were briefly reviewed.

7. Implementing digital filters in the fixed-point DSP system requires scaling filter coefficients so that the filters are in Q-15 format, and input scaling for the adder so that overflow during the MAC operations can be avoided.

8. The floating-point processor is easy to code using floating-point arithmetic and develops the prototype quickly. However, it is not efficient in terms of the number of instructions it has to complete compared with the fixed-point processor.

9. The fixed-point processor using fixed-point arithmetic takes much effort to code. But it offers the least number of instructions for the CPU to execute.

Bibliography

Dahnoun, N. (2000). *Digital Signal Processing Implementation Using the TMS320C6000TM DSP Platform*. Englewood Cliffs, NJ: Prentice Hall.

Embree, P. M. (1995). *C Algorithms for Real-Time DSP*. Upper Saddle River, NJ: Prentice Hall.

Ifeachor, E. C., & Jervis, B. W. (2002). *Digital Signal Processing: A Practical Approach* (2nd ed.). Upper Saddle River, NJ: Prentice Hall.

Kehtarnavaz, N., & Simsek, B. (2000). *C6X-Based Digital Signal Processing*. Upper Saddle River, NJ: Prentice Hall.

Sorensen, H. V., & Chen, J. P. (1997). *A Digital Signal Processing Laboratory Using TMS320C30*. Upper Saddle River, NJ: Prentice Hall.

Texas Instruments (1991). *TMS320C3x User's Guide*. Dallas, TX: Author.

———. (1998). *TMS320C6x CPU and Instruction Set Reference Guide*. Literature ID# SPRU 189C. Dallas, TX: Author.

———. (2001). *Code Composer Studio: Getting Started Guide*. Dallas, TX: Author.

van der Vegte, J. (2002). *Fundamentals of Digital Signal Processing*. Upper Saddle River, NJ: Prentice Hall.

Wikipedia. (2007). Harvard Mark I. Retrieved March 14, 2007, from http://en.wikipedia.org/wiki/Harvard_Mark_I

CHAPTER 9

Code Optimization and Resource Partitioning

David Katz
Rick Gentile

DSP is all about algorithms. These algorithms place heavy burdens on systems that often have limited cost and power budgets. Thus, DSP code usually requires heavy optimization and careful use of memory and other resources. In the early days of DSP, there was only one way to meet these goals: Use DMA to bring data into internal SRAM (caches were rare) and then process it in hand-coded assembly (many early DSPs only had an assembler!).

Today, DSP algorithms are implemented on many different classes of hardware, often with caches and multiple processing cores. You can program in high-level languages like C, low-level assembly, or a mixture of both. This presents developers with a much more complex set of decisions.

In this chapter, David Katz and Rick Gentile show the modern programmer how to manage code optimization and system resource partitioning. Katz and Gentile do a good job exploring the topic, and bring up important issues that every DSP programmer should understand. This chapter focuses on the Blackfin processor from Analog Devices, but their ideas apply to most processors used for DSP.

The chapter starts with a look at event handling and generation. These are hugely important topics because most DSP systems require fast real-time responses. Katz and Gentile then show how to choose between C/C++, assembly, or a mixture of both. With this out of the way, they present guidelines for efficient programming, including tips on helping compilers produce efficient code. Finally, they go into some detail on memory management. They show when to use cache over DMA (and vice versa), how to use a memory management unit (MMU), and they review the physics of data movement. This last point is important, because data movement can create major bottlenecks and can boost system power significantly.

There's a lot to digest in this chapter, but it is all crucial information. Woe to any developer who starts coding without a firm grasp of these concepts!

—Kenton Williston

9.1 Introduction

In an ideal situation, we can select an embedded processor for our application that provides maximum performance for minimum extra development effort. In this utopian environment, we could code everything in a high-level language like C, we wouldn't need an intimate knowledge of our chosen device, it wouldn't matter where we placed our data and code, we wouldn't need to devise any data movement subsystem, the performance of external devices wouldn't matter… In short, everything would just work.

Alas, this is only the stuff of dreams and marketing presentations. The reality is, as embedded processors evolve in performance and flexibility, their complexity also increases. Depending on the time-to-market for your application, you will have to walk a fine line to reach your performance targets. The key is to find the right balance between getting the application to work and achieving optimum performance. Knowing when the performance is "good enough" rather than optimal can mean getting your product out on time versus missing a market window.

In this chapter, we want to explain some important aspects of processor architectures that can make a real difference in designing a successful multimedia system. Once you understand the basic mechanics of how the various architectural sections behave, you will be able to gauge where to focus your efforts, rather than embark on the noble yet unwieldy goal of becoming an expert on all aspects of your chosen processor.

Here, we'll explore in detail some Blackfin processor architectural constructs. Again, keep in mind that much of our discussion generalizes to other processor families from different vendors as well.

We will begin with what should be key focal points in any complex application: interrupt and exception handling and response times.

9.2 Event Generation and Handling

Nothing in an application should make you think "performance" more than event management. If you have used a microprocessor, you know that "events" encompass two categories: interrupts and exceptions. An interrupt is an event that happens asynchronous to processor execution. For example, when a peripheral completes a transfer, it can generate an interrupt to alert the processor that data is ready for processing.

Exceptions, on the other hand, occur synchronously to program execution. An exception occurs based on the instruction about to be executed. The change of flow due to an exception occurs prior to the offending instruction actually being executed. Later in this chapter, we'll describe the most widely used exception handler in an embedded processor—the handler that manages pages describing memory attributes. Now, however, we will focus on interrupts rather than exceptions, because managing interrupts plays such a critical role in achieving peak performance.

9.2.1 System Interrupts

System level interrupts (those that are generated by peripherals) are handled in two stages—first in the system domain, and then in the core domain. Once the system interrupt controller (SIC) acknowledges an interrupt request from a peripheral, it compares the peripheral's assigned priority to all current activity from other peripherals to decide when to service this particular interrupt request. The most important peripherals in an application should be mapped to the highest priority levels. In general, the highest-bandwidth peripherals need the highest priority. One "exception" to this rule (pardon the pun!) is where an external processor or supervisory circuit uses a nonmaskable interrupt (NMI) to indicate the occurrence of an important event, such as powering down.

When the SIC is ready, it passes the interrupt request information to the core event controller (CEC), which handles all types of events, not just interrupts. Every interrupt from the SIC maps into a priority level at the CEC that regulates how to service interrupts with respect to one another, as Figure 9.1 shows. The CEC checks the "vector" assignment for the current interrupt request, to find the address of the appropriate interrupt service routine (ISR). Finally, it loads this address into the processor's execution pipeline to start executing the ISR.

There are two key interrupt-related questions you need to ask when building your system. The first is, "How long does the processor take to respond to an interrupt?" The second is, "How long can any given task afford to wait when an interrupt comes in?"

The answers to these questions will determine what your processor can actually perform within an interrupt or exception handler.

For the purposes of this discussion, we define interrupt response time as the number of cycles it takes from when the interrupt is generated at the source (including the time it takes for the current instruction to finish executing) to the time that the first instruction is executed in the interrupt service routine. In our experience, the most common method software engineers use to evaluate this interval for themselves is to set up a programmable flag to generate an interrupt when its pin is triggered by an externally generated pulse. The first instruction in the interrupt service routine then performs a write to a different flag pin. The resulting time

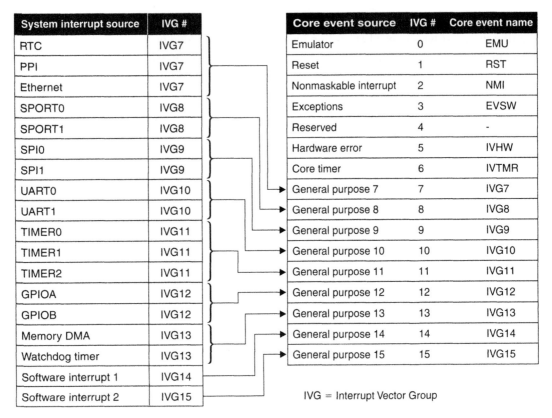

System interrupt source	IVG #
RTC	IVG7
PPI	IVG7
Ethernet	IVG7
SPORT0	IVG8
SPORT1	IVG8
SPI0	IVG9
SPI1	IVG9
UART0	IVG10
UART1	IVG10
TIMER0	IVG11
TIMER1	IVG11
TIMER2	IVG11
GPIOA	IVG12
GPIOB	IVG12
Memory DMA	IVG13
Watchdog timer	IVG13
Software interrupt 1	IVG14
Software interrupt 2	IVG15

Core event source	IVG #	Core event name
Emulator	0	EMU
Reset	1	RST
Nonmaskable interrupt	2	NMI
Exceptions	3	EVSW
Reserved	4	-
Hardware error	5	IVHW
Core timer	6	IVTMR
General purpose 7	7	IVG7
General purpose 8	8	IVG8
General purpose 9	9	IVG9
General purpose 10	10	IVG10
General purpose 11	11	IVG11
General purpose 12	12	IVG12
General purpose 13	13	IVG13
General purpose 14	14	IVG14
General purpose 15	15	IVG15

IVG = Interrupt Vector Group

Figure 9.1: Sample system-to-core interrupt mapping

difference is then measured on an oscilloscope. This method only provides a rough idea of the time taken to service interrupts, including the time required to latch an interrupt at the peripheral, propagate the interrupt through to the core, and then vector the core to the first instruction in the interrupt service routine. Thus, it is important to run a benchmark that more closely simulates the profile of your end application.

Once the processor is running code in an ISR, other higher priority interrupts are held off until the return address associated with the current interrupt is saved off to the stack. This is an important point, because even if you designate all other interrupt channels as higher priority than the currently serviced interrupt, these other channels will all be held off until you save the return address to the stack. The mechanism to re-enable interrupts kicks in automatically when you save the return address. When you program in C, any register the ISR uses will automatically be saved to the stack. Before exiting the ISR, the registers are restored from the stack. This also happens automatically, but depending on where your stack is located and how many registers are involved, saving and restoring data to the stack can take a significant amount of cycles.

Interrupt service routines often perform some type of processing. For example, when a line of video data arrives into its destination buffer, the ISR might run code to filter or downsample it. For this case, when the handler does the work, other interrupts are held off (provided that nesting is disabled) until the processor services the current interrupt.

When an operating system or kernel is used, however, the most common technique is to service the interrupt as soon as possible, release a semaphore, and perhaps make a call to a callback function, which then does the actual processing. The semaphore in this context provides a way to signal other tasks that it is okay to continue or to assume control over some resource.

For example, we can allocate a semaphore to a routine in shared memory. To prevent more than one task from accessing the routine, one task takes the semaphore while it is using the routine, and the other task has to wait until the semaphore has been relinquished before it can use the routine. A Callback Manager can optionally assist with this activity by allocating a callback function to each interrupt. This adds a protocol layer on top of the lowest layer of application code, but in turn it allows the processor to exit the ISR as soon as possible and return to a lower-priority task. Once the ISR is exited, the intended processing can occur without holding off new interrupts.

We already mentioned that a higher-priority interrupt can break into an existing ISR once you save the return address to the stack. However, some processors (like Blackfin) also support self-nesting of core interrupts, where an interrupt of one priority level can interrupt an ISR of the same level, once the return address is saved. This feature can be useful for building a simple scheduler or kernel that uses low-priority software-generated interrupts to preempt an ISR and allow the processing of ongoing tasks.

There are two additional performance-related issues to consider when you plan out your interrupt usage. The first is the placement of your ISR code. For interrupts that run most frequently, every attempt should be made to locate these in L1 instruction memory. On Blackfin processors, this strategy allows single-cycle access time. Moreover, if the processor were in the midst of a multi-cycle fetch from external memory, the fetch would be interrupted, and the processor would vector to the ISR code.

Keep in mind that before you re-enable higher priority interrupts, you have to save more than just the return address to the stack. Any register used inside the current ISR must also be saved. This is one reason why the stack should be located in the fastest available memory in your system. An L1 "scratchpad" memory bank, usually smaller in size than the other L1 data banks, can be used to hold the stack. This allows the fastest context switching when taking an interrupt.

9.3 Programming Methodology

It's nice not to have to be an expert in your chosen processor, but even if you program in a high-level language, it's important to understand certain things about the architecture for which you're writing code.

One mandatory task when undertaking a signal-processing-intensive project is deciding what kind of programming methodology to use. The choice is usually between assembly language and a high-level language (HLL) like C or C++. This decision revolves around many factors, so it's important to understand the benefits and drawbacks each approach entails.

The obvious benefits of C/C++ include modularity, portability and reusability. Not only do the majority of embedded programmers have experience with one of these high-level languages, but also a huge code base exists that can be ported from an existing processor domain to a new processor in a relatively straightforward manner. Because assembly language is architecture-specific, reuse is typically restricted to devices in the same processor family. Also, within a development team it is often desirable to have various teams coding different system modules, and an HLL allows these cross-functional teams to be processor-agnostic.

One reason assembly has been difficult to program is its focus on actual data flow between the processor register sets, computational units and memories. In C/C++, this manipulation occurs at a much more abstract level through the use of variables and function/procedure calls, making the code easier to follow and maintain.

The C/C++ compilers available today are quite resourceful, and they do a great job of compiling the HLL code into tight assembly code. One common mistake happens when programmers try to "outsmart" the compiler. In trying to make it easier for the compiler, they in fact make things more difficult! It's often best to just let the optimizing compiler do its job. However, the fact remains that compiler performance is tuned to a specific set of features that the tool developer considered most important. Therefore, it cannot exceed handcrafted assembly code performance in all situations.

The bottom line is that developers use assembly language only when it is necessary to optimize important processing-intensive code blocks for efficient execution. Compiler features can do a very good job, but nothing beats thoughtful, direct control of your application data flow and computation.

9.4 Architectural Features for Efficient Programming

In order to achieve high performance media processing capability, you must understand the types of core processor structures that can help optimize performance. These include the following capabilities:

- Multiple operations per cycle

- Hardware loop constructs

- Specialized addressing modes

- Interlocked instruction pipelines

These features can make an enormous difference in computational efficiency. Let's discuss each one in turn.

9.4.1 Multiple Operations per Cycle

Processors are often benchmarked by how many millions of instructions they can execute per second (MIPS). However, for today's processors, this can be misleading because of the confusion surrounding what actually constitutes an instruction. For example, multi-issue instructions, which were once reserved for use in higher-cost parallel processors, are now also available in low-cost, fixed-point processors. In addition to performing multiple ALU/MAC operations each core processor cycle, additional data loads and stores can be completed in the same cycle. This type of construct has obvious advantages in code density and execution time.

An example of a Blackfin multi-operation instruction is shown in Figure 9.2. In addition to two separate MAC operations, a data fetch and data store (or two data fetches) can also be accomplished in the same processor clock cycle. Correspondingly, each address can be updated in the same cycle that all of the other activities are occurring.

9.4.2 Hardware Loop Constructs

Looping is a critical feature in real-time processing algorithms. There are two key looping-related features that can improve performance on a wide variety of algorithms: *zero-overhead hardware loops* and *hardware loop buffers*.

Zero-overhead loops allow programmers to initialize loops simply by setting up a count value and defining the loop bounds. The processor will continue to execute this loop until the count has been reached. In contrast, a software implementation would add overhead that would cut into the real-time processing budget.

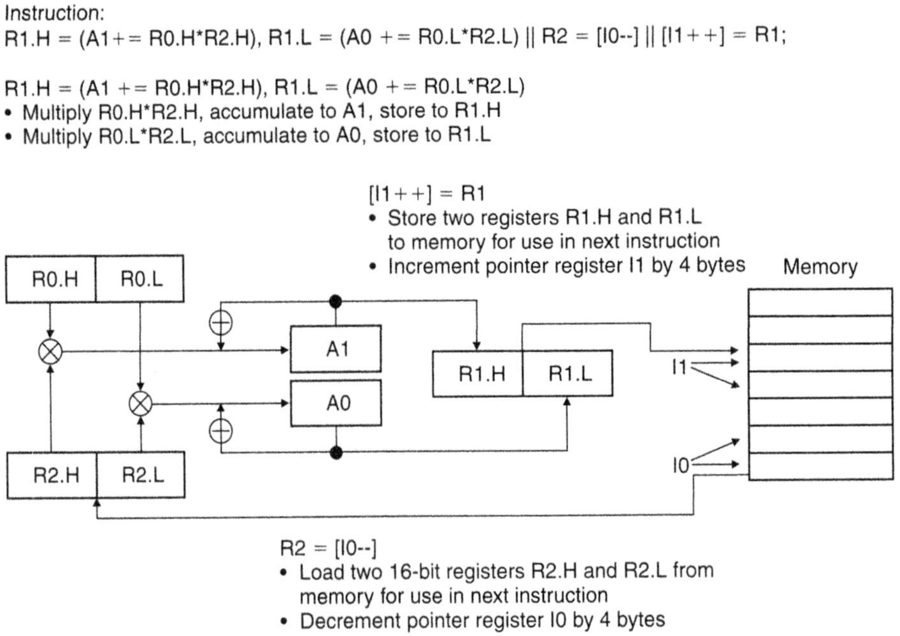

Instruction:
R1.H = (A1+= R0.H*R2.H), R1.L = (A0 += R0.L*R2.L) || R2 = [I0--] || [I1++] = R1;

R1.H = (A1 += R0.H*R2.H), R1.L = (A0 += R0.L*R2.L)
• Multiply R0.H*R2.H, accumulate to A1, store to R1.H
• Multiply R0.L*R2.L, accumulate to A0, store to R1.L

[I1++] = R1
• Store two registers R1.H and R1.L
  to memory for use in next instruction
• Increment pointer register I1 by 4 bytes

R2 = [I0--]
• Load two 16-bit registers R2.H and R2.L from
  memory for use in next instruction
• Decrement pointer register I0 by 4 bytes

Figure 9.2: Example of single-cycle, multi-issue instruction

Many processors offer zero-overhead loops, but hardware loop buffers, which are less common, can really add increased performance in looping constructs. They act as a kind of cache for instructions being executed in the loop. For example, after the first time through a loop, the instructions can be kept in the loop buffer, eliminating the need to re-fetch the same code each time through the loop. This can produce a significant savings in cycles by keeping several loop instructions in a buffer where they can be accessed in a single cycle. The use of the hardware loop construct comes at no cost to the HLL programmer, since the compiler should automatically use hardware looping instead of conditional jumps.

Let's look at some examples to illustrate the concepts we've just discussed.

9.4.3 Specialized Addressing Modes

9.4.3.1 Byte Addressability

Allowing the processor to access multiple data words in a single cycle requires substantial flexibility in address generation. In addition to the more signal-processing-centric access sizes along 16-and 32-bit boundaries, byte addressing is required for the most efficient processing. This is important for multimedia processing because many video-based systems operate on

Example 9.1 Dot Product

The dot product, or scalar product, is an operation useful in measuring orthogonality of two vectors. It's also a fundamental operator in digital filter computations. Most C programmers should be familiar with the following implementation of a dot product:

```
short dot(const short a[],const short b[],int size) {
```

/* Note: It is important to declare the input buffer arrays as const, because this gives the compiler a guarantee that neither "a" nor "b" will be modified by the function. */

```
        int i;
        int output=0;

        for(i=0; i<size; i++) {
                output += (a[i] * b[i]);
        }
        return output;
}
```

Below is the main portion of the equivalent assembly code:

```
        /* P0=Loop Count, P1 & I0 hold starting addresses of a & b
            arrays */
        A1=A0=0;    /* A0 & A1 are accumulators */
        LSETUP (loop1, loop1) LC0=P0 ; /* Set up hardware loop
            starting and ending at label loop1 */
        loop1: A1 += R1.H * R0.H , A0 += R1.L * R0.L || R1=[ P1 ++ ]
            || R0=[ I0 ++ ] ;
```

The following points illustrate how a processor's architectural features can facilitate this tight coding.

Hardware loop buffers and loop counters eliminate the need for a jump instruction at the end of each iteration. Since a dot product is a summation of products, it is implemented in a loop. Some processors use a JUMP instruction at the end of each iteration in order to process the next iteration of the loop. This contrasts with the assembly program above, which shows the LSETUP instruction as the only instruction needed to implement a loop.

Multi-issue instructions allow computation and two data accesses with pointer updates in the same cycle. In each iteration, the values a[i] and b[i] must be read, then multiplied, and finally written back to the running summation in the variable output. On many microcontroller platforms, this effectively amounts to four instructions. The last line of the assembly code shows that all of these operations can be executed in one cycle.

Parallel ALU operations allow two 16-bit instructions to be executed simultaneously. The assembly code shows two accumulator units (A0 and A1) used in each iteration. This reduces the number of iterations by 50%, effectively halving the original execution time.

8-bit data. When memory accesses are restricted to a single boundary, the processor may spend extra cycles to mask off relevant bits.

9.4.3.2 Circular Buffering

Another beneficial addressing capability is *circular buffering*. For maximum efficiency, this feature must be supported directly by the processor, with no special management overhead. Circular buffering allows a programmer to define buffers in memory and stride through them automatically. Once the buffer is set up, no special software interaction is required to navigate through the data. The address generator handles nonunity strides and, more importantly, handles the "wrap-around" feature illustrated in Figure 9.3. Without this automated address

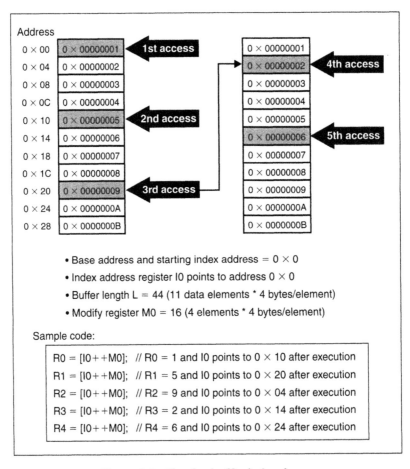

Figure 9.3: Circular buffer in hardware

generation, the programmer would have to manually keep track of buffer pointer positions, thus wasting valuable processing cycles.

Many optimizing compilers will automatically use hardware circular buffering when they encounter array addressing with a modulus operator.

9.4.3.3 Bit Reversal

An essential addressing mode for efficient signal-processing operations such as the FFT and DCT is bit reversal. Just as the name implies, bit reversal involves reversing the bits

---

Example 9.2 Single-Sample FIR

The finite impulse response filter is a very common filter structure equivalent to the convolution operator. A straightforward C implementation follows:

```
// sample the signal into a circular buffer
x[cur]= sampling_function ();
cur =(cur+1)%TAPS; // advance cur pointer in circular fashion

// perform the multiply-addition
y = 0;
for (k=0; k<TAPS; k++) {
    y += h[k] * x[(cur+k)%TAPS];
}
```

The essential part of an FIR kernel written in assembly is shown below.

```
/* the samples are stored in the R0 register, while the
        coefficients are stored in the R1 register */
LSETUP (loop_begin, loop_end) LC0=P0; /* loop counter set to
        traverse the filter */
loop_begin: A1+=R0.H*R1.L, A0+=R0.L*R1.L || R0.L=[I0++] ;
        /* perform MAC and fetch next data */
loop_end: A1+=R0.L*R1.H, A0+=R0.H*R1.H || R0.H=[I0++] ||
    R1=[I1++];
    /* perform MAC and fetch next data */
```

In the C code snippet, the % (modulus) operator provides a mechanism for circular buffering. As shown in the assembly kernel, this modulus operator does not get translated into an additional instruction inside the loop. Instead, the Data Address Generator registers I0 and I1 are configured outside the loop to automatically wrap around to the beginning upon hitting the buffer boundary.

---

in a binary address. That is, the least significant bits are swapped in position with the most significant bits. The data ordering required by a radix-2 butterfly is in "bit-reversed" order, so bit-reversed indices are used to combine FFT stages. It is possible to calculate these bit-reversed indices in software, but this is very inefficient. An example of bit reversal address flow is shown in Figure 9.4.

Example 9.3 FFT

A fast Fourier transform is an integral part of many signal-processing algorithms. One of its peculiarities is that if the input vector is in sequential time order, the output comes out in bit-reversed order. Most traditional general-purpose processors require the programmer to implement a separate routine to unscramble the bit-reversed output. On a media processor, bit reversal is often designed into the addressing engine.

Allowing the hardware to automatically bit-reverse the output of an FFT algorithm relieves the programmer from writing additional utilities, and thus improves performance.

Address LSB	Input buffer	Bit-reversed buffer	Address LSB
000	0 × 00000000	0 × 00000000	000
001	0 × 00000001	0 × 00000004	100
010	0 × 00000002	0 × 00000002	010
011	0 × 00000003	0 × 00000006	110
100	0 × 00000004	0 × 00000001	001
101	0 × 00000005	0 × 00000005	101
110	0 × 00000006	0 × 00000003	011
111	0 × 00000007	0 × 00000007	111

Sample code:

```
LSETUP (start, end) LC0 = P0;           //Loop count = 8
start: R0 = [I0] || I0 += M0 (BREV);    // I0 points to input buffer, automatically incremented in
                                        //bit-reversed progression
end: [I2++] = R0;                       // I2 points to bit-reversed buffer
```

Figure 9.4: Bit reversal in hardware

Since bit reversal is very specific to algorithms like fast Fourier transforms and discrete Fourier transforms, it is difficult for any HLL compiler to employ hardware bit reversal. For this reason, comprehensive knowledge of the underlying architecture and assembly language are key to fully utilizing this addressing mode.

9.4.4 Interlocked Instruction Pipelines

As processors increase in speed, it is necessary to add stages to the processing pipeline. For instances where a high-level language is the sole programming language, the compiler is responsible for dealing with instruction scheduling to maximize performance through the pipeline. That said, the following information is important to understand even if you're programming in C.

On older processor architectures, pipelines are usually not interlocked. On these architectures, executing certain combinations of neighboring instructions can yield incorrect results. Interlocked pipelines like the one in Figure 9.5, on the other hand, make assembly

```
IF1-3: Instruction Fetch
DC: Decode
AC: Address Calculation
EX1-4: Execution
WB: Writeback
```

	Pipeline stage									
	IF1	IF2	IF3	DC	AC	EX1	EX2	EX3	EX4	WB
1	Inst1									
2	Inst2	Inst1								
3	Inst3	Inst2	Inst1							
4	Inst4	Inst3	Inst2	Inst1						
5	Inst5	Inst4	Inst3	Inst2	Inst1					
6	Branch	Inst5	Inst4	Inst3	Inst2	Inst1				
7	Stall	Branch	Inst5	Inst4	Inst3	Inst2	Inst1			
8	Stall	Stall	Branch	Inst5	Inst4	Inst3	Inst2	Inst1		
9	Stall	Stall	Stall	Branch	Inst5	Inst4	Inst3	Inst2	Inst1	
10	Stall	Stall	Stall	Stall	Branch	Inst5	Inst4	Inst3	Inst2	Inst1

(Time axis labeled "Time" along left side)

Figure 9.5: Example of interlocked pipeline architecture with stalls inserted

programming (as well as the life of compiler engineers) easier by automatically inserting stalls when necessary. This prevents the assembly programmer from scheduling instructions in a way that will produce inaccurate results. It should be noted that, even if the pipeline is interlocked, instruction rearrangement can still yield optimization improvements by eliminating unnecessary stalls.

Let's take a look at stalls in more detail. Stalls will show up for one of four reasons:

1. The instruction in question may itself take more than one cycle to execute. When this is the case, there isn't anything you can do to eliminate the stall. For example, a 32-bit integer multiply might take three core-clock cycles to execute on a 16-bit processor. This will cause a "bubble" in two pipeline stages for a three-cycle instruction.

2. The second case involves the location of one instruction in the pipeline with respect to an instruction that follows it. For example, in some instructions, a stall may exist because the result of the first instruction is used as an operand of the following instruction. When this happens and you are programming in assembly, it is often possible to move the instruction so that the stall is not in the critical path of execution.

   Here are some simple examples on Blackfin processors that demonstrate these concepts.

   Register Transfer/Multiply latencies (One stall, due to R0 being used in the multiply):

   ```
   R0=R4; /* load R0 with contents of R4 */
   <STALL>
   R2.H=R1.L * R0.H; /* R0 is used as an operand */
   ```

   In this example, any instruction that does not change the value of the operands can be placed in-between the two instructions to hide the stall.

   When we load a pointer register and try to use the content in the next instruction, there is a latency of three stalls:

   ```
   P3=[SP++]; /* Pointer register loaded from stack */
   <STALL>
   <STALL>
   <STALL>
   R0=P3; /* Use contents of P3 after it gets its value from earlier
           instruction */
   ```

3. The third case involves a change of flow. While a deeper pipeline allows increased clock speeds, any time a change of flow occurs, a portion of the pipeline is flushed, and this

consumes core-clock cycles. The branching latency associated with a change of flow varies based on the pipeline depth. Blackfin's 10-stage pipeline yields the following latencies:

Instruction flow dependencies (Static Prediction):

Correctly predicted branch	(4 stalls)
Incorrectly predicted branch	(8 stalls)
Unconditional branch	(8 stalls)
"Drop-through" conditional branch	(0 stalls)

The term "predicted" is used to describe what the sequencer does as instructions that will complete ten core-clock cycles later enter the pipeline. You can see that when the sequencer does not take a branch, and in effect "drops through" to the next instruction after the conditional one, there are no added cycles. When an unconditional branch occurs, the maximum number of stalls occurs (eight cycles). When the processor predicts that a branch occurs and it actually is taken, the number of stalls is four. In the case where it predicted no branch, but one is actually taken, it mirrors the case of an unconditional branch.

One more note here. The maximum number of stalls is eight, while the depth of the pipeline is ten. This shows that the branching logic in an architecture does not implicitly have to match the full size of the pipeline.

4. The last case involves a conflict when the processor is accessing the same memory space as another resource (or simply fetching data from memory other than L1). For instance, a core fetch from SDRAM will take multiple core-clock cycles. As another example, if the processor and a DMA channel are trying to access the same memory bank, stalls will occur until the resource is available to the lower-priority process.

9.5 Compiler Considerations for Efficient Programming

Since the compiler's foremost task is to create correct code, there are cases where the optimizer is too conservative. In these cases, providing the compiler with extra information (through pragmas, built-in keywords, or command-line switches) will help it create more optimized code.

In general, compilers can't make assumptions about what an application is doing. This is why pragmas exist—to let the compiler know it is okay to make certain assumptions. For example, a pragma can instruct the compiler that variables in a loop are aligned and that they are not referenced to the same memory location. This extra information allows the compiler to optimize more aggressively, because the programmer has made a guarantee dictated by the pragma.

In general, a four-step process can be used to optimize an application consisting primarily of HLL code:

1. Compile with an HLL-optimizing compiler.

2. Profile the resulting code to determine the "hot spots" that consume the most processing bandwidth.

3. Update HLL code with pragmas, built-in keywords, and compiler switches to speed up the "hot spots."

4. Replace HLL procedures/functions with assembly routines in places where the optimizer did not meet the timing budget.

For maximum efficiency, it is always a good idea to inspect the most frequently executed compiler-generated assembly code to make a judgment on whether the code could be more vectorized. Sometimes, the HLL program can be changed to help the compiler produce faster code through more use of multi-issue instructions. If this still fails to produce code that is fast enough, then it is up to the assembly programmer to fine-tune the code line-by-line to keep all available hardware resources from idling.

9.5.1 Choosing Data Types

It is important to remember how the standard data types available in C actually map to the architecture you are using. For Blackfin processors, each type is shown in Table 9.1.

Table 9.1: C data types and their mapping to Blackfin registers

C type	Blackfin equivalent
char	8-bit signed
unsigned char	8-bit unsigned
short	16-bit signed integer
unsigned short	16-bit unsigned integer
Int	32-bit signed integer
unsigned int	32-bit unsigned integer
long	32-bit signed integer
unsigned long	32-bit unsigned integer

The float (32-bit), double (32-bit), long long (64-bit) and unsigned long long (64-bit) formats are not supported natively by the processor, but these can be emulated.

9.5.2 Arrays versus Pointers

We are often asked whether it is better to use arrays to represent data buffers in C, or whether pointers are better. Compiler performance engineers always point out that arrays are easier to analyze. Consider the example:

```
void array_example (int a[],int b[],int sum[],int n)
{
      int i;
      for (i=0; i<n; ++i)
      sum[i]=a[i]+b[i];
}
```

Even though we chose a simple example, the point is that these constructs are very easy to follow.

Now let's look at the same function using pointers. With pointers, the code is "closer" to the processor's native language.

```
void pointer_example (int a[],int b[],int sum[],int n){
      int i;
      for (i = 0; i < n; ++i)
            *out++ = *a++ + *b++ ;
}
```

Which produces the most efficient code? Actually, there is usually very little difference. It is best to start by using the array notation because it is easier to read. An array format can be better for "alias" analysis in helping to ensure there is no overlap between elements in a buffer. If performance is not adequate with arrays (for instance, in the case of tight inner loops), pointers may be more useful.

9.5.3 Division

Fixed-point processors often do not support division natively. Instead, they offer division primitives in the instruction set, and these help accelerate division.

The "cost" of division depends on the range of the inputs. There are two possibilities: You can use division primitives where the result and divisor each fit into 16 bits. On Blackfin processors, this results in an operation of ~40 cycles. For more precise, bitwise 32-bit division, the result is ~10× more cycles.

If possible, it is best to avoid division, because of the additional overhead it entails. Consider the example:

if (X/Y > A/B)

This can easily be rewritten as:

if (X * B > A * Y)

to eliminate the division.

Keep in mind that the compiler does not know anything about the data precision in your application. For example, in the context of the above equation rewrite, two 12-bit inputs are "safe," because the result of the multiplication will be 24 bits maximum. This quick check will indicate when you can take a shortcut, and when you have to use actual division.

9.5.4 Loops

We already discussed hardware looping constructs. Here we'll talk about software looping in C. We will attempt to summarize what you can do to ensure best performance for your application.

1. Try to keep loops short. Large loop bodies are usually more complex and difficult to optimize. Additionally, they may require register data to be stored in memory, decreasing code density and execution performance.

2. Avoid loop-carried dependencies. These occur when computations in the present iteration depend on values from previous iterations. Dependencies prevent the compiler from taking advantage of loop overlapping (i.e., nested loops).

3. Avoid manually unrolling loops. This confuses the compiler and cheats it out of a job at which it typically excels.

4. Don't execute loads and stores from a noncurrent iteration while doing computations in the current loop iteration. This introduces loop-carried dependencies. This means avoiding loop array writes of the form:

    ```
    for (i = 0; i < n; ++i)
      a[i] = b[i] * a[c[i]]; /* has array dependency*/
    ```

5. Make sure that inner loops iterate more than outer loops, since most optimizers focus on inner loop performance.

6. Avoid conditional code in loops. Large control-flow latencies may occur if the compiler needs to generate conditional jumps.

   As an example,

   ```
   for {
        if { ….. } else {…..}
        }
   ```

   should be replaced, if possible, by:

   ```
   if {
        for   {…..}
     } else {
        for   {…..}
               }
   ```

7. Don't place function calls in loops. This prevents the compiler from using hardware loop constructs, as we described earlier in this chapter.

8. Try to avoid using variables to specify stride values. The compiler may need to use division to figure out the number of loop iterations required, and you now know why this is not desirable!

9.5.5 Data Buffers

It is important to think about how data is represented in your system. It's better to pre-arrange the data in anticipation of "wider" data fetches—that is, data fetches that optimize the amount of data accessed with each fetch. Let's look at an example that represents complex data.

One approach that may seem intuitive is:

```
short Real_Part[ N ];
short Imaginary_Part [ N ];
```

While this is perfectly adequate, data will be fetched in two separate 16-bit accesses. It is often better to arrange the array in one of the following ways:

```
short Complex [ N*2 ];
```

or

```
long Complex [ N ];
```

Here, the data can be fetched via one 32-bit load and used whenever it's needed. This single fetch is faster than the previous approach.

On a related note, a common performance-degrading buffer layout involves constructing a 2D array with a column of pointers to `malloc`'d rows of data. While this allows complete flexibility in row and column size and storage, it may inhibit a compiler's ability to optimize, because the compiler no longer knows if one row follows another, and therefore it can see no constant offset between the rows.

9.5.6 Intrinsics and In-lining

It is difficult for compilers to solve all of your problems automatically and consistently. This is why you should, if possible, avail yourself of "in-line" assembly instructions and intrinsics.

In-lining allows you to insert an assembly instruction into your C code directly. Sometimes this is unavoidable, so you should probably learn how to in-line for the compiler you're using.

In addition to in-lining, most compilers support intrinsics, and their optimizers fully understand intrinsics and their effects. The Blackfin compiler supports a comprehensive array of 16-bit intrinsic functions, which must be programmed explicitly. Below is a simple example of an intrinsic that multiplies two 16-bit values.

```
#include <fract.h>
fract32 fdot(fract16 *x, fract16 *y, int n)
{
        fract32 sum=0;
        int i;
        for (i = 0; i < n; i++)
                sum = add_fr1x32(sum, mult_fr1x32(x[i], y[i]));
        return sum;
}
```

Here are some other operations that can be accomplished through intrinsics:

- Align operations

- Packing operations

- Disaligned loads

- Unpacking

- Quad 8-bit add/subtract

- Dual 16-bit add/clip

- Quad 8-bit average

- Accumulator extract with addition

- Subtract/absolute value/accumulate

The intrinsics that perform the above functions allow the compiler to take advantage of video-specific instructions that improve performance but that are difficult for a compiler to use natively.

When should you use in-lining, and when should you use intrinsics? Well, you really don't have to choose between the two. Rather, it is important to understand the results of using both, so that they become tools in your programming arsenal. With regard to in-lining of assembly instructions, look for an option where you can include in the in-lining construct the registers you will be "touching" in the assembly instruction. Without this information, the compiler will invariably spend more cycles, because it's limited in the assumptions it can make and therefore has to take steps that can result in lower performance. With intrinsics, the compiler can use its knowledge to improve the code it generates on both sides of the intrinsic code. In addition, the fact that the intrinsic exists means someone who knows the compiler and architecture very well has already translated a common function to an optimized code section.

9.5.7 Volatile Data

The `volatile` data type is essential for peripheral-related registers and interrupt-related data.

Some variables may be accessed by resources not visible to the compiler. For example, they may be accessed by interrupt routines, or they may be set or read by peripherals.

The `volatile` attribute forces all operations with that variable to occur exactly as written in the code. This means that a variable is read from memory each time it is needed, and it's written back to memory each time it's modified. The exact order of events is preserved. Missing a `volatile` qualifier is the largest single cause of trouble when engineers port from one C-based processor to another. Architectures that don't require `volatile` for hardware-related accesses probably treat all accesses as volatile by default and thus may perform at a lower performance level than those that require you to state this explicitly. When a C program works with optimization turned off but doesn't work with optimization on, a missing `volatile` qualifier is usually the culprit.

9.6 System and Core Synchronization

Earlier we discussed the importance of an interlocked pipeline, but we also need to discuss the implications of the pipeline on the different operating domains of a processor.

On Blackfin devices, there are two synchronization instructions that help manage the relationship between when the core and the peripherals complete specific instructions or sequences. While these instructions are very straightforward, they are sometimes used more than necessary. The CSYNC instruction prevents any other instructions from entering the pipeline until all pending core activities have completed. The SSYNC behaves in a similar manner, except that it holds off new instructions until all pending system actions have completed. The performance impact from a CSYNC is measured in multiple CCLK cycles, while the impact of an SSYNC is measured in multiple SCLKs. When either of these instructions is used too often, performance will suffer needlessly.

So when do you need these instructions? We'll find out in a minute. But first we need to talk about memory transaction ordering.

9.6.1 Load/Store Synchronization

Many embedded processors support the concept of a Load/Store data access mechanism. What does this mean, and how does it impact your application? "Load/Store" refers to the characteristic in an architecture where memory operations (loads and stores) are intentionally separated from the arithmetic functions that use the results of fetches from memory operations. The separation is made because memory operations, especially instructions that access off-chip memory or I/O devices, take multiple cycles to complete and would normally halt the processor, preventing an instruction execution rate of one instruction per core-clock cycle. To avoid this situation, data is brought into a data register from a source memory location, and once it is in the register, it can be fed into a computation unit.

In write operations, the "store" instruction is considered complete as soon as it executes, even though many clock cycles may occur before the data is actually written to an external memory or I/O location. This arrangement allows the processor to execute one instruction per clock cycle, and it implies that the synchronization between when writes complete and when subsequent instructions execute is not guaranteed. This synchronization is considered unimportant in the context of most memory operations. With the presence of a write buffer that sits between the processor and external memory, multiple writes can, in fact, be made without stalling the processor.

For example, consider the case where we write a simple code sequence consisting of a single write to L3 memory surrounded by five NOP ("no operation") instructions. Measuring the cycle count of this sequence running from L1 memory shows that it takes six cycles to execute. Now let's add another write to L3 memory and measure the cycle count again. We will see the cycle count increase by one cycle each time, until we reach the limits of the write buffer, at which point it will increase substantially until the write buffer is drained.

9.6.2 Ordering

The relaxation of synchronization between memory accesses and their surrounding instructions is referred to as "weak ordering" of loads and stores. Weak ordering implies that the timing of the actual completion of the memory operations—even the order in which these events occur—may not align with how they appear in the sequence of a program's source code.

In a system with weak ordering, only the following items are guaranteed:

- Load operations will complete before a subsequent instruction uses the returned data.

- Load operations using previously written data will use the updated values, even if they haven't yet propagated out to memory.

- Store operations will eventually propagate to their ultimate destination.

Because of weak ordering, the memory system is allowed to prioritize reads over writes. In this case, a write that is queued anywhere in the pipeline, but not completed, may be deferred by a subsequent read operation, and the read is allowed to be completed before the write. Reads are prioritized over writes because the read operation has a dependent operation waiting on its completion, whereas the processor considers the write operation complete, and the write does not stall the pipeline if it takes more cycles to propagate the value out to memory.

For most applications, this behavior will greatly improve performance. Consider the case where we are writing to some variable in external memory. If the processor performs a write to one location followed by a read from a different location, we would prefer to have the read complete before the write.

This ordering provides significant performance advantages in the operation of most memory instructions. However, it can cause side effects—when writing to or reading from non-memory locations such as I/O device registers, the order of how read and write operations complete is often significant.

For example, a read of a status register may depend on a write to a control register. If the address in either case is the same, the read would return a value from the write buffer rather than from the actual I/O device register, and the order of the read and write at the register may be reversed. Both of these outcomes could cause undesirable side effects. To prevent these occurrences in code that requires precise (strong) ordering of load and store operations, synchronization instructions like `CSYNC` or `SSYNC` should be used.

The CSYNC instruction ensures all pending core operations have completed and the core buffer (between the processor core and the L1 memories) has been flushed before proceeding to the next instruction. Pending core operations may include any pending interrupts, speculative states (such as branch predictions) and exceptions. A CSYNC is typically required after writing to a control register that is in the core domain. It ensures that whatever action you wanted to happen by writing to the register takes place before you execute the next instruction.

The SSYNC instruction does everything the CSYNC does, and more. As with CSYNC, it ensures all pending operations have to be completed between the processor core and the L1 memories. SSYNC further ensures completion of all operations between the processor core, external memory and the system peripherals. There are many cases where this is important, but the best example is when an interrupt condition needs to be cleared at a peripheral before an interrupt service routine (ISR) is exited. Somewhere in the ISR, a write is made to a peripheral register to "clear" and, in effect, acknowledge the interrupt. Because of differing clock domains between the core and system portions of the processor, the SSYNC ensures the peripheral clears the interrupt before exiting the ISR. If the ISR were exited before the interrupt was cleared, the processor might jump right back into the ISR.

Load operations from memory do not change the state of the memory value itself. Consequently, issuing a speculative memory-read operation for a subsequent load instruction usually has no undesirable side effect. In some code sequences, such as a conditional branch instruction followed by a load, performance may be improved by speculatively issuing the read request to the memory system before the conditional branch is resolved.

For example,

```
IF CC JUMP away_from_here
R0 = [P2];
...
away_from_here:
```

If the branch is taken, then the load is flushed from the pipeline, and any results that are in the process of being returned can be ignored. Conversely, if the branch is not taken, the memory will have returned the correct value earlier than if the operation were stalled until the branch condition was resolved.

However, this could cause an undesirable side effect for a peripheral that returns sequential data from a FIFO or from a register that changes value based on the number of reads that are requested. To avoid this effect, use an SSYNC instruction to guarantee the correct behavior between read operations.

Store operations never access memory speculatively, because this could cause modification of a memory value before it is determined whether the instruction should have executed.

9.6.3 Atomic Operations

We have already introduced several ways to use semaphores in a system. While there are many ways to implement a semaphore, using atomic operations is preferable, because they provide noninterruptible memory operations in support of semaphores between tasks.

The Blackfin processor provides a single atomic operation: TESTSET. The TESTSET instruction loads an indirectly addressed memory word, tests whether the low byte is zero, and then sets the most significant bit of the low memory byte without affecting any other bits. If the byte is originally zero, the instruction sets a status bit. If the byte is originally nonzero, the instruction clears the status bit. The sequence of this memory transaction is atomic— hardware bus locking insures that no other memory operation can occur between the test and set portions of this instruction. The TESTSET instruction can be interrupted by the core. If this happens, the TESTSET instruction is executed again upon return from the interrupt. Without something like this TESTSET facility, it is difficult to ensure true protection when more than one entity (for example, two cores in a dual-core device) vies for a shared resource.

9.7 Memory Architecture—The Need for Management

In this section we will discuss how to best use memory in your application.

9.7.1 Memory Access Trade-offs

Embedded media processors usually have a small amount of fast, on-chip memory, whereas microcontrollers usually have access to large external memories. A hierarchical memory architecture combines the best of both approaches, providing several tiers of memory with different performance levels. For applications that require the most determinism, on-chip SRAM can be accessed in a single core-clock cycle. Systems with larger code sizes can utilize bigger, higher-latency on-chip and off-chip memories.

On its own, this hierarchy is only part of the answer, since most complex programs today are large enough to require external memory, and this would dictate an unacceptably slow execution speed. As a result, programmers would be forced to manually move key code in and out of internal SRAM. However, by adding data and instruction caches into the architecture, external memory becomes much more manageable. The cache reduces the manual movement of instructions and data into the processor core, thus greatly simplifying the programming model.

Figure 9.6 demonstrates a typical memory configuration where instructions are brought in from external memory as they are needed. Instruction cache usually operates with some type of least recently used (LRU) algorithm, insuring that instructions that run more often get replaced less often. The figure also illustrates that having the ability to configure some on-chip data memory as cache and some as SRAM can optimize performance. DMA controllers can feed the core directly, while data from tables can be brought into the data cache as they are needed.

When porting existing applications to a new processor, "out-of-the-box" performance is important. As we saw earlier, there are many features compilers exploit that require minimal developer involvement. Yet, there are many other techniques that, with a little extra effort by the programmer, can have a big impact on system performance.

Proper memory configuration and data placement always pays big dividends in improving system performance. On high-performance media processors, there are typically three paths into

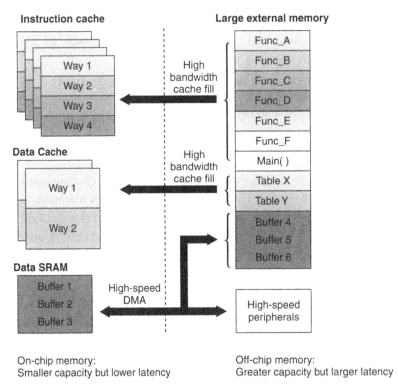

Figure 9.6: Typical memory configuration

a memory bank. This allows the core to make multiple accesses in a single clock cycle (e.g., a load and store, or two loads). By laying out an intelligent data flow, a developer can avoid conflicts created when the core processor and DMA vie for access to the same memory bank.

9.7.2 Instruction Memory Management—To Cache or To DMA?

Maximum performance is only realized when code runs from internal L1 memory. Of course, the ideal embedded processor would have an unlimited amount of L1 memory, but this is not practical. Therefore, programmers must consider several alternatives to take advantage of the L1 memory that exists in the processor, while optimizing memory and data flows for their particular system. Let's examine some of these scenarios.

The first, and most straightforward, situation is when the target application code fits entirely into L1 instruction memory. For this case, there are no special actions required, other than for the programmer to map the application code directly to this memory space. It thus becomes intuitive that media processors must excel in code density at the architectural level.

In the second scenario, a caching mechanism is used to allow programmers access to larger, less expensive external memories. The cache serves as a way to automatically bring code into L1 instruction memory as needed. The key advantage of this process is that the programmer does not have to manage the movement of code into and out of the cache. This method is best when the code being executed is somewhat linear in nature. For nonlinear code, cache lines may be replaced too often to allow any real performance improvement.

The instruction cache really performs two roles. For one, it helps pre-fetch instructions from external memory in a more efficient manner. That is, when a cache miss occurs, a cache-line fill will fetch the desired instruction, along with the other instructions contained within the cache line. This ensures that, by the time the first instruction in the line has been executed, the instructions that immediately follow have also been fetched. In addition, since caches usually operate with an LRU algorithm, instructions that run most often tend to be retained in cache.

Some strict real-time programmers tend not to trust cache to obtain the best system performance. Their argument is that if a set of instructions is not in cache when needed for execution, performance will degrade. Taking advantage of cache-locking mechanisms can offset this issue. Once the critical instructions are loaded into cache, the cache lines can be locked, and thus not replaced. This gives programmers the ability to keep what they need in cache and to let the caching mechanism manage less-critical instructions.

In a final scenario, code can be moved into and out of L1 memory using a DMA channel that is independent of the processor core. While the core is operating on one section of memory, the DMA is bringing in the section to be executed next. This scheme is commonly referred to as an overlay technique.

While overlaying code into L1 instruction memory via DMA provides more determinism than caching it, the tradeoff comes in the form of increased programmer involvement. In other words, the programmer needs to map out an overlay strategy and configure the DMA channels appropriately. Still, the performance payoff for a well-planned approach can be well worth the extra effort.

9.7.3 Data Memory Management

The data memory architecture of an embedded media processor is just as important to the overall system performance as the instruction clock speed. Because multiple data transfers take place simultaneously in a multimedia application, the bus structure must support both core and DMA accesses to all areas of internal and external memory. It is critical that arbitration between the DMA controller and the processor core be handled automatically, or performance will be greatly reduced. Core-to-DMA interaction should only be required to set up the DMA controller, and then again to respond to interrupts when data is ready to be processed.

A processor performs data fetches as part of its basic functionality. While this is typically the least efficient mechanism for transferring data to or from off-chip memory, it provides the simplest programming model. A small, fast scratchpad memory is sometimes available as part of L1 data memory, but for larger, off-chip buffers, access time will suffer if the core must fetch everything from external memory. Not only will it take multiple cycles to fetch the data, but the core will also be busy doing the fetches. It is important to consider how the core processor handles reads and writes. As we detailed above, Blackfin processors possess a multi-slot write buffer that can allow the core to proceed with subsequent instructions before all posted writes have completed. For example, in the following code sample, if the pointer register P0 points to an address in external memory and P1 points to an address in internal memory, line 50 will be executed before R0 (from line 46) is written to external memory:

```
...
Line 45: R0 =R1+R2;
Line 46: [P0]=R0; /* Write the value contained in R0 to slower
    external memory */
Line 47: R3=0x0 (z);
Line 48: R4=0x0 (z);
Line 49: R5=0x0 (z);
Line 50: [P1]=R0; /* Write the value contained in R0 to faster
    internal memory */
```

In applications where large data stores constantly move into and out of external DRAM, relying on core accesses creates a difficult situation. While core fetches are inevitably needed at times, DMA should be used for large data transfers, in order to preserve performance.

9.7.3.1 What about Data Cache?

The flexibility of the DMA controller is a double-edged sword. When a large C/C++ application is ported between processors, a programmer is sometimes hesitant to integrate DMA functionality into already-working code. This is where data cache can be very useful, bringing data into L1 memory for the fastest processing. The data cache is attractive because it acts like a mini-DMA, but with minimal interaction on the programmer's part.

Because of the nature of cache-line fills, data cache is most useful when the processor operates on consecutive data locations in external memory. This is because the cache doesn't just store the immediate data currently being processed; instead, it prefetches data in a region contiguous to the current data. In other words, the cache mechanism assumes there's a good chance that the current data word is part of a block of neighboring data about to be processed. For multimedia streams, this is a reasonable conjecture.

Since data buffers usually originate from external peripherals, operating with data cache is not always as easy as with instruction cache. This is due to the fact that coherency must be managed manually in "non-snooping" caches. "Non-snooping" means that the cache is not aware of when data changes in source memory unless it makes the change directly. For these caches, the data buffer must be invalidated before making any attempt to access the new data. In the context of a C-based application, this type of data is "volatile." This situation is shown in Figure 9.7.

In the general case, when the value of a variable stored in cache is different from its value in the source memory, this can mean that the cache line is "dirty" and still needs to be written back to memory. This concept does not apply for volatile data. Rather, in this case the cache line may be "clean," but the source memory may have changed without the knowledge of the core processor. In this scenario, before the core can safely access a volatile variable in data cache, it must invalidate (but not flush!) the affected cache line.

This can be performed in one of two ways. The cache tag associated with the cache line can be directly written, *or* a "Cache Invalidate" instruction can be executed to invalidate the target memory address. Both techniques can be used interchangeably, but the direct method is usually a better option when a large data buffer is present (e.g., one greater in size than the data cache size). The Invalidate instruction is always preferable when the buffer size is smaller than the size of the cache. This is true even when a loop is required, since the

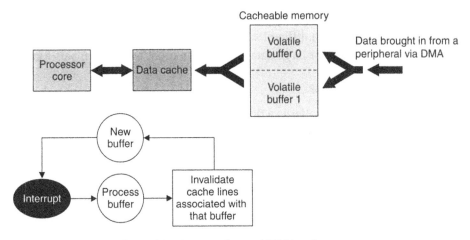

Figure 9.7: Data cache and DMA coherency

Invalidate instruction usually increments by the size of each cache line instead of by the more typical 1-, 2- or 4-byte increment of normal addressing modes.

From a performance perspective, this use of data cache cuts down on improvement gains, in that data has to be brought into cache each time a new buffer arrives. In this case, the benefit of caching is derived solely from the pre-fetch nature of a cache-line fill. Recall that the prime benefit of cache is that the data is present the second time through the loop.

One more important point about volatile variables, regardless of whether or not they are cached—if they are shared by both the core processor and the DMA controller, the programmer must implement some type of semaphore for safe operation. In sum, it is best to keep volatiles out of data cache altogether.

9.7.4 System Guidelines for Choosing Between DMA and Cache

Let's consider three widely used system configurations to shed some light on which approach works best for different system classifications.

9.7.4.1 Instruction Cache, Data DMA

This is perhaps the most popular system model, because media processors are often architected with this usage profile in mind. Caching the code alleviates complex instruction flow management, assuming the application can afford this luxury. This works well when the system has no hard real-time constraints, so that a cache miss would not wreak havoc on the timing of tightly coupled events (for example, video refresh or audio/video synchronization).

Also, in cases where processor performance far outstrips processing demand, caching instructions is often a safe path to follow, since cache misses are then less likely to cause bottlenecks. Although it might seem unusual to consider that an "oversized" processor would ever be used in practice, consider the case of a portable media player that can decode and play both compressed video and audio. In its audio-only mode, its performance requirements will be only a fraction of its needs during video playback. Therefore, the instruction/data management mechanism could be different in each mode.

Managing data through DMA is the natural choice for most multimedia applications, because these usually involve manipulating large buffers of compressed and uncompressed video, graphics and audio. Except in cases where the data is quasi-static (for instance, a graphics icon constantly displayed on a screen), caching these buffers makes little sense, since the data changes rapidly and constantly. Furthermore, as discussed above, there are usually multiple data buffers moving around the chip at one time—unprocessed blocks headed for conditioning, partly conditioned sections headed for temporary storage, and completely processed segments destined for external display or storage. DMA is the logical management tool for these buffers, since it allows the core to operate on them without having to worry about how to move them around.

9.7.4.2 Instruction Cache, Data DMA/Cache

This approach is similar to the one we just described, except in this case part of L1 data memory is partitioned as cache, and the rest is left as SRAM for DMA access. This structure is very useful for handling algorithms that involve a lot of static coefficients or lookup tables. For example, storing a sine/cosine table in data cache facilitates quick computation of FFTs. Or, quantization tables could be cached to expedite JPEG encoding or decoding.

Keep in mind that this approach involves an inherent tradeoff. While the application gains single-cycle access to commonly used constants and tables, it relinquishes the equivalent amount of L1 data SRAM, thus limiting the buffer size available for single-cycle access to data. A useful way to evaluate this tradeoff is to try alternate scenarios (Data DMA/Cache versus only DMA) in a Statistical Profiler (offered in many development tools suites) to determine the percentage of time spent in code blocks under each circumstance.

9.7.4.3 Instruction DMA, Data DMA

In this scenario, data and code dependencies are so tightly intertwined that the developer must manually schedule when instruction and data segments move through the chip. In such hard real-time systems, determinism is mandatory, and thus cache isn't ideal.

Although this approach requires more planning, the reward is a deterministic system where code is always present before the data needed to execute it, and no data blocks are lost via buffer overruns. Because DMA processes can link together without core involvement, the start of a new process guarantees that the last one has finished, so that the data or code movement is verified to have happened. This is the most efficient way to synchronize data and instruction blocks.

The Instruction/Data DMA combination is also noteworthy for another reason. It provides a convenient way to test code and data flows in a system during emulation and debug. The programmer can then make adjustments or highlight "trouble spots" in the system configuration.

An example of a system that might require DMA for both instructions and data is a video encoder/decoder. Certainly, video and its associated audio need to be deterministic for a satisfactory user experience. If the DMA signaled an interrupt to the core after each complete buffer transfer, this could introduce significant latency into the system, since the interrupt would need to compete in priority with other events. What's more, the context switch at the beginning and end of an interrupt service routine would consume several core processor cycles. All of these factors interfere with the primary objective of keeping the system deterministic.

Figures 9.8 and 9.9 provide guidance in choosing between cache and DMA for instructions and data, as well as how to navigate the tradeoff between using cache and using SRAM, based on the guidelines we discussed previously.

As a real-world illustration of these flowchart choices, Tables 9.2 and 9.3 provide actual benchmarks for G.729 and GSM AMR algorithms running on a Blackfin processor under various cache and DMA scenarios. You can see that the best performance can be obtained when a balance is achieved between cache and SRAM.

In short, there is no single answer as to whether cache or DMA should be the mechanism of choice for code and data movement in a given multimedia system. However, once developers are aware of the tradeoffs involved, they should settle into the "middle ground," the perfect optimization point for their system.

9.7.5 Memory Management Unit (MMU)

An MMU in a processor controls the way memory is set up and accessed in a system. The most basic capabilities of an MMU provides for memory protection, and when cache is used,

Instruction cache vs code overlay decision flow

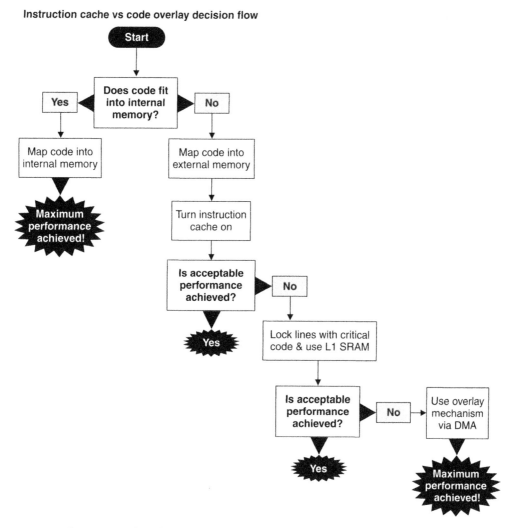

Figure 9.8: Checklist for choosing between instruction cache and DMA

it also determines whether or not a memory page is cacheable. Explicitly using the MMU is usually optional, because you can default to the standard memory properties on your processor.

On Blackfin processors, the MMU contains a set of registers that can define the properties of a given memory space. Using something called *cacheability protection look-aside buffers*

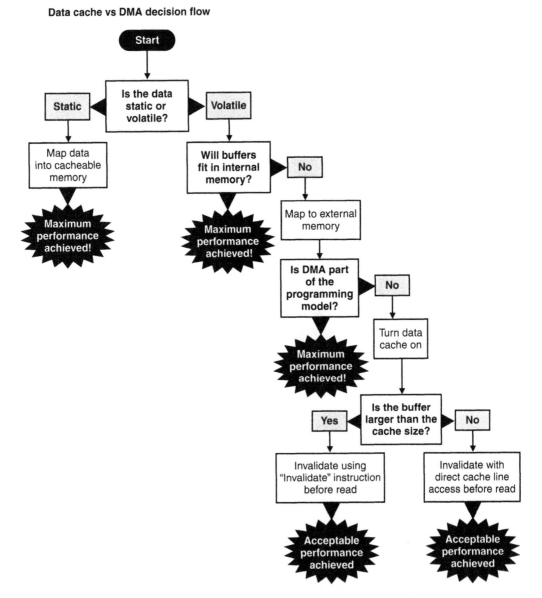

Figure 9.9: Checklist for choosing between data cache and DMA

(CPLBs), you can define parameters such as whether or not a memory page is cacheable, and whether or not a memory space can be accessed. Because the 32-bit-addressable external memory space is so large, it is likely that CPLBs will have to be swapped in and out of the MMU registers.

Table 9.2: Benchmarks (relative cycles per frame) for G.729A algorithm with cache enabled

	L1 banks configured as SRAM		L1 banks configured as cache			Cache + SRAM
	All L2	L1	Code only	Code + DataA	Code + DataB	DataA cache, DataB SRAM
Coder	1.00	0.24	0.70	0.21	0.21	0.21
Decoder	1.00	0.19	0.80	0.20	0.19	0.19

Table 9.3: Benchmarks (relative cycles per frame) for GSM AMR algorithm with cache enabled

	L1 banks configured as SRAM		L1 banks configured as cache			Cache + SRAM
	All L2	L1	Code	Code + DataA	Code + DataB	DataA cache, DataB SRAM
Coder	1.00	0.34	0.74	0.20	0.20	0.20
Decoder	1.00	0.42	0.75	0.23	0.23	0.23

9.7.5.1 CPLB Management

Because the amount of memory in an application can greatly exceed the number of available CPLBs, it may be necessary to use a CPLB manager. If so, it's important to tackle some issues that could otherwise lead to performance degradation. First, whenever CPLBs are enabled, any access to a location without a valid CPLB will result in an exception being executed prior to the instruction completing. In the exception handler, the code must free up a CPLB and re-allocate it to the location about to be accessed. When the processor returns from the exception handler, the instruction that generated the exception then executes.

If you take this exception too often, it will impact performance, because every time you take an exception, you have to save off the resources used in your exception handler. The processor then has to execute code to re-program the CPLB. One way to alleviate this problem is to profile the code and data access patterns. Since the CPLBs can be "locked," you can protect the most frequently used CPLBs from repeated page swaps.

Another performance consideration involves the search method for finding new page information. For example, a "nonexistent CPLB" exception handler only knows the address where an access was attempted. This information must be used to find the corresponding address "range" that needs to be swapped into a valid page. By locking the most frequently used pages and setting up a sensible search based on your memory access usage (for instructions and/or data), exception-handling cycles can be amortized across thousands of accesses.

9.7.5.2 Memory Translation

A given MMU may also provide memory translation capabilities, enabling what's known as *virtual memory*. This feature is controlled in a manner that is analogous to memory protection. Instead of CPLBs, *translation look-aside buffers* (TLBs) are used to describe physical memory space. There are two main ways in which memory translation is used in an application. As a holdover from older systems that had limited memory resources, operating systems would have to swap code in and out of a memory space from which execution could take place.

A more common use on today's embedded systems still relates to operating system support. In this case, all software applications run thinking they are at the same physical memory space, when, of course, they are not. On processors that support memory translation, operating systems can use this feature to have the MMU translate the actual physical memory address to the same virtual address based on which specific task is running. This translation is done transparently, without the software application getting involved.

9.8 Physics of Data Movement

So far, we've seen that the compiler and assembler provide a bunch of ways to maximize performance on code segments in your system. Using of cache and DMA provide the next level for potential optimization. We will now review the third tier of optimization in your system—it's a matter of physics.

Understanding the "physics" of data movement in a system is a required step at the start of any project. Determining if the desired throughput is even possible for an application can yield big performance savings without much initial investment.

For multimedia applications, on-chip memory is almost always insufficient for storing entire video frames. Therefore, the system must usually rely on L3 DRAM to support relatively fast access to large buffers. The processor interface to off-chip memory constitutes a major factor in designing efficient media frameworks, because access patterns to external memory must be well planned in order to guarantee optimal data throughput. There are several high-level steps that can ensure that data flows smoothly through memory in any system. Some of these are discussed below and play a key role in the design of system frameworks.

1. Grouping Like Transfers to Minimize Memory Bus Turnarounds

Accesses to external memory are most efficient when they are made in the same direction (e.g., consecutive reads or consecutive writes). For example, when accessing off-chip synchronous memory, 16 reads followed by 16 writes is always completed sooner than 16 individual read/write sequences. This is because a write followed by a read incurs latency. Random accesses to external memory generate a high probability of bus turnarounds. This added latency can easily halve available bandwidth. Therefore, it is important to take advantage of the ability to control the number of transfers in a given direction. This can be done either automatically (as we'll see here) or by manually scheduling your data movements.

A DMA channel garners access according to its priority, signified on Blackfin processors by its channel number. Higher priority channels are granted access to the DMA bus(es) first. Because of this, you should always assign higher priority DMA channels to peripherals with the highest data rates or with requirements for lowest latency.

To this end, MemDMA streams are always lower in priority than peripheral DMA activity. This is due to the fact that with Memory DMA, no external devices will be held off or starved of data. Since a Memory DMA channel requests access to the DMA bus as long as the channel is active, efficient use of any time slots unused by a peripheral DMA are applied to MemDMA transfers. By default, when more than one MemDMA stream is enabled and ready, only the highest priority MemDMA stream is granted.

When it is desirable for the MemDMA streams to share the available DMA bus bandwidth, however, the DMA controller can be programmed to select each stream in turn for a fixed number of transfers.

This "Direction Control" facility is an important consideration in optimizing use of DMA resources on each DMA bus. By grouping same-direction transfers together, it provides a way to manage how frequently the transfer direction changes on the DMA buses. This is a handy way to perform a first level of optimization without real-time processor intervention. More importantly, there's no need to manually schedule bursts into the DMA streams.

When direction control features are used, the DMA controller preferentially grants data transfers on the DMA or memory buses that are going in the same read/write direction as in the previous transfer, until either the direction control counter times out, or until traffic stops or changes direction on its own. When the direction counter reaches zero, the DMA controller changes its preference to the opposite flow direction.

In this case, reversing direction wastes no bus cycles other than any physical bus turnaround delay time. This type of traffic control represents a tradeoff of increased latency for improved utilization (efficiency). Higher block transfer values might increase the length of time each request waits for its grant, but they can dramatically improve the maximum attainable bandwidth in congested systems, often to above 90%.

Here's an example that puts these concepts into some perspective.

As a rule of thumb, it is best to maximize same-direction contiguous transfers during moderate system activity. For the most taxing system flows, however, it is best to select a block transfer value in the middle of the range to ensure no one peripheral gets locked out of accesses to external memory. This is especially crucial when at least two high-bandwidth peripherals (like PPIs) are used in the system.

In addition to using direction control, transfers among MDMA streams can be alternated in a "round-robin" fashion on the bus as the application requires. With this type of arbitration, the first DMA process is granted access to the DMA bus for some number of cycles, followed by the second DMA process, and then back to the first. The channels alternate in this pattern until all of the data is transferred. This capability is most useful on dual-core processors (for example, when both core processors have tasks that are awaiting a data stream transfer).

Example 9.4

First, we set up a memory DMA from L1 to L3 memory, using 16-bit transfers, that takes about 1100 system clock (SCLK) cycles to move 1024 16-bit words.

We then begin a transfer from a different bank of external memory to the video port (PPI). Using 16-bit unpacking in the PPI, we continuously feed an NTSC video encoder with 8-bit data. Since the PPI sends out an 8-bit quantity at a 27 MHz rate, the DMA bus bandwidth required for the PPI transfer is roughly 13.5 M transfers/second.

When we measure the time it takes to complete the same 1024-word MemDMA transfer with the PPI transferring simultaneously, it now takes three times as long.

Why is this? It's because the PPI DMA activity takes priority over the MemDMA channel transactions. Every time the PPI is ready for its next sample, the bus effectively reverses direction. This translates into cycles that are lost both at the external memory interface and on the various internal DMA buses.

When we enable Direction Control, the performance increases because there are fewer bus turn-arounds.

Without this "round-robin" feature, the first set of DMA transfers will occur, and the second DMA process will be held off until the first one completes. Round-robin prioritization can help insure that both transfer streams will complete back-to-back.

Another thing to note: using DMA and/or cache will always help performance because these types of transactions transfer large data blocks in the same direction. For example, a DMA transfer typically moves a large data buffer from one location to another. Similarly, a cache-line fill moves a set of consecutive memory locations into the device, by utilizing block transfers in the same direction.

Buffering data bound for L3 in on-chip memory serves many important roles. For one, the processor core can access on-chip buffers for pre-processing functions with much lower latency than it can by going off-chip for the same accesses. This leads to a direct increase in system performance. Moreover, buffering this data in on-chip memory allows more efficient peripheral DMA access to this data. For instance, transferring a video frame on-the-fly through a video port and into L3 memory creates a situation where other peripherals might be locked out from accessing the data they need, because the video transfer is a high-priority process. However, by transferring lines incrementally from the video port into L1 or L2 memory, a Memory DMA stream can be initiated that will quietly transfer this data into L3 as a low-priority process, allowing system peripherals access to the needed data.

2. Understanding Core and DMA SDRAM Accesses

Consider that on a Blackfin processor, core reads from L1 memory take one *core*-clock cycle, whereas core reads from SDRAM consume eight *system* clock cycles. Based on typical CCLK/SCLK ratios, this could mean that eight SCLK cycles equate to 40 CCLKs. Incidentally, these eight SCLKs reduce to only one SCLK by using a DMA controller in a burst mode instead of direct core accesses.

There is another point to make on this topic. For processors that have multiple data fetch units, it is better to use a dual-fetch instruction instead of back-to-back fetches. On Blackfin processors with a 32-bit external bus, a dual-fetch instruction with two 32-bit fetches takes nine SCLKs (eight for the first fetch and one for the second). Back-to-back fetches in separate instructions take 16 SCLKs (eight for each). The difference is that, in the first case, the request for the second fetch in the single instruction is pipelined, so it has a head start.

Similarly, when the external bus is 16 bits in width, it is better to use a 32-bit access rather than two 16-bit fetches. For example, when the data is in consecutive locations, the 32-bit fetch takes nine SCLKs (eight for the first 16 bits and one for the second). Two 16-bit fetches take 16 SCLKs (eight for each).

3. Keeping SDRAM Rows Open and Performing Multiple Passes on Data

Each access to SDRAM can take several SCLK cycles, especially if the required SDRAM row has not yet been activated. Once a row is active, it is possible to read data from an entire row without reopening that row on every access. In other words, it is possible to access any location in memory on every SCLK cycle, as long as those locations are within the same row in SDRAM. Multiple SDRAM clock cycles are needed to close a row, and therefore constant row closures can severely restrict SDRAM throughput. Just to put this into perspective, an SDRAM page miss can take 20–50 CCLK cycles, depending on the SDRAM type.

Applications should take advantage of open SDRAM banks by placing data buffers appropriately and managing accesses whenever possible. Blackfin processors, as an example, keep track of up to four open SDRAM rows at a time, so as to reduce the setup time—and thus increase throughput—for subsequent accesses to the same row within an open bank. For example, in a system with one row open, row activation latency would greatly reduce the overall performance. With four rows open at one time, on the other hand, row activation latency can be amortized over hundreds of accesses. Let's look at an example that illustrates the impact this SDRAM row management can have on memory access bandwidth:

Figure 9.10 shows two different scenarios of data and code mapped to a single *external* SDRAM bank. In the first case, all of the code and data buffers in external memory fit in a single bank, but because the access patterns of each code and data line are random, almost every access involves the activation of a new row. In the second case, even though the access patterns are randomly interspersed between code and data accesses, each set of accesses has a high probability of being within the same row. For example, even when an instruction fetch occurs immediately before and after a data access, two rows are kept open and no additional row activation cycles are incurred.

When we ran an MPEG-4 encoder from external memory (with both code and data in SDRAM), we gained a 6.5% performance improvement by properly spreading out the code and data in external memory.

4. Optimizing the System Clock Settings and Ensuring Refresh Rates are Tuned for the Speed at Which SDRAM Runs

External DRAM requires periodic refreshes to ensure that the data stored in memory retains its proper value. Accesses by the core processor or DMA engine are held off until an in-process refresh cycle has completed. If the refresh occurs too frequently, the processor can't access SDRAM as often, and throughput to SDRAM decreases as a result.

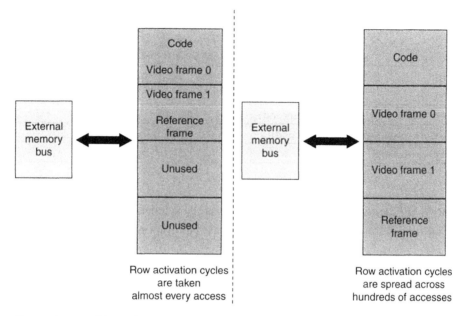

Figure 9.10: Taking advantage of code and data partitioning in external memory

On the Blackfin processor, the SDRAM Refresh Rate Control register provides a flexible mechanism for specifying the Auto-Refresh timing. Since the clock frequency supplied to the SDRAM can vary, this register implements a programmable refresh counter. This counter coordinates the supplied clock rate with the SDRAM device's required refresh rate.

Once the desired delay (in number of SDRAM clock cycles) between consecutive refresh counter time-outs is specified, a subsequent refresh counter time-out triggers an Auto-Refresh command to all external SDRAM devices.

Not only should you take care not to refresh SDRAM too often, but also be sure you're refreshing it often enough. Otherwise, stored data will start to decay because the SDRAM controller will not be able to keep corresponding memory cells refreshed.

Table 9.4 shows the impact of running with the best clock values and optimal refresh rates. Just in case you were wondering, RGB, CYMK and YIQ are imaging/video formats. Conversion between the formats involves basic linear transformation that is common in video-based systems. Table 9.4 illustrates that the performance degradation can be significant with a nonoptimal refresh rate, depending on your actual access patterns. In this example, CCLK is reduced to run with an increased SCLK to illustrate this point. Doing this improves performance for this algorithm because the code fits into L1 memory and the data is partially

Table 9.4: Using the optimal refresh rate

	Sub-optimal SDRAM refresh rate		Optimal SDRAM refresh rate	
CCLK (MHz)	594 MHz	526 MHz	526 MHz	
SCLK (MHz)	119 MHz	132 MHz	132 MHz	
RGB to CMYK Conversion (iterations per second)	226	244	250	
RGB to YIQ Conversion (iterations per second)	266	276	282	Total
Cumulative Improvement		5%	2%	7%

in L3 memory. By increasing the SCLK rate, data can be fetched faster. What's more, by setting the optimal refresh rate, performance increases a bit more.

5. Exploiting Priority and Arbitration Schemes between System Resources

Another important consideration is the priority and arbitration schemes that regulate how processor subsystems behave with respect to one another. For instance, on Blackfin processors, the core has priority over DMA accesses, by default, for transactions involving L3 memory that arrive at the same time. This means that if a core read from L3 occurs at the same time a DMA controller requests a read from L3, the core will win, and its read will be completed first.

Let's look at a scenario that can cause trouble in a real-time system. When the processor has priority over the DMA controller on accesses to a shared resource like L3 memory, it can lock out a DMA channel that also may be trying to access the memory. Consider the case where the processor executes a tight loop that involves fetching data from external memory. DMA activity will be held off until the processor loop has completed. It's not only a loop with a read embedded inside that can cause trouble. Activities like cache line fills or nonlinear code execution from L3 memory can also cause problems because they can result in a series of uninterruptible accesses.

There is always a temptation to rely on core accesses (instead of DMA) at early stages in a project, for a number of reasons. The first is that this mimics the way data is accessed on a typical prototype system. The second is that you don't always want to dig into the internal workings of DMA functionality and performance. However, with the core and DMA arbitration flexibility, using the memory DMA controller to bring data into and out of internal memory gives you more control of your destiny early on in a project.

Bibliography

Analog Devices, Inc., ADSP-BF533 Blackfin Processor Hardware Reference, Rev 3.0, September 2004.

Analog Devices, Inc., VisualDSP++ 4.1C/C++ Compiler and Library Manual for Blackfin Processors.

Heath, S. (2003). Embedded Systems Design (Second edition). Elsevier (Newnes).

CHAPTER 10

Testing and Debugging DSP Systems

Robert Oshana

While not the sexiest part of embedded design, test and debug is crucial for a product's success. It's also one of the most vexing parts. In survey after survey, developers list test and debug as their biggest source of project delays. In the 2008 Tech Insights survey, for example, 38% of engineers reported that debugging was their biggest concern. This was followed closely by worries over the related issue of code complexity, a concern that vexed 26% of respondents. The responding engineers' greatest concern for future projects? Debugging tools and software integration.

The anxiety over the future is unsurprising. Test and debug isn't getting any easier. With design complexity on the rise, particularly in multi-core designs, makers of debug hardware and software tools face an uphill battle. DSP applications can be particularly difficult to debug, as DSP engineers are often called upon to do more low-level optimization—and therefore are more likely to introduce changes that will break the code.

In this chapter, Robert Oshana tackles this complicated subject, giving us a broad overview of the state of the art of test and debug. He starts with the old methods like printf statements and shows how they're broken. He then goes into some detail on JTAG, the most commonly used debug technology. JTAG is built into most of today's processors, and is a technology most DSP engineers will use at some point in their careers. He then delves into many important topics such as emulation, trace, high-speed data collection and visualization, compiler and linker dependencies, real-time embedded software testing techniques, and common DSP algorithm bugs. By the time you get to the end of the chapter, you should have a good handle on the key techniques for DSP debugging.

—Kenton Williston

10.1 Multicore System-on-a-Chip

Designing and building embedded systems is a difficult task, given the inherent scarcity of resources in embedded systems (processing power, memory, throughput, battery life, and cost). Various trade-offs are made between these resources when designing an embedded system.

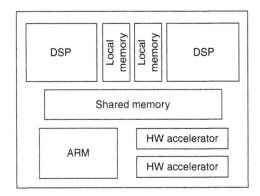

Figure 10.1: Block diagram of a DSP SoC

Modern embedded systems are using devices with multiple processing units manufactured on a single chip, creating a sort of multicore system-on-a-chip (SoC), which can increase the processing power and throughput of the system while at the same time increasing the battery life and reducing the overall cost. One example of a DSP-based SoC is shown in Figure 10.1. Multicore approaches keep hardware design in the low frequency range (each individual processor can run at a lower speed, which reduces overall power consumption as well as heat generation), offering significant price, performance, and flexibility (in software design and partitioning) over higher speed single-core designs.

There are several characteristics of SoC that we will discuss (see Jerraya). I will use an example processor to demonstrate these characteristics and how they are deployed in an existing SoC.

1. *Customized to the application* – Like embedded systems in general, SoCs are customized to an application space. As an example, I will reference the video application space. A suitable block diagram showing the flow of an embedded video application space is shown in Figure 10.2. This system consists of input capture, real-time signal processing, and output display components. As a system there are multiple technologies associated with building a flexible system including analog formats, video converters, digital formats, and digital processing. An SoC processor will incorporate a system of components; processing elements, peripherals, memories, I/O, and so forth to implement a system such as that shown in Figure 10.2. An example of an SoC processor that implements a digital video system is shown in Figure 10.3. This processor consists of various components to input, process, and output digital video information. More about the details of this in a moment.

2. *SoCs improve power/performance ratio* – Large processors running at high frequencies consume more power, and are more expensive to cool. Several smaller processors

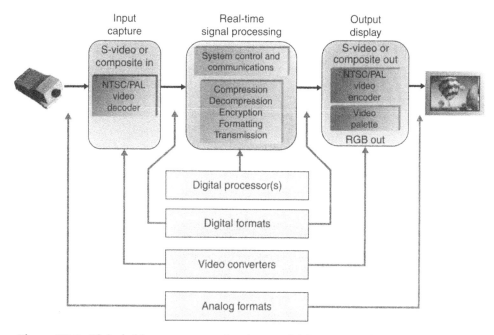

Figure 10.2: Digital video system application model (courtesy of Texas Instruments)

running at a lower frequency can perform the same amount of work without consuming as much energy and power. In Figure 10.1, the ARM processor, the two DSPs, and the hardware accelerators can run a large signal processing application efficiently by properly partitioning the application across these four different processing elements.

3. *Many apps require programmability* – SoCs contain multiple programmable processing elements. These are required for a number of reasons:

   - *New technology* – Programmability supports upgradeability and changeability easier than nonprogrammable devices. For example, as new video codec technology is developed, the algorithms to support these new standards can be implemented on a programmable processing element easily. New features are also easier to add.

   - *Support for multiple standards and algorithms* – Some digital video applications require support for multiple video standards, resolutions, and quality. Its easier to implement these on a programmable system.

   - *Full algorithm control* – A programmable system provides the designer the ability to customize and/or optimize a specific algorithm as necessary which provides the application developer more control over differentiation of the application.

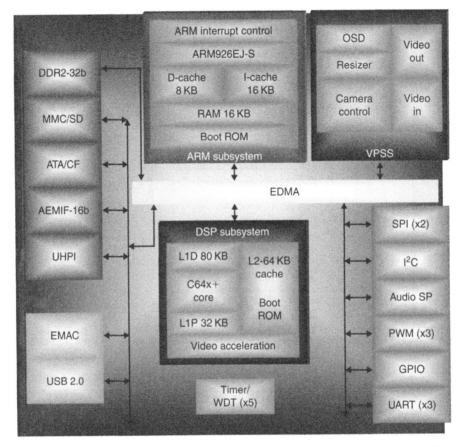

Figure 10.3: A SoC processor customized for Digital Video Systems (courtesy of Texas Instruments)

- *Software reuse in future systems* – By developing digital video software as components, these can be reuse/repackaged as building blocks for future systems as necessary.

4. *Constraints such as real-time, power, cost* – There are many constraints in real-time embedded systems. Many of these constraints are met by customizing to the application.

5. *Special instructions* – SoCs have special CPU instructions to speed up the application. As an example, the SoC in Figure 10.3 contains special instructions on the DSP to accelerate operations such as:

- 32-bit multiply instructions for extended precision computation

- Expanded arithmetic functions to support FFT and DCT algorithms

- Improve complex multiplications

- Double dot product instructions for improving throughput of FIR loops

- Parallel packing Instructions

- Enhanced Galois Field Multiply

Each of these instructions accelerate the processing of certain digital video algorithms. Of course, compiler support is necessary to schedule these instructions, so the tools become an important part of the entire system as well.

6. *Extensible* – Many SoCs are extensible in ways such as word size and cache size. Special tooling is also made available to analyze systems as these system parameters are changes.

7. *Hardware acceleration* – There are several benefits to using hardware acceleration in an SoC. The primary reason is better cost/performance ratio. Fast processors are costly. By partitioning into several smaller processing elements, cost can be reduced in the overall system. Smaller processing elements also consume less power and can actually be better at implementing real-time systems as the dedicated units can respond more efficiently to external events.

Hardware accelerators are useful in applications that have algorithmic functions that do not map to a CPU architecture well. For example, algorithms that require a lot of bit manipulation require a lot of registers. A traditional CPU register model may not be suited to efficiently execute these algorithms. A specialized hardware accelerator can be built that performs bit manipulation efficiently which sits beside the CPU and used by the CPU for bit manipulation operations. Highly responsive I/O operations are another area where a dedicated accelerator with an attached I/O peripheral will perform better. Finally, applications that are required to process streams of data, such as many wireless and multimedia applications, do not map well to the traditional CPU architecture, especially those that implement caching systems. Since each streaming data element may have a limited lifetime, processing will require the constant thrashing of cache for new data elements. A specialized hardware accelerator with special fetch logic can be implemented to provide dedicated support to these data streams.

Hardware acceleration is used on SoCs as a way to efficiently execute classes of algorithms. The use of accelerators if possible can lower overall system power, since these accelerators are customized to the class of processing and, therefore, perform these calculations very efficiently. The SoC in Figure 10.3 has hardware acceleration support.

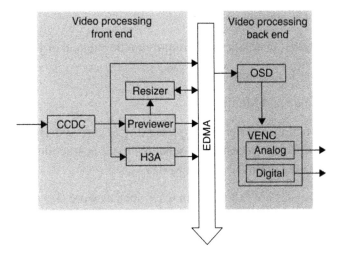

Figure 10.4: Block diagram of the video processing subsystem acceleration module of the SoC in Figure 10.3 (courtesy of Texas Instruments)

In particular, the video processing subsystem (VPSS) as well as the Video Acceleration block within the DSP subsystem are examples of hardware acceleration blocks used to efficiently process video algorithms. Figure 10.4 shows a block diagram of one of the VPSS. This hardware accelerator contains:

A front end module containing:

- CCDC (charge coupled device)

- Previewer

- Resizer (accepts data from the previewer or from external memory and resizes from ¼× to 4×)

and a back end module containing:

- Color space conversion

- DACS

- Digital output

- On-screen display

This VPSS processing element eases the overall DSP/ARM loading through hardware acceleration. An example application using the VPSS is shown in Figure 10.5.

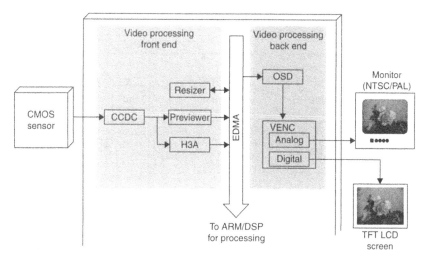

**Figure 10.5: A video phone example using the VPSS acceleration module
(courtesy of Texas Instruments)**

8. *Heterogeneous memory systems* – Many SoC devices contain separate memories for the different processing elements. This provides a performance boost because of lower latencies on memory accesses, as well as lower power from reduced bus arbitration and switching.

This programmable coprocessor is optimized for imaging and video applications. Specifically, this accelerator is optimized to perform operations such as filtering, scaling, matrix multiplication, addition, subtraction, summing absolute differences, and other related computations.

Much of the computation is specified in the form of commands which operate on arrays of streaming data. A simple set of APIs can be used to make processing calls into this accelerator. In that sense, a single command can drive hundreds or thousands of cycles.

As discussed previously, accelerators are used to perform computations that do not map efficiently to a CPU. The accelerator in Figure 10.6 is an example of an accelerator that performs efficient operations using parallel computation. This accelerator has an 8-parallel multiply accumulate (MAC) engine which significantly accelerates classes of signal processing algorithms that requires this type of parallel computation.

Examples include:

- JPEG encode and decode
- MPEG-1/2/4 encode and decode
- H.263 encode and decode

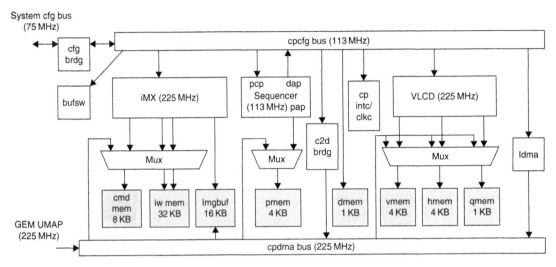

Figure 10.6: A hardware accelerator example; video and imaging coprocessor (courtesy of Texas Instruments)

- WMV9 decode

- H.264 baseline profile decode

The variable length code/decode (VLCD) module in this accelerator supports the following fundamental operations very efficiently;

- Quantization and inverse quantization (Q/IQ)

- Variable length coding and decoding (VLC/VLD)

- Huffman tables

- Zigzag scan flexibility

The design of this block is such that it operates on a macroblock of data at a time (max 6 8 × 8 blocks, 4:2:0 format). Before starting to encode or decode a bitstream, the proper registers and memory in the VLCD module must first be initialized by the application software.

This hardware accelerator also contains a block called a *sequencer* which is really just a 16-bit microprocessor targeted for simple control, address calculation, and loop control functions. This simple processing element offloads the sequential operations from the DSP. The application developer can program this sequencer to coordinate the operations among the other accelerator elements including the iMX, VLCD, System DMA, and the DSP.

The sequencer code is compiled using a simple macro using support tools, and is linked with the DSP code to be later loaded by the CPU at run time.

One of the other driving factors for the development of SoC technology is the fact that there is an increasing demand for programmable performance. For many applications, performance requirements are increasing faster than the ability of a single CPU to keep pace. The allocation of performance, and thus response time, for complex real-time systems is often easier with multiple CPUs. And dedicated CPUs in peripherals or special accelerators can offload low-level functionality from a main CPU, allowing it to focus on higher-level functions.

10.2 Software Architecture for SoC

Software development for SoC involves partitioning the application among the various processing elements based on the most efficient computational model. This can require a lot of trial and error to establish the proper partitioning. At a high level the SoC partitioning algorithm is as follows:

- Place the state machine software (those algorithms that provide application control, sequencing, user interface control, event driven software, and so on) on a RISC processor such as an ARM.

- Place the signal processing software on the DSP, taking advantage of the application specific architecture that a DSP offers for signal processing functions.

- Place high rate, computationally intensive algorithms in hardware accelerators, if they exist and if they are customized to the specific algorithm of consideration.

As an example, consider the software partitioning shown in Figure 10.7. This SoC model contains a general-purpose processor (GPP), a DSP, and hardware acceleration. The GPP contains a chip support library which is a set of low level peripheral APIs that provide efficient access to the device peripherals, a general-purpose operating system, an algorithmic abstraction layer and a set of API for and application and user interface layer. The DSP contains a similar chip support library, a DSP centric kernel, a set of DSP specific algorithms and interfaces to higher level application software. The hardware accelerator contains a set of APIs for the programmer to access and some very specific algorithms mapped to the acceleration. The application programmer is responsible for the overall partitioning of the system and the mapping of the algorithms to the respective processing elements. Some vendors may provide a "black box" solution to one or more of these processing elements, including the DSP and the hardware accelerators. This provides another level of abstraction

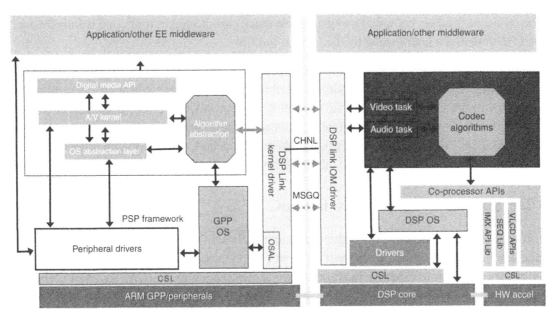

Figure 10.7: Software architecture for SoC (courtesy of Texas Instruments)

to the application developer who does not need to know the details of some of the underlying algorithms. Other system developers may want access to these low level algorithms, so there is normally flexibility in the programming model for these systems, depending on the amount of customization and tailoring required.

Communication in an SoC is primarily established by means of software. The communication interface between the DSP and the ARM in Figure 10.7, for example, is realized by defining memory locations in the DSP data space as registers. The ARM gains read/write access to these registers through a host interface. Both processors can asynchronously issue commands to each other, no one masters the other. The command sequence is purely sequential; the ARM cannot issue a new command unless the DSP has sent a "command complete" acknowledgement.

There exist two register pairs to establish the two-way asynchronous communication between ARM and DSP, one register pair is for the sending commands to ARM, and the other register pair is for the sending commands to DSP. Each register pair has:

- a command register, which is used pass commands to ARM or DSP;

- a command complete register, which is used to return the status of execution of the command;

- each command can pass up to 30 words of command parameters;

- also, each command execution can return up to 30 words of command return parameters.

An ARM to DSP command sequence is as follows:

- ARM writes a command to the command register

- ARM writes number of parameters to number register

- ARM writes command parameters into the command parameter space

- ARM issues a nonmaskable interrupt to the DSP

- DSP reads the command

- DSP reads the command parameters

- DSP executes the command

- DSP clears the command register

- DSP writes result parameters into the result parameter space

- DSP writes "command complete" register

- DSP issues HINT interrupt to ARM

The DSP to ARM command sequence is as follows:

- DSP writes command to command register

- DSP writes number of parameters to number register

- DSP writes command parameters into the command parameter space

- DSP issues an HINT interrupt to the DSP

- ARM reads the command

- ARM reads the command parameters

- ARM executes DSP command

- ARM clears the command register

- ARM writes result parameters into the result parameter space

- ARM writes "command complete" register

- ARM sends an INT0 interrupt to the DSP

Communication between the ARM and the DSP is usually accomplished using a set of communication APIs. Below is an example of a set of communication APIs between a general-purpose processor (in this case an ARM) and a DSP. The detailed software implementation for these APIs is given at the end of the chapter.

```
#define        ARM_DSP_COMM_AREA_START_ADDR 0X80
               Start DSP address for ARM-DSP.
#define        ARM_DSP_COMM_AREA_END_ADDR 0xFF
               End DSP address for ARM-DSP.
#define        ARM_DSP_DSPCR (ARM_DSP_COMM_AREA_START_ADDR)
               ARM to DSP, parameters and command from ARM.
#define        ARM_DSP_DSPCCR (ARM_DSP_COMM_AREA_START_ADDR+32)
               ARM to DSP, return values and completion code from
               DSP.
#define        ARM_DSP_ARMCR (ARM_DSP_COMM_AREA_START_ADDR+64)
               DSP to ARM, parameters and command from DSP.
#define        ARM_DSP_ARMCCR (ARM_DSP_COMM_AREA_START_ADDR+96)
               DSP to ARM, return values and completion code from
               ARM.
#define        DSP_CMD_MASK (Uint16)0X0FFF
               Command mask for DSP.
#define        DSP_CMD_COMPLETE (Uint16)0X4000
               ARM-DSP command complete, from DSP.
#define        DSP_CMD_OK (Uint16)0X0000
               ARM-DSP valid command.
#define        DSP_CMD_INVALID_CMD (Uint16)0X1000
               ARM-DSP invalid command.
#define        DSP_CMD_INVALID_PARAM (Uint16)0X2000
               ARM-DSP invalid parameters.
```

10.2.1 Functions

STATUS	ARMDSP_sendDspCmd (Uint16 cmd, Uint16 *cmdParams, Uint16 nParams)
	Send command, parameters from ARM to DSP.
STATUS	ARMDSP_getDspReply (Uint16 *status, Uint16 *retParams, Uint16 nParams)
	Get command execution status, return parameters sent by DSP to ARM.
STATUS	ARMDSP_getArmCmd (Uint16 *cmd, Uint16 *cmdParams, Uint16 nParams)
	Get command, parameters sent by DSP to ARM.
STATUS	ARMDSP_sendArmReply (Uint16 status, Uint16 *retParams, Uint16 nParams)
	Send command execution status, return parameters from ARM to DSP.
STATUS	ARMDSP_clearReg ()
	Clear ARM-DSP communication area.

10.3 SoC System Boot Sequence

Normally, the boot image for DSP is part of the ARM boot image. There could be many different boot images for the DSP for the different tasks DSP needs to execute. The sequence starts with the ARM downloading the image related to the specific task to be executed by the DSP. ARM resets then the DSP (via a control register) and then brings the DSP out of reset. At this stage the DSP begins execution at a predefined location, usually in ROM. The ROM code at this address initializes the DSP internal registers and places the DSP into an idle mode. At this point ARM downloads the DSP code by using a host port interface. After it completes downloading the DSP image, the ARM can send an interrupt to the DSP, which wakes it up from the idle mode, vectors to a start location and begins running the application code loaded by the ARM. The DSP boot sequence is given below:

- ARM resets DSP and then brings it out of reset.

- DSP gets out of reset and load its program counter (PC) register with a start address.

- The ROM code in this location branches the DSP to an initialization routine address.

- A DSP status register is initialized to move the vector table to a dedicated location, all the interrupts are disabled except for a dedicated unmaskable interrupt and the DSP is set to an mode.

- While DSP is in its mode, the ARM loads the DSP Program/Data memory with the DSP code/data.

- When the ARM finishes downloading the DSP code, it wakes up DSP from the mode by asserting an interrupt signal.

- The DSP then branches to a start address where the new interrupt vector table is located. The ARM should have loaded this location with at least a branch to the start code.

10.4 Tools Support for SoC

SoC, and heterogeneous processors in general, require more sophisticated tools support. A SoC may contain several programmable debuggable processing elements that require tools support for code generation, debug access and visibility, and real-time data analysis. A general model for this is shown in Figure 10.8. A SoC processor will have several processing elements such as an ARM and DSP. Each of these processing elements will

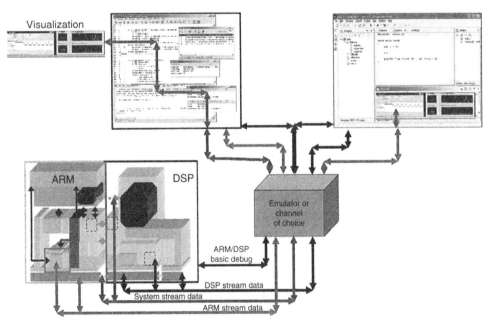

Figure 10.8: An SoC tools environment (courtesy of Texas Instruments)

require a development environment that includes mechanisms to extract, process, and display debug and instrumentation streams of data, mechanisms to peak and poke at memory and control execution of the programmable element, and tools to generate, link, and build executable images for the programmable elements.

SoC tools environments also contain support for monitoring the detailed status of each of the processing elements. As shown in Figure 10.9, detailed status reporting and control of the processing elements in an SoC allows the developer to gain visibility into the execution profile of the system. Also, since power-sensitive SoC devices may power down some or all of the device as the application executes, it is useful to also understand the power profile of the application. This can also be obtained using the proper analysis tools.

10.5 A Video Processing Example of SoC

Video processing is a good example of a commercial application requiring a system on a chip solution. Video processing applications are computationally intensive and demand a lot

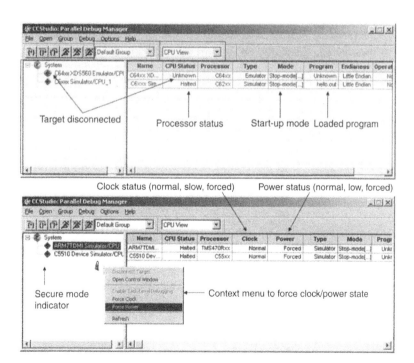

Figure 10.9: Tools support provides visibility into the status of each of the SoC processing elements (courtesy of Texas Instruments)

of MIPS to maintain the data throughput required for these applications. Some of the very compute-intensive algorithms in these applications include:

- Image pipe processing and video stabilization

- Compression and decompression

- Color conversion

- Watermarking and various forms of encryption

To perform a 30 frame per second MPEG-4 algorithm can take as much as 2500 MIPS depending on the resolution of the video.

The Audio channel processing is not as demanding but still requires enough overall MIPS to perform audio compression and decompression, equalization and sample rate conversion.

As these applications become even more complex and demanding (for example new compression technologies are still being invented), these SoC will need to support not just one but several different compression standards. SoCs for video applications include dedicated instruction set accelerators to improve performance. The SoC programming model and peripheral mix allows for the flexibility to support several formats of these standards efficiently.

For example the DM320 SoC processor in Figure 10.10 has an on chip SIMD engine (called iMX) dedicated to video processing. This hardware accelerator can perform the common video processing algorithms (Discrete Cosine Transform (DCT), IDCT, Motion Estimation, Motion Correlation to name a few).

The VLCD (variable length coding/decoding) processor is built to support variable length encoding and decoding as well as quantization of standards such as JPEG, H.263, MPEG-1/2/4 video compression standards.

As you can see from the figure, an SoC solution contains appropriate acceleration mechanisms, specialized instruction sets, hardware coprocessors, etc. that provide efficient execution of the important algorithms in DSP applications. We discussed an example of video processing but you will find the same mechanisms supporting other applications such as wireless basestation and cellular handset.

The code listings below implement the ARM-side APIs that talk to the DSP Controller module that manages the ARM/DSP interface across the DSP's Host Port Interface. These APIs are used to boot and reset the DSP and load the DSP code from the ARM, since the DSP can only execute code from internal memory that ARM loads.

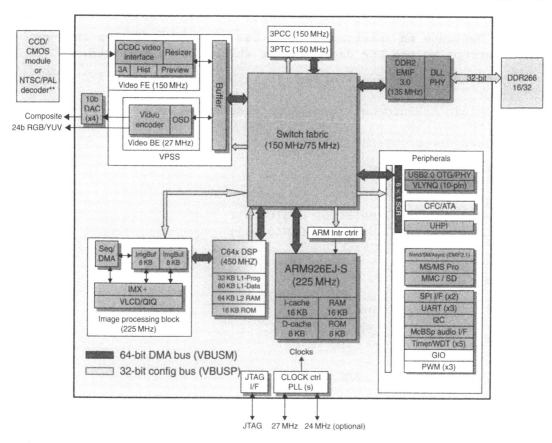

Figure 10.10: A SoC designed for video and image processing using a RISC device (ARM926) and a DSP (courtesy of Texas Instruments)

```
/**
     \ DSP Control Related APIs
*/
static STATUS DSPC_hpiAddrValidate(Uint32 dspAddr, Uint8 read);
/**
     \ Reset the DSP, Resets the DSP by toggling the DRST bit of
HPIB Control Register. \n
*/
STATUS DSPC_reset() {
     DSPC_FSET( HPIBCTL, DRST, 0);
     DSPC_FSET( HPIBCTL, DRST, 1);
     return E_PASS;
}
```

```
/**

      \ Generate an Interrupt to the DSP. Generates either INT0 or
NMI interrupt to the DSP depending on which one is specified.

      \param int ID DSP interrupt ID : INT0 -interrupt 0 NMI -NMI
interrupt
      \return if success, \c E_PASS, else error code
*/
STATUS DSPC_strobeINT(DSP_INT_ID intID) {

STATUS status = E_PASS;
    switch(intID){
      case INT0:
            DSPC_FSET( HPIBCTL, DINT0, 0);
            DSPC_FSET( HPIBCTL, DINT0, 1);
         status = E_PASS;
         break;
      case NMI:
            DSPC_FSET( HPIBCTL, HPNMI, 0);
            DSPC_FSET( HPIBCTL, HPNMI, 1);
            status = E_PASS;
            break;
      default:
            status = E_INVALID_INPUT;
            break;
    }
    return (status);
}
/**

      \ Assert the hold signal to the DSP
*/
STATUS DSPC_assertHOLD() {
      DSPC_FSET( HPIBCTL, DHOLD, 0);
      return E_PASS;
}
/**

      \ Release the hold signal that was asserted to the DSP
*/
STATUS DSPC_releaseHOLD() {
      DSPC_FSET( HPIBCTL, DHOLD, 1);
      return E_PASS;
}
/**

      \ Check if HOLD acknowledge signal received from DSP
```

```
*/
DM_BOOL DSPC_checkHOLDACK() {
      return((DM_BOOL)( DSPC_FGET( HPIBSTAT, HOLDA ) == 0 ? DM_TRUE :
DM_ FALSE));
}

/**
      \ Enable/Disable byte swapping when transferring data over HPI
interface
      \param enable Byte swap, DM_TRUE: enable, DM_FALSE: disable
*/
STATUS DSPC_byteSwap(DM_BOOL enable) {
      DSPC_FSET( HPIBCTL, EXCHG, ((enable == DM_TRUE) ? 1 : 0));
      return E_PASS;
}
/**
      \ Enable/Disable HPI interface
      \param enable HPI interface, DM_TRUE: enable, DM_FALSE: disable
*/
STATUS DSPC_hpiEnable(DM_BOOL enable) {
      DSPC_FSET( HPIBCTL, EXCHG, ((enable == DM_TRUE) ? 1 : 0));
      return E_PASS;
}
/**
      \ Get HPI interface status register HPIBSTAT
      \return register HPIBSTAT (0×30602)
*/
Uint16 DSPC_hpiStatus() {
      return DSPC_RGET( HPIBSTAT );
}
/**
      \ Write data from ARM address space to DSP address space
      Memory map in DSP address space is as follows:
      \code

      Address      Address Access   Description
      Start        End
      0×60         0×7F R/W         DSP specific memory area (32W)
      0×80         0×7FFF R/W       DSP on-chip RAM, mapped on
both program and data space (~32KW)
      0×8000       0×BFFF R/W       DSP on-chip RAM, mapped on
data space only (16KW)
```

```
      0X1C000    0X1FFFF R/W       DSP on-chip RAM, mapped on program
space only (16KW)
\endcode
      \param address Absolute address in ARM address space, must be
                 16-bit aligned
      \param size    Size of data to be written, in units of 16-bit
                 words
      \param dspAddr Absolute address in DSP address space, 0X0 ..
                 0X1FFFF
      \return if success, \c E_PASS, else error code
*/
STATUS DSPC_writeData(Uint16 *address, Uint32 size, Uint32 dspAddr) {
      if(size==0)
            return E_PASS;
      if((Uint32)address & 0X1 )
            return E_INVALID_INPUT;
      if( DSPC_hpiAddrValidate(dspAddr, 0) != E_PASS )
            return E_INVALID_INPUT;
      {
          Uint16 *hpiAddr;
          Uint16 *armAddr;
          hpiAddr=(Uint16*)HPI_DSP_START_ADDR;
          armAddr=(Uint16*)address;

          if(((dspAddr >= 0X10000) && (dspAddr<0X18000)) ||
      (dspAddr >= 0X1C000 ))
          {
               hpiAddr += (dspAddr-0X10000);
          }else if((dspAddr >= 0X0060)&&(dspAddr<0XC000)){
               hpiAddr += dspAddr;
          }else {
               hpiAddr=(Uint16*)COP_SHARED_MEM_START_ADDR;
               hpiAddr += (dspAddr-0XC000);
          }
          while(size--)
               *hpiAddr++=*armAddr++;
      }
      return E_PASS;
}
/**

      \ Read data from DSP address space to ARM address space
      Memory map in DSP address space is as follows:
      \code
```

```
      Address         Address Access       Description
      Start           End
      0×60            0×7F R/W             DSP specific memory area (32W)
      0×80            0×7FFF R/W           DSP on-chip RAM, mapped on both
                                           program and data space (~32KW)
      0×8000          0×BFFF    R/W        DSP on-chip RAM, mapped on data
                                           space only (16KW)
      0×1C000         0×1FFFF R/W          DSP on-chip RAM, mapped on
                                           program space only (16KW)
\endcode
      \param address                      Absolute address in ARM address
                                          space, must be 16-bit aligned
      \param size                         Size of data to be read, in units
                                          of 16-bit words
      \param dspAddr                      Absolute address in DSP address
                                          space, 0×0 .. 0×1FFFF
      \return if success, \c E_PASS, else error code
*/
STATUS DSPC_readData(Uint16 *address, Uint32 size, Uint32 dspAddr) {
      if(size==0)
           return E_PASS;
      if((Uint32)address & 0×1 )
           return E_INVALID_INPUT;
      if( DSPC_hpiAddrValidate(dspAddr, 1) != E_PASS )
           return E_INVALID_INPUT;
      {
           Uint16 *hpiAddr;
           Uint16 *armAddr;

           hpiAddr=(Uint16*)HPI_DSP_START_ADDR;
           armAddr=(Uint16*)address;

           if(((dspAddr >= 0×10000) && (dspAddr<0×18000)) ||
           (dspAddr >+ 0×1C000 ))
           {
                hpiAddr += (dspAddr - 0×10000);
           }else if((dspAddr >= 0×0060) && (dspAddr<;0×C000)){
                hpiAddr += dspAddr;
           }else {
                hpiAddr=(Uint16*)COP_SHARED_MEM_START_ADDR;
                hpiAddr += (dspAddr - 0×C000);
           }
      while(size--)
                **armAddr++=hpiAddr++;
```

```
        }
        return E_PASS;
}
/**

        \ Similar to DSPC_writeData(), except that after writing it
verifies the contents written to the DSP memory
        Memory map in DSP address space is as follows:
        \code
        Address         Address Access      Description
        Start           End
        0×60            0×7F R/W            DSP specific memory area (32W)
        0×80            0×7FFF R/W          DSP on-chip RAM, mapped on both
                                           program and data
                                           space
                                           (~32KW)
        0×8000          0×BFFF R/W          DSP on-chip RAM, mapped on data
                                           space only (16KW)
        0×1C000         0×1FFFF R/W         DSP on-chip RAM, mapped on
                                           program space o
\endcode
        \param address          Absolute address in ARM address space,
                                must be 16-bit aligned
        \param size             Size of data to be written, in units of
                                16-bit words
        \param dspAddr          Absolute address in DSP address space,
                                0×0 .. 0×1FFFF
        \param retryCount       Number of times to retry in case of
                                failure in writing data to DSP
                                memory
        \return if success,     \c E_PASS, else error code
*/
STATUS DSPC_writeDataVerify(Uint16 *address, Uint32 size, Uint32
dspAddr, Uint16 retryCount) {
        if(size==0)
            return E_PASS;

        if((Uint32)address & 0×1 )
            return E_INVALID_INPUT;

        if( DSPC_hpiAddrValidate(dspAddr, 0) != E_PASS )
            return E_INVALID_INPUT;
        {
            volatile Uint16 *hpiAddr;
            volatile Uint16 *armAddr;
```

```
        hpiAddr=(Uint16*)HPI_DSP_START_ADDR;
        armAddr=(Uint16*)address;
        if((((dspAddr >= 0×10000) && (dspAddr<0×18000)) ||
        (dspAddr >= 0×1C000 ))
        {
                hpiAddr += (dspAddr-0×10000);
        }else if((dspAddr >= 0×0060) && (dspAddr<0×C000)){
                hpiAddr += dspAddr;
        }else {
                hpiAddr=(Uint16*)COP_SHARED_MEM_START_ADDR;
                hpiAddr += (dspAddr-0×C000);
        }
        {
                Uint16 i;
                volatile DM_BOOL error;

                while(size--) {
                        error=(DM_BOOL)DM_TRUE;
                        for(i=0;i<retryCount;i++) {
                                *hpiAddr=*armAddr;
                                if(*hpiAddr==*armAddr) {
                                        error=(DM_BOOL)DM_FALSE;
                                        break;
                                }
                        }
                        if(error==DM_TRUE)
                                return E_DEVICE;
                        hpiAddr++;
                        armAddr++;
                }
        }
    }
    return E_PASS;
}
/**

    \ Download code to DSP memory
    \param pCode code to be dowloaded
    \see DSPCODESOURCE
*/
STATUS DSPC_loadCode(const DSPCODESOURCE* pCode) {
    if ( pCode == NULL || pCode->size == 0 )
        return E_INVALID_INPUT;

        // reset DSP
```

```
        DSPC_reset();

        // download the code to DSP memory
        while ( pCode->size != 0 ) {
                Uint16 nRetry=5;
                if( DSPC_writeDataVerify((Uint16 *)pCode->code, pCode-
>size, pCode->address, nRetry) != E_PASS )
                        return E_DEVICE;
                pCode++;
        }
        // let DSP go
        DSPC_strobeINT(INT0);
        return E_PASS;
}
static STATUS DSPC_hpiAddrValidate(Uint32 dspAddr, Uint8 read) {
// even if dspAddr <= 0X80 allow write
                if(dspAddr >= 0X60 && dspAddr <= 0XFFFF )
                        return E_PASS;
                if(dspAddr >= 0X10000 && dspAddr <= 0X17FFF )
                        return E_PASS;
                if(dspAddr >= 0X1c000 && dspAddr <= 0X1FFFF )
                        return E_PASS;
                return E_INVALID_INPUT;
}
/**

        \ ARM-DSP Communication APIs
*/
/*
/**

        \ Send command, parameters from ARM to DSP
        This routine also triggers the NMI interrupt to DSP
        \param cmd command to be sent to DSP
        \param cmdParams pointer to paramters
        \param nParams number of parameters to be sent 0..30, \n
if \c nParams<30, then remaining ARM-DSP register set is filled with 0's
        \return if success, \c E_PASS, else error code
*/
STATUS ARMDSP_sendDspCmd(Uint16 cmd, Uint16* cmdParams, Uint16
nParams) {
        DSPC_writeData( &cmd, 1, ARM_DSP_COMM_AREA_START_ADDR);
        DSPC_writeData( &nParams, 1, ARM_DSP_COMM_AREA_START_ADDR+1);
        DSPC_writeData( cmdParams, nParams,
ARM_DSP_COMM_AREA_START_ADDR+2);
```

```
        DSPC_strobeINT(NMI);
        return E_PASS;
}
/**
        \ Get command execution status, return parameters sent by DSP
to ARM
        \param      status      command status received from DSP
        \param      retParams   pointer to return paramters
        \param      nParams     number of parameters to be fetched
                                from ARM-DSP communication area, 0..30
        \return if success, \c E_PASS, else error code
*/
STATUS ARMDSP_getDspReply( Uint16* status, Uint16* retParams, Uint16
nParams)
{
        DSPC_readData( status, 1, ARM_DSP_COMM_AREA_START_ADDR+32);
        DSPC_readData( retParams, nParams,
ARM_DSP_COMM_AREA_START_ADDR+34);
        return E_PASS;
}
/**

        \ Get command, parameters sent by DSP to ARM
        \param cmd command received from DSP
        \param cmdParams        pointer to paramters
        \param nParams          number of parameters to be fetched from
                                ARM-DSP communication area, 0..30
        \return if success, \c E_PASS, else error code
*/
STATUS ARMDSP_getArmCmd( Uint16* cmd, Uint16* cmdParams, Uint16
nParams) {
        DSPC_readData( cmd, 1, ARM_DSP_COMM_AREA_START_ADDR+64);
        DSPC_readData( cmdParams, nParams,
ARM_DSP_COMM_AREA_START_ADDR+66);
        return E_PASS;
}
/**

        \ Send command execution status, return parameters from ARM to
DSP
        This routine also triggers the NMI interrupt to DSP
        \param status       command execution status to be sent to DSP
        \param retPrm        pointer to return paramters
        \param nParams        number of parameters to be sent 0..30, \n
if \c nParams<30, then remaining ARM-DSP register set is filled with 0's
```

```
        \return if success, \c E_PASS, else error code
*/
STATUS ARMDSP_sendArmReply( Uint16 status, Uint16* retParams, Uint16
nParams )
{
        DSPC_writeData( &status, 1, ARM_DSP_COMM_AREA_START_ADDR+96);
        DSPC_writeData( retParams, nParams,
ARM_DSP_COMM_AREA_START_ADDR+98);
        DSPC_strobeINT(INT0);
        return E_PASS;
}
/**

        \ Clear ARM-DSP communication area
        \return if success, \c E_PASS, else error code
*/
STATUS ARMDSP_clearReg() {
        Uint16 nullArray[128];
        memset((char*)nullArray, 0, 256);
        if(DSPC_writeData(nullArray, 128,ARM_DSP_COMM_AREA_START_ADDR)
!= E_PASS )
                return E_DEVICE;
        return E_PASS;

}
```

Bibliography

Jerraya A, Tenhunen H, Wolf W. Multiprocessor systems-on-chips. *IEEE Computer,* July, 36.

Embedded Software in Real-Time Signal Processing Systems: Design Technologies. *Proceedings of the IEEE*, 85(3), March 1997.

A Software/Hardware Co-design Methodology for Embedded Microprocessor Core Design. *IEEE* 1999.

Component-Based Design Approach for Multicore SoCs, 2002, ACM.

A Customizable Embedded SoC Platform Architecture, *IEEE IWSOC'04, International Workshop on System-on-Chip for Real-Time Applications.*

Hellestrand G. (May 2004). How virtual prototypes aid SoC hardware design. *EEdesign.com.*

Edwards C. (June 2000). Panel Weighs Hardware, Software Design Options. *EETUK.com.*

Zeidman, B. (July 2005). Back to the Basics: Programmable SoCs. *Embedded.com.*

Wolf, W., (2001). *Computers as Components*. Boston: Morgan Kaufmann.

Index

Printed and bound by CPI Group (UK) Ltd, Croydon, CR0 4YY

03/10/2024

01040340-0013